DATE DUE

JAN 3 1 2015	

DEMCO, INC. 38-2931

Soybean Utilization

SOYBEAN UTILIZATION

Harry E. Snyder

Professor, Food Science Department
University of Arkansas
Fayetteville, Arkansas

T. W. Kwon

Principal Research Scientist
Division of Biological Science and Engineering
Korea Advanced Institute of Science and Technology
Seoul, Korea

An Book
Published by Van Nostrand Reinhold Company
New York

An AVI Book
(AVI is an imprint of Van Nostrand Reinhold Company Inc.)
Copyright © 1987 by Van Nostrand Reinhold Company Inc.

Library of Congress Catalog Card Number LC 87-10688

ISBN 0-442-28216-6

Printed in the United States of America

Van Nostrand Reinhold Company Inc.
115 Fifth Avenue
New York, New York 10003

Van Nostrand Reinhold Company Limited
Molly Millars Lane
Wokingham, Berkshire RG11 2PY, England

Van Nostrand Reinhold
480 La Trobe Street
Melbourne, Victoria 3000, Australia

Macmillan of Canada
Division of Canada Publishing Corporation
164 Commander Boulevard
Agincourt, Ontario M1S 3C7, Canada

16 15 14 13 12 11 10 9 8 7 6 5 4 3 2 1

Library of Congress Cataloging-in-Publication Data

Snyder, Harry E.
  Soybean utilization.

  "An AVI book."
  Bibliography: p.
  Includes index.
  1. Soybean products. I. Kwon, T. W. II. Title.
TP438.S6S68   1987   664'.64   87-10688
ISBN 0-442-28216-6

Contents

Preface

An author should have a good reason before releasing another book to the general public or to students. We think that we have two. The first is that we both do research and teach a course dealing with soybean utilization, and we feel that both the students and instructors have suffered from lack of a suitable textbook.

Second, there is a lot of emphasis in graduate programs in Food Science on developing courses of a highly specialized nature based on scientific disciplines rather than on food commodities. For example, the material we have covered in this text might be split into several courses on the chemistry of lipids, the chemistry of proteins, the unit processees of oil extraction, the nutritive aspects of oil seed products, and the food technology of oil seed products. These courses would treat many more commodities and chemical compounds than we have covered here, thus they would have achieved breadth in a different sense.

Our use of the soybean as the focus for the book allows us to pull together information on the physical and physiological properties, the biochemistry, the toxicology, and the food quality aspects of one particular commodity and thus achieve perspective that otherwise might be lacking. We also are convinced that for the purposes of a textbook it is somewhat unfair to subject the student to limited treatises by experts and to expect the student to put all of that material into the correct relational concept. We do not claim to have succeeded in our attempts, but we think the attempt is important.

Of course the attempt owes a great deal to others. Our wives should be mentioned first for having sacrificed time to this effort but still having provided invaluable help and encouragement. Others who have supplied particularly useful graphs, figures, information, or perspective are Craig Bair, Judson Harper, Bryan Hill, Lothar Leistner, George Reinbold, Eugene Sander, Frederic Senti, Hwa L. Wang, and Walter Wolf. We thank them all.

We are indebted to Sally Rader and Roxanne Rackerby for the drawing of many of the figures.

Finally we are making a plea to those who make use of this book to supply us with information or points of view that differ with those expressed herein. We know that there will be errors, for which we alone are responsible, and we will appreciate the opportunity to correct those. This effort should be viewed as a first try at a comprehensive view of soybean utilization. In subsequent editions, there will be opportunities not only to correct errors, but to expand the perspective and thus to achieve more completely our objective of providing the serious student with a consistent and correct picture of how we make use of soybeans.

1

Production, Marketing, and Sources of Information

INTRODUCTION

There are two distinct stories to tell about soybean utilization. One is a long and complicated tale of the use of these beans as a food source in the Orient. The other is swift and dramatic: the tremendous increase in production and use since World War II in the West. We shall attempt to tell something of both stories, with the emphasis on the current situation.

The soybean (*Glycine max* (L.) Merrill family Leguminosae) undoubtedly originated in the Orient, probably in China. One of the mysteries about soybeans is why there has been so little direct use of them as food: very few cultures have simply cooked and eaten soybeans as a substantial part of their diet. This may be because (1) soybeans have a high satiety value due to their relatively high oil content; (2) they contain oligosaccharides, which can cause gastric distress; and (3) they contain a trypsin inhibitor that may cause difficulties for humans, but

Table 1.1

Changes in World Soybean Production by Selected Countries from 1935 to 1984 (1000 MT)[a]

Country	1935–1939	1945–1949	1950–1954	1955–1959	1960–1964	1965–1969	1970–1974	1975–1979	1980–1984
United States	1,505	5,596	7,994	12,963	17,696	26,752	34,256	39,536	51,826
China	9,615	8,217	8,840	9,215	7,447	6,541	7,817	7,600	8,900
Japan	335	199	443	442	334	182	124	159	206
South Korea	—	133	130	147	152	199	246	295	242
Indonesia	261	261	290	373	396	394	519	597	608
Brazil	—	—	93	123	285	675	5,009	11,645	14,600
World	12,438	14,775	18,315	23,958	27,266	36,202	50,322	73,741	85,250

[a]Source: American Soybean Association (1981, 1984).

this may not be the case in a mixed diet. Regardless of the reason, the fact is that soybeans are inevitably processed in some way into food products. In the West the main products are a purified oil and a defatted meal. In the Orient, the main products are also oil and meal, but a fascinating variety of nonfermented and fermented soy foods are used.

Although the role of soybeans as food in the Orient must have been decisive over centuries, the amounts being used currently are not as great as one might expect. Wang *et al.* (1979) have reported an annual consumption of 15–20 kg per person in China in all forms. Consumption in the United States is about 110 kg per person annually, but most of this is due to the use of soybeans as animal feed.

Soybeans were first introduced into the United States in the early 1800s but remained a minor curiosity until the twentieth century, when some farmers started to grow them as a hay crop. It was not until after 1945 that their value as a supplier of feed and food oil was recognized and exploited. Table 1.1 shows the situation with respect to production of soybeans from the 1930s to the present for the world and for several countries.

Table 1.1 outlines the spectacular growth of soybean production in the United States. Smith and Circle noted in 1978 that the U.S. production had increased from about 110,000 MT (metric tons) in 1922 to 32 million MT in 1971, an average yearly increase of about 680,000 MT. They commented that such a tremendous increase in production would not likely be duplicated by soybeans or any other crop. Yet, in the next decade production of soybeans almost doubled, going to 62 million MT—an average yearly increase of about 3 million MT. Brazil is also showing the capacity to increase soybean production rapidly. In contrast, China, the dominant producing country in the 1930s and 1940s, has demonstrated little ability to increase production during recent years.

In comparison to the major grains produced in the world, soybean production is not impressive. For the 1981/1982 production year, worldwide production was

Table 1.2
World Production of Selected Vegetable Oils
(1000 MT)[a]

	1984/1985	1985/1986
Soybean	13,300	13,640
Cottonseed	3,870	3,430
Peanut	3,100	3,150
Sunflower	6,080	6,380
Rapeseed	5,630	6,250
Palm	7,040	8,290
Cocount	2,690	3,330
Olive	1,580	1,480

[a]Source: Anon. (1986).

approximately 450 million MT of wheat, 770 million MT of coarse grain, mainly maize (corn), and 410 million MT of rough rice, but only 87 million MT of soybeans.

The increased production of soybeans in recent years has led to dominance of soybean oil among the various vegetable oils available for food use. Table 1.2 shows world production of the major vegetable oils. The recent trends are up for soybean, rapeseed, coconut, and palm oils, whereas the others are relatively constant in production. Soybean oil is clearly the predominant vegetable oil in the world, but there are many different sources of fats and oils, and soybean oil makes up approximately 20% of the total fats and oils available worldwide.

The spectacular increases in production of soybeans already noted are due to two valuable components—the food oil and the feed meal. One metric ton of soybeans yields about 183 kg of oil and 800 kg of meal. When soybeans were first processed, the oil was the valuable component and the meal was considered a by-product. From 1950 to 1960 the value of oil and meal from a single unit of soybeans was roughly the same. However, since 1960 the demand throughout the world for protein foods and thus for good protein feeds has been high, and the meal has become the more valuable component (American Soybean Association 1981). Currently (1986), meal is worth about $150/MT and oil about $400/MT, and so the value in a metric ton of soybeans would be $120 for meal and $70 for oil.

We intend to concentrate in this book on the processing and utilization of soybeans for food and feed, but the agricultural production and marketing of soybeans are essential parts of the total picture and often influence processing and utilization. For example, condition of the beans at harvest will influence the quality of oil obtained. Also, the marketing channel and number of transfers in and out of railway cars, ships, etc. will influence the ease with which the oil can be degummed, because transfers cause beans to split, and oil from split beans is difficult to degum. Therefore, we shall consider some of the more important aspects of agricultural production and marketing in brief.

AGRICULTURAL PRODUCTION

The agricultural production of soybeans involves many complicated management decisions by the producer that affect yield and profit. There are also many factors beyond control such as temperature and rainfall, although increased utilization of irrigation is an important factor in yield and profit. The knowledge that has been generated in the past 40 years about the physiology of the soybean has been extremely helpful to the soybean producer. In particular, knowledge about maturity groups, breeding, diseases and pests, and nitrogen assimilation makes the management tasks of the soybean producer considerably easier.

Maturity Groups.   When the soybean was first introduced to the United States and farmers were growing it primarily for hay, observant farmers and researchers noted that maturity of the crop varied widely. These observations led eventually to the understanding that cultivars (cultivated varieties) differed in their time of flowering, and that the time of flowering was controlled by the dark period. As nights lengthened, an unknown mechanism triggered the onset of flowering.

As a result of this knowledge, cultivars have been divided into 12 maturity groups. Those adapted for flowering at the highest latitudes are labeled 00. As the latitude decreases, maturity groups 0 through X can be produced in their respective zones. In the northern latitudes where nights change rapidly as the year proceeds, the bands of latitude for maturity groups are narrow (Fig. 1.1). In the south where day length changes are much less pronounced, maturity groups are suitable for wider latitude bands.

The maturity groups were specifically designed to aid U.S. farmers and may not strictly apply to other sections of the world, in which altitude and temperature will be additional important factors.

If a maturity group adapted for northern latitudes is planted farther south, it will encounter longer nights earlier in its growth cycle. As a result, the plant will flower and set pods before it has reached its full vegetative growth. This could reduce the yield. In the converse situation, if a southern maturity group is planted in northern latitudes, the night length conducive for flowering will be reached late in the growing season. Such a cultivar may then encounter early frost before the seeds are mature, and again yield would be decreased.

The understanding and use of maturity groups have played an important role in the increased yield of soybeans in the United States. Another important activity responsible for increased yields is the breeding of soybeans.

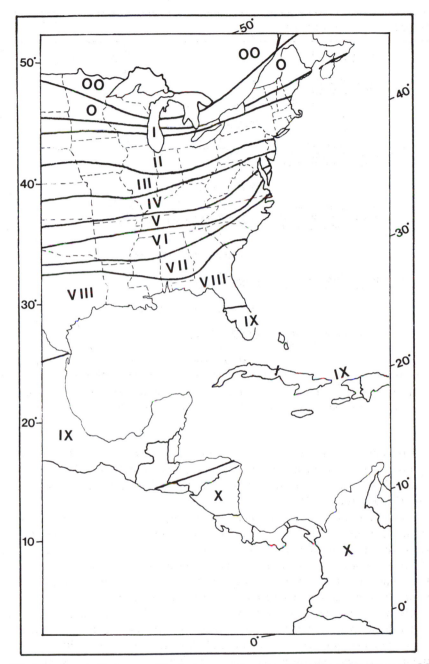

Fig. 1.1. Geographical zones of the American continent, to which soybean maturity groups OO through X are best adapted. Source: Whigman and Minor (1978). Copyright by Academic Press, Inc.

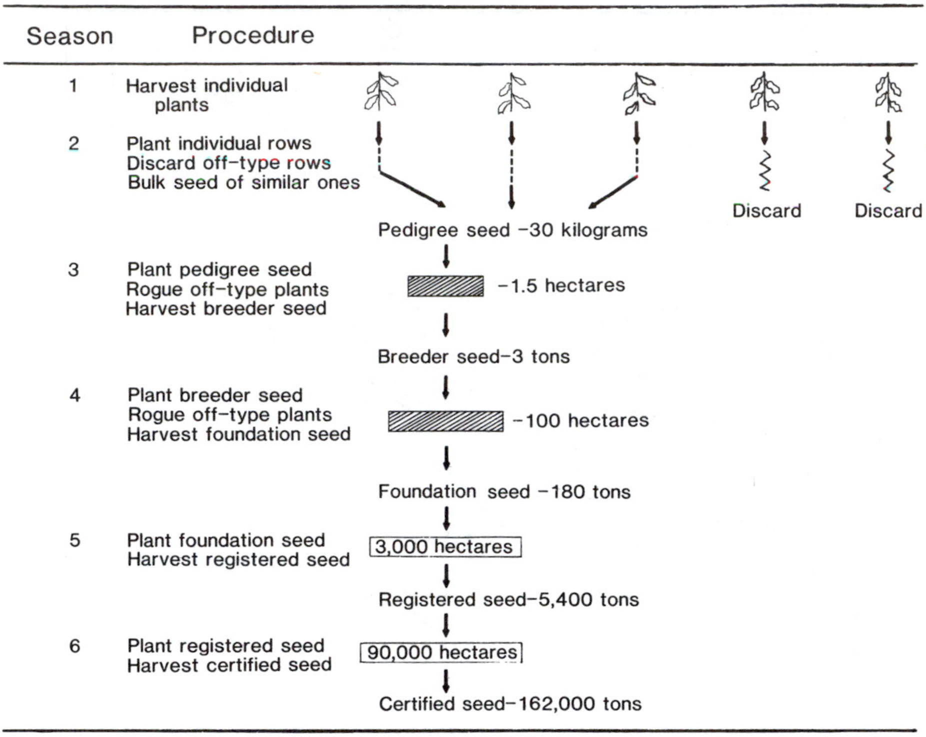

Season	Procedure
1	Harvest individual plants
2	Plant individual rows Discard off-type rows Bulk seed of similar ones

Discard   Discard

Pedigree seed −30 kilograms

| 3 | Plant pedigree seed
Rogue off-type plants
Harvest breeder seed |

−1.5 hectares

Breeder seed−3 tons

| 4 | Plant breeder seed
Rogue off-type plants
Harvest foundation seed |

−100 hectares

Foundation seed −180 tons

| 5 | Plant foundation seed
Harvest registered seed |

3,000 hectares

Registered seed−5,400 tons

| 6 | Plant registered seed
Harvest certified seed |

90,000 hectares

Certified seed−162,000 tons

Fig. 1.2. Representative steps needed between the original cross and certified seed availability for a new soybean variety. Source: Fehr (1978). Copyright by Academic Press, Inc.

Breeding. As a source of genes with which to work, soybean breeders have approximately 5000 plant introductions from other countries maintained at Urbana, Illinois and at Stoneville, Mississippi, by the U.S. Department of Agriculture (USDA) (Fehr 1978).

The flowers of soybeans are small and inconspicuous. The female portion of the flower matures before the pollen-producing anthers. Normally the flowers are self-fertilizing and natural crossing has been estimated to be less than 0.5% (Caviness 1966). If the soybean breeder selects the flower at the proper time for female maturity but male immaturity, the flower can be fertilized with selected pollen from the chosen male parent and can produce hybrid seed.

The new seed must be grown through many generations to produce pure lines that will continue to breed true. Also many generations are needed to produce sufficient quantity of seed for commercial introduction of a new cultivar. It may require six to ten years from the time of the initial cross until a new cultivar is ready for distribution to growers, but soybean breeders have the option of obtain-

	Percentage of individuals					
Generation	TT	Tt	tt			
				Parent 1 Gray pubescence tt	X	Parent 2 Brown pubescence TT
F_1	0	100	0		Tt	Brown pubescence
F_2	25	50	25	tt	Tt	TT
F_3	37.5	25	37.5	tt	Tt	TT
F_4	43.75	12.5	43.75	tt	Tt	TT
F_5	46.875	6.25	46.875	tt	Tt	TT

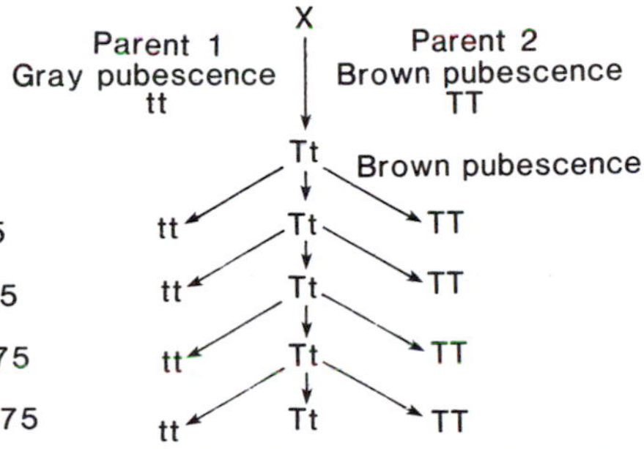

Fig. 1.3. The segregation of a single gene into true breeding lines with continued self pollination. Source: Fehr (1978). Copyright by Academic Press, Inc.

ing exclusive rights to the sale and distribution of their new cultivar. Figure 1.2 shows the buildup of seed tonnage and area planted with successive seasons.

Some characteristics of soybeans such as flower color, seed coat color, and time of flowering are qualitative in nature: that is, they are traits that are distinctly different and are controlled by one or a few genes. Other traits are quantitative: they are not distinctly different, but form part of a continuous progression and are controlled by a large number of genes. Traits such as yield, plant height, and protein and oil percentage are quantitative.

To achieve improvements in either qualitative or quantitative characteristics, one must select at least one parent plant with the desired characteristic. When the cross is made to obtain hybrid seed, the new seed is going to be heterozygous for many characteristics and has to be grown for many generations to obtain true breeding lines. Figure 1.3 shows how successive generations for one characteristic, pubescence color, lead to higher and higher percentages of homozygous or true breeding lines.

Breeding for resistance to disease may involve selection of plants with specific resistance to a particular disease organism or selection of plants with general resistance. General resistance means the plants are not immune to the disease organism under laboratory conditions, but may survive under field conditions. Both strategies have advantages and disadvantages. Specific resistance may be controlled by a single or a few major genes and may be easy to select for, but the plants will be resistant to only a specific disease organism and may lack resistance to a mutant strain. Plants with general disease resistance may show some susceptibility to disease organisms but may be able to withstand attacks by mutants.

Diseases and Pests.   Disease resistance can be accomplished by breeding in some instances, but breeding resistant cultivars is not the only answer for soybean disease problems. Bacteria, fungi, and viruses may cause soybean diseases as well as soil nematodes. Some of the more serious soybean diseases are phytophthora rot caused by *Phytophthora megasperma*, a fungus; brown stem rot caused by *Phialophora gregata*, a fungus; bacterial pustule caused by *Xanthomonas phaseoli*; soybean rust caused by *Phakospora pachyrhizi*, a fungus; and soybean mosaic caused by the soybean virus (Whigham and Minor 1978).

Parasitic soil nematodes can be a serious problem in soybean production, particularly the cyst nematode *Heterodera glycines*, and the root-knot nematode *Meloidogyne* spp. Yield loss due to nematodes has been estimated at 10%. Damage to roots by nematodes can allow entry of fungal disease organisms, thus compounding the problem.

All parts of the plant can be subject to disease, and over 100 disease organisms are known to attack soybeans. In addition to selecting seed bred for disease resistance, the producer may combat disease problems by rotating crops and by proper handling of crop residues, but often there is little control available. Disease severity can vary widely depending on environmental conditions.

Pests such as birds, rodents, insects, and weeds have the ability to decrease soybean yields greatly. Birds and rodents are most often a problem immediately after planting since the seeds or seedlings are suitable food for them. The seed produced by the soybean plant is not subject to such attack since the pod surrounding the seeds is an effective barrier.

Insects will feed on all parts of the soybean plant and are most serious pests in tropical and subtropical areas. The soybean plant does have considerable ability to compensate for insect damage to vegetative structures but is more susceptible to attack directly on seed pods. Control is by natural predators of insects and chemical pesticides.

Weeds can seriously reduce the yield of soybeans since they compete for water, sunlight, and soil nutrients. Several tools are available to the producer to control weeds such as cultivation, row spacing, choice of cultivar, and chemical herbicides, but the situation is complicated by weather, type of weeds, number of weed seeds, and amount of fertilizer used.

Nitrogen Assimilation.   The nitrogen assimilation necessary for soybean growth and for production of the high protein content of seeds is due to two systems. Early in its growth, the soybean plant makes use of nitrate reductase to convert soil nitrate into reduced nitrogen compounds needed for amino acid synthesis. As the soybean plant matures it begins to make use of a second system by which nitrogen gas in the air is converted to reduced nitrogen, that is, nitrogen fixation.

The two systems are not independent. If there is sufficient nitrate in the soil, the ability of the soybean to form nodules and to begin nitrogen fixation is inhibited. Also, if the soybean plant is actively fixing nitrogen, it will not respond favorably to added nitrogen fertilizer. Because of the interdependence of the two systems, it is difficult to state the contribution of each to the nitrogen assimilation of the plant.

Nodule formation begins soon after the plant is established and is caused by *Rhizobium japonicum*, a normally occurring soil bacterium. The soybean and other legumes are specific for the species of *Rhizobium* that can form root nodules, but the mechanism for establishing this specificity is not known.

The soybean producer can have some influence on nitrogen fixation by purchasing and spreading a soil inoculant consisting of *R. japonicum* cultures. This practice is probably most effective in those fields where soybeans have not been grown for a long time, but is less effective where soybeans have been grown and the numbers of *R. japonica* in the soil are sufficiently high.

Both nitrogen assimilation systems make use of energy and photosynthate generated by the photosynthesis system of the soybean. The two systems are equally efficient in nitrogen assimilation.

From this brief introduction to some of the problems faced in soybean production, we see that it is obviously a complicated management task. To choose the land, the seed, the irrigation, the pesticides, the fertilization, and the cultivation such that a successful crop is produced is a risky business. Equally important to the success of the producer is how he chooses to market his soybeans.

MARKETING

All of those decisions and practices involved in buying and selling soybeans and in moving soybeans through the channels from producer to consumer make up marketing.

One of the main factors going into a producer's decision on what crop to plant is the price of that crop. Since the price generally reflects the supply and demand situation, the price becomes an effective production regulator. If the price is up, meaning strong demand and weak supply, the producer will tend to plant that crop, and the supply situation will be corrected.

Also, the price of soybeans may help to relieve the pressure of harvest time. If many producers sell at harvest, the price will drop, and producers will tend to hold their soybeans in storage either on the farm or at a local elevator until the price improves. Thus soybeans will be marketed throughout the year.

The producer has several options in selling his crop. He may be part of a cooperative marketing structure in which the producers in effect own the processing facility and pay themselves a fair price. The producer may sell to a local elevator but minimize risk and profit by making use of the futures market. The producer also may store his soybeans until a better price is offered. Most producers make use of several options with each years crop.

When the crop is sold by the producer, it is usually graded at that time, and the price may be adjusted based on the grade. Grades for soybeans are similar to grades for corn and wheat. The important criteria for soybeans are test weight (weight per bushel), foreign material or dockage, splits, and color.

Grading of soybeans and of soybean products is considered in detail in Chapter 11.

Grades are useful to promote fair trading, so that buyers and sellers can know and agree upon the quality of the soybeans being traded. It would be useful if grades also reflected the end value of the soybeans, their protein and oil content, but for soybeans, grades have nothing to say about this (Updaw and Nichols 1979). Proposals to include oil and protein content in the grading system have been made, but changes depend upon an accurate and rapid analysis for oil and protein and upon agreement by the buyers and sellers that such a change in grading would be beneficial. We are probably closer to having a suitable analytical system based on infrared reflectance than we are to having agreement that a change would be beneficial. With so much money at stake, people are reluctant to change the grading system.

The processors buy their soybeans from producers or local elevators, and the price they pay reflects the demand for oil and meal. Processors may produce and sell crude oil only. That would be a crushing operation (making soybean flakes and extracting the oil) that may or may not include degumming. Processors also may extract, refine, and market oil ready for the retail market. With respect to meal, processors may sell either 44 or 50% protein meal to feed formulators, or they may formulate the feeds or actually feed and market animals. The trend in processing is toward doing all of the processing that needs to be done (vertical integration) rather than one part only. The marketing options for the processor are multiple, depending on the kinds of oil and meal products produced.

More than half of the soybeans marketed in the United States are exported, and this fact makes the transportation systems—trucks, barges, rail, and ocean shipping—an important part of the soybean marketing system. For local transportation from the producer to the local elevator, trucks are most useful. For export through Gulf of Mexico ports or through the Great Lakes ports, barges are generally the cheapest transportation. For export through the Atlantic coast ports or for export to Canada or Mexico, rail systems are most frequently used. With the recent increase in exports of soybeans and other grains from the United States, huge demands have been made on transportation systems, and consequently one of the important factors in marketing soybeans is the availability of transportation at a suitable price.

When soybeans are exported, governments frequently become involved in the transaction. If the U.S. government has a trade agreement with another government, then the purchase of soybeans will be monitored, although the selling would be through a private or cooperative grain company. In countries with centrally planned economies, state trading organizations exist for the purpose of importing grain and then reselling to other organizations within the country.

The processors who purchase imported soybeans may market the several products they produce much the same as U.S. processors. Often the processor of imported soybeans will export the products, thus further complicating the flow of soybeans from producer to consumer.

This brief description of marketing soybeans emphasizes the systems used in the West. Similar buying, selling, and processing procedures exist in the East along with more traditional practices. The traditional practice is production of small quantities of soybeans for home consumption. In this instance no marketing would be involved, and the products made (soy sauce, soy curd, soymilk, soy paste, tempeh, and other foods) would be for home consumption.

Oriental soybean producers may grow soybeans on a small scale as a cash crop and sell them to middlemen (either cooperatives or private dealers). There may be several transactions through middlemen before the soybeans are sold to processors of indigenous products. These soybean processors would then retail their products to those who had no opportunity to produce their own. Some of these retail products have high-quality identification with consumers (for example, locally produced Japanese soy sauce) and will continue to be produced from locally grown soybeans with special processing techniques.

SOURCES OF INFORMATION

As an aid to the student and as an introduction to the literature, we list and comment on the several sources of information that were useful to us in preparation of this book. We mention here only the books and kinds of journals that we found helpful. For specific journal references, these sources and later sections of this book should be consulted.

Sources Covering All Aspects of Utilization

In 1951 a major monograph on soybeans appeared "Soybeans and Soybean Products" (Markley 1951). It was edited by K. S. Markley and contained 25 chapters in two volumes written by authorities in their respective fields. There is an introductory chapter on history and agricultural production and a chapter on world trade, but most of the book is concerned with the processing of soybeans and the chemistry, physical properties, and nutritive value of soybeans and products from soybeans. There is a wealth of detail in these two volumes, but the difficulty for a student is to know what is still valid and what has become obsolete in the intervening years.

"Soybeans as a Food Source" (Wolf and Cowan 1975) appeared first as a review in 1971 but was later updated and printed as a separate monograph. This is a well-written, short explanation of the western processes for obtaining oil and meal from soybeans. Also, the use of the basic soy products in various foods is emphasized. Only mention is made of the traditional Oriental products. There is considerable emphasis on proteins, protein products, and functional properties of soy proteins in other foods.

In 1984 "The Book of Soybeans" by Watanabe and Kishi appeared in English. The emphasis is on oriental use of soybean based foods, and about half of this book is devoted to cooking of soybean foods including recipes. There is a short introductory section on production and properties of soybeans, and some material is included on the western products such as defatted flours, protein concentrates and isolates. The strength of this book is detailed information on how to prepare specific soybean foods.

One might expect to find a wealth of information written in Oriental languages about soybeans, but there seems to be relatively little at least in the contemporary scientific literature. The only book of which we are aware that covers the broad field of soybean utilization is "Soybean Foods" by Watanabe *et al.* (1971). This book, written in Japanese, begins with a short chapter on soybean varieties and maturity groups used in Japan. Next is a discussion of physical properties such as structure, color, size, and microstructure of soybeans. Under chemical properties, they treat the chemical composition of soybeans with emphasis on the proteins. There is a chapter on soybean grading (both for imports and for domestic production) followed by a chapter on the special nutritional features of soybeans that need to be considered—the antinutritional factors.

The remainder of the book is concerned with food uses of soybeans in Japan. There are chapters on soymilk and on soy curds (there are several curd products such as frozen, fried, and conventional); on fermented soy foods; on other soy foods (toasted flours used as coatings for rice cakes, sprouts, and tempeh); on quality and food uses of defatted flours; on new food applications such as concentrates, isolates, and textured products; and a final chapter on recommendations by the United Nations and other international bodies on uses of soybean products to help with world food problems. This book is now dated, but it is factually correct, and it is the best book on soybean utilization that we are aware of in an Oriental language.

A second book, "Science of Soybean, Soy Curd and Soy Sprouts" (Kim 1982) is written in Korean and is devoted mainly to soy curd and soy sprouts with a general introduction about soybeans. In the introduction, the author covers world production of soybeans, use as food, and nutritional components. The chapter on soy curd deals with different kinds of curd and their manufacturing procedures as well as the nutritional components and how they are distributed during the manufacturing process. Use of the curd in school luncheon programs and lists of the manufacturers throughout the world complete this chapter. In the final chapter, manufacturing procedures for soy sprouts and nutritional components of the sprouts are covered.

Agricultural Production of Soybeans

For a brief, well-written, and informative introduction to the problems and solutions of soybean production, the book edited by Norman (1978) is recommended. It is multiauthored and suffers somewhat from overlapping subjects and uneven treatment, but the student who knows little of soybean physiology, breeding, and production problems will find the book useful. Although the title includes the subject of utilization, only a cursory chapter is included on this aspect of soybeans.

For a different perspective on soybean production, there is "Soybean of Japan," edited by the Japanese Ministry of Agriculture and Forestry (Anon. 1977). This Japanese language compilation has chapters on production of soybeans; supply and demand; trade, marketing and price; processed foods; research and development; and regulations and laws on soybean production and trade. The book is unique for the insight it gives into the rules and regulations used by the Japanese government to control soybean production, marketing, and utilization. There are government incentives for production and for maintenance of reserves. The regulations on imports, duties, and inspection are spelled out, and research subsidies and institutes for production of soybeans in Japan are delineated.

Three research conferences on soybean production have been held recently: at the University of Illinois in 1975 (Hill 1976), at the University of North Carolina in 1979 (Corbin 1980), and at Iowa State University in 1984 (Shibles 1985). These conferences emphasized current research on production, but they included interesting papers on aspects of marketing and utilization.

Soybean Protein

While information about soybean protein can be found in most of the material written about soybeans, there are three publications devoted to discussions of soybean protein. Two of these are the result of international conferences cosponsored by the American Soybean Association and by the American Oil Chemists' Society (Anon. 1974, 1979). The first conference was held in Munich in 1973. It was divided into seven sessions: World Protein Markets; Soy Protein Products, Their Production and Properties; Legal and Regulatory Aspects of Soy Utilization in Foods; Utilization of Soy Proteins in Foods: (A) Meat and Bakery Products, (B) International Products; Nutritional Aspects of Soy Protein Foods; and Future Developments and Prospects. In a summary of the conference it was mentioned that a main thrust was the knowledge and techniques needed to use soy protein directly in human foods. It was also acknowledged that the progress would be slow and that regulatory agencies need to develop rules to encourage rather than discourage soy protein use.

The second conference was held in Amsterdam in 1978. This time the emphasis was on vegetable food proteins and the material was divided into 11 major categories: Protein Nutrition; Economics of Vegetable Protein; Current Developments in Protein Food Regulations; Characteristics of Protein Ingredients; Vegetable Proteins in Cereals, Snacks and Bakery Products; Vegetable Proteins in Confectionary Products; Vegetable Proteins in Fermented Foods and Other Products; Vegetable Proteins in Dairy Products; Marketing Requirements and Experiences; and Advances in New Vegetable Proteins. Although the titles stress vegetable proteins, most of the information is about soy proteins. The emphasis, as in the 1973 conference, is on direct use of proteins in human diets rather than as animal feed.

Probably the best single source of information about soy protein is the book first published in 1972 and with minor revision in 1978, "Soybeans: Chemistry and Technology, Vol. 1, Proteins" (Smith and Circle 1978), now out of print. This book is multiauthored and covers a wide range of topics including chemistry, nutrition, processing, and products of soybeans with the emphasis on proteins but including information on other constituents. This book is now dated, meaning that the reader may have some doubts about what material is out of date and what is still valid. The book is well indexed and remains a valuable reference work. Its shortcomings as a textbook are due to overlapping of subjects covered by different authors and no discussion of the processing and chemistry of the oil component of soybeans.

Altschul and Wilcke (1985) have edited "Seed Storage Proteins" as the final volume in their series in new protein foods. Although not devoted entirely to soybean proteins, this book does concentrate on them and covers chemical structure, genetic engineering, functional properties, chemical modification, nutritional value, and a chapter each on isolated soy protein and soy protein concentrate.

Soybean Oil

Recently, a detailed, extensive, well-written handbook on soy oil appeared (Erickson et al. 1980). This is the first book of its kind, although reports from recent conferences on soy processing (Anon. 1976, 1981, 1983) cover many of the same topics. There are 26 chapters written by 17 authorities in their respective specialty areas. There are chapters on a comparison of soy oil with other oils, nonfood uses, world and U.S. markets, environmental concerns, and sources of information. The bulk of the book, however, tells the story of the chemical and physical properties of soy oil, the industrial process of producing an oil or shortening for food use and the quality and use of the various oil products. This is a valuable book (and an exceptional bargain) for any serious student of soybean utilization. We shall be covering many of the same topics in this text, but in much less detail.

Weiss (1983) has recently updated his book "Food Oils and Their Uses." This is a valuable source of information on foods in which oil is an important component such as mayonnaise, salad dressing, margarine, and shortening. The book also contains useful sections on chemical and physical properties of fats and oils, common sources, processing methods, and various additives.

There is a new edition of "Bailey's Industrial Oil and Fat Products" (Swern 1979, 1982; Applewhite 1985) as a three-volume compendium on chemistry, processing, and products of oils. This has been a valuable reference work for many years and will continue to be. It is particularly useful for comparisons of chemical compositions, processing techniques, and products derived from the many raw materials yielding important fats and oils.

Nutrition and Health

The soybean has been acclaimed (or accused?) as a health food. While it is often difficult to separate fact from fiction in the health food area, one book that is useful in this regard is "Soy Protein and Human Nutrition" (Wilcke *et al.*, 1979). The emphasis is on measurement of protein nutritive value in humans and on soy protein isolate as a nutrient for humans. In addition there is discussion of plant protein in general as an answer to human nutritional needs, of the antinutritional factors that may come into play, and of the regulatory obstacles that need to be overcome.

At a different level but still of importance in today's world are a number of books extolling the nutritional virtues of the soybean. One Japanese author makes the point that many of the advanced nations have populations that are suffering from degenerative diseases. He is concerned that the Japanese may be moving in this same direction, and he sees the soybean as a reasonable dietary supplement to avoid excessive meat consumption.

In a somewhat similar vein, a number of books have been published recently describing how to make traditional Oriental soy foods on a semi-industrial scale. This small-scale production of soy products for sale is linked with a philosophy that there is some inherent benefit from the use of soy foods. Either soy foods are more healthful than the foods they replace, or soy foods avoid some disadvantage associated with consumption of animal products. Our purpose in calling attention to such a philosophy is neither to condemn nor to advocate it, but to recognize that it exists, and to point out that soy foods play an important role for the people who follow this line of thinking.

Miscellaneous

A few sources do not fit easily into the classification given above. One such source is the Soya Bluebook (formerly the Soybean Bluebook) published annually by the American Soybean Association (1981, 1984). This little book contains a wealth of information about organizations that are concerned with soybeans, about companies that buy and sell soybean products or the equipment to manufacture them, about soybean production statistics in the world and in the United States, and about standard terms and grading standards for soybeans.

The National Soybean Processors Association regularly publishes a set of rules for trading soybean oil and meal. In addition to contractual information, there are standards for various soybean products along with references to accepted analytical techniques (National Soybean Processors Association 1984).

Another source that we have found useful is a study conducted for the U.S. Agency for International Development (USAID) by the Northern Regional Research Center of USDA (Wang *et al.* 1979). This study was a worldwide survey of the use of soybeans in traditional foods. It includes descriptions of the foods, methods of preparation, and often the amounts consumed. It is undoubtedly the most recent and best survey of the use of soy products throughout the world.

Journals

As a source of up-to-date research information on soy products and processes, the Journal of the American Oil Chemists' Society is the best. This is partly because this organization has cosponsored and published the results of five international conferences on soy utilization. In addition to the conference reports, it regularly accepts and publishes research articles on soy oil and soy protein and on the processes used to produce such products. It also frequently publishes news items on the current status of soybean, oil, and protein supplies, and markets.

Other research journals (and their publishers) that are important for information on soybeans are:

Journal of Food Science and Food Technology—Institute of Food Technologists.

Cereal Chemistry and Cereal Foods World—American Association of Cereal Chemists.

Lipids—American Oil Chemists' Society.

Oil Mill Gazetteer—International Oil Mill Superintendents.

Fette, Seifen, Anstrichmittel—German Society for Lipid Science, Germany.

Zeitschrift fur̈ Lebensmittel—Untersuchung und Forschung—contains information on soy products chemistry and food use; Springer-Verlag, Germany.

Fleisch—contains information on use of soy products in meat products; VEB, Germany.

Die Fleischwirtschaft—contains information on the use of soy products in meat products; Federal Association of German Meat Industry, Germany.

Oil World—contains statistical information on world markets for all oils and fats; ISTA Mielke Co., Germany.

Revue Françqise des Corps Gras—Institute of Fats, France.

Oleagineux—an international review journal on fats and oils; Institute for Research on Fats and Oils, France.

La Revista Italiana delle Sostanze Grasse—industrial and experimental information on fats and oils; Experimental Station for the Fats and Oil Industry, Italy.

Grasas y Aceites—Institute of Fats and Their Derivatives, Spain.

Yukagaku—Japan Oil Chemists' Society, Japan.

Journal of the Japanese Society of Food Science and Technology—contains information on both soy oil and soy protein, Japan.

Agricultural and Biological Chemistry—Agricultural Chemical Society of Japan, Japan.

Korean Soybean Digest—Korean Soybean Society, Korea.

REFERENCES

Altschul, A. M., and H. W. Wilcke (1985). "New Protein Foods," Vol. 5, Seed Storage Proteins. Academic Press, New York.

American Soybean Association (1981). "Soya Bluebook 81." American Soybean Association, St. Louis, MO.

American Soybean Association (1984). "'84 Soya Bluebook." American Soybean Association, St. Louis, MO.

Anon. (1974). *Proc. World Soy Protein Conf., J. Am. Oil Chem. Soc.* **51:**51A–216A.

Anon. (1976). *Proc. World Conf. Oilseed Vegetable Oil Processing Technol., J. Am. Oil Chem. Soc.* **53:**221–461.

Anon. (1977). "Soybean of Japan." Jikyusha, Tokyo (in Japanese).

Anon. (1979). *Proc. World Conf. Vegetable Food Proteins, J. Am. Oil Chem. Soc.* **56:**99–484.

Anon. (1981). *Proc. World Conf. Soya Processing Utilization, J. Am. Oil Chem. Soc.* **58:**121–539.

Anon. (1983). *Proc. World Conf. Oilseed Edible Oil Processing, J. Am. Oil Chem. Soc.* **60:**185–478.

Anon. (1986). World fats and oils report. *J. Am. Oil Chem. Soc.* **63:**944.

Applewhite, T. H. (1985). "Bailey's Industrial Oil and Fat Products," Vol. 3. Wiley, New York.

Caviness, C. E. (1966). Estimates of natural cross-pollination in Jackson soybeans in Arkansas. *Crop. Sci.* **6:**211.

Corbin, F. T., ed. (1980). "World Soybean Research Conference II: Proceedings." Westview Press, Boulder, CO.

Erickson, D. R., E. H. Pryde, O. L. Brekke, T. L. Mounts, and R. A. Falb (1980). "Handbook of Soy Oil Processing and Utilization." American Soybean Association, St. Louis, MO,.and American Oil Chemists Society, Champaign, IL.

Fehr, W. R. (1978). Breeding. *In* "Soybean Physiology, Agronomy, and Utilization" (A. G. Norman, ed.). Academic Press, New York.

Hill, L. D. (1976). "World Soybean Research. Proceedings of the World Soybean Research Conference." Interstate Printers and Publishers, Inc., Danville, IL.

Kim, K. H. (1982). "Science of Soybean, Soy Curd and Soy Sprouts." Korea Advanced Institute of Science and Technology, Seoul.

Markley, K. S. (1951). "Soybeans and Soybean Products," Vols. 1 and 2. Wiley (Interscience), New York.

National Soybean Processors Association (1984). "1983–84 Yearbook and Trading Rules." National Soybean Processors Association, Washington D.C.

Norman, A. G. (1978). "Soybean Physiology, Agronomy, and Utilization." Academic Press, New York.

Shibles, R. (1985). "World Soybean Research Conference III: Proceedings." Westview Press, Boulder, CO.

Smith, A. K., and S. J. Circle, eds. (1978). "Soybeans: Chemistry and Technology," Vol. 1, Proteins. AVI Publ. Co., Westport, CT.

Swern, D. (1979). "Bailey's Industrial Oil and Fat Products," Vol. 1. Wiley, New York.

Swern, D. (1982). "Bailey's Industrial Oil and Fat Products," Vol. 2. Wiley, New York.

Updaw, N. J., and Nichols, T. E., Jr. (1979). Pricing soybeans on the basis of chemical constituents. *In* " World Soybean Research Conference II: Proceedings" (F. T. Corbin, ed.). Westview Press, Boulder, CO.

Wang, H. L., G. C. Mustakas, W. J. Wolf, L. C. Wang, C. W. Hesseltine, and E. B. Bagley (1979). "Soybeans as Human Food—Unprocessed and Simply Processed." *Util. Res. Rep.* **5**, USDA, NRRC, Peoria IL.

Watanabe, D. J., and A. Kishi (1984). "The Book of Soybeans." Japan Publications Inc., Tokyo and New York.

Watanabe, D. J., H. O. Ebine, and D. O. Ohda (1971). "Soybean Foods." Kohrin Shoin, Tokyo (in Japanese).

Weiss, T. J. (1983). "Food Oils and Their Uses," 2nd ed. AVI Publ. Co., Westport, CT.

Whigham, D. K., and H. C. Minor (1978). Agronomic characteristics and environmental stress. *In* "Soybean Physiology, Agronomy, and Utilization" (A. G. Norman, ed.). Academic Press Inc., New York.

Wilcke, H. L., D. T. Hopkins, and D. H. Waggle (1979). "Soy Protein and Human Nutrition." Academic Press, New York.

Wolf, W. J., and J. C. Cowan (1975). "Soybeans as a Food Source," rev. ed. CRC Press, Cleveland, OH.

2

Morphology and Composition

MORPHOLOGY

Dry soybeans are close to spherical in shape and vary considerably in size. The size will vary with the growing conditions and the variety. Some so-called vegetable varieties, which are cultivated for cooking and eating, are larger than the usual field varieties. The first features that are identifiable in soybeans are those associated with the seed coat.

Seed Coat

The outermost layer of the soybean is the seed coat or testa. The seed coat has its origin in the integuments of the ovule, the female portion of the flower. Therefore, genetically the seed coat is the same as the maternal plant and differs genetically from the embryo. Identifiable features of the seed coat are the hilum, the point of attachment of the bean to the pod; the micropyle, the opening through which the germ tube grew to reach the embryo sac; and the chalaza, a small groove at the end of the hilum opposite the micropyle. Figure 2.1 diagrams these structural features of the seed coat, but if possible, the student should actually observe these features in several varieties of beans.

The color of the seed coat can vary widely. Soybean classes for grading in the

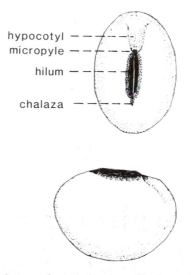

hypocotyl
micropyle
hilum
chalaza

Fig. 2.1.   Morphological features of a typical soybean. The hypocotyl is covered by the seed coat. Source: Wolf and Cowan (1975).

United States are based on color and include yellow, green, brown, and black beans. (Although green, brown, and black soybeans are classified by the grading standards, the standards discriminate against them. For example, grade 1 may have a maximum of only 1% brown or black beans.) All soybeans are green early in maturity due to chlorophyll. As the beans mature, chlorophyll disappears and residual flavonoid pigments predominate. In some varieties chlorophyll is not lost, and hence they have green seed coats. The cotyledons are invariably light yellow unless soybeans are harvested before full maturity. Then the chlorophyll becomes a problem in oil refining and requires special bleaching procedures.

Often the hilum is a different color than the remainder of the seed coat. The color is useful mainly for variety identification, although in some countries black soybeans are cooked and mixed with white rice for an interesting visual effect. A dark-colored hilum may be a defect in products such as soymilk or soy curd in which small pieces of the hilum may remain and appear as dark foreign material. A purple discoloration of the seed coat can be caused by a fungal infestation of the beans during development or storage.

The seed coat makes up about 9% of the soybean by dry weight. Seed coat is mainly composed of cellulosic type materials, but it does contain about 9% protein. There is very little oil in the seed coat, and so normally hulls are removed from the beans before going to the extractor. They are added back to the defatted meal for some animal feeding purposes.

The seed coat is well fitted to the cotyledons and is difficult to remove from an intact, dry soybean. However, if the soybean is cracked into several pieces, the

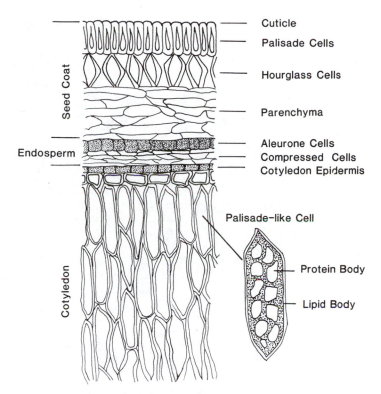

Cuticle

Palisade Cells

Hourglass Cells

Parenchyma

Aleurone Cells
Compressed Cells
Cotyledon Epidermis

Seed Coat

Endosperm

Cotyledon

Palisade-like Cell

Protein Body

Lipid Body

Fig. 2.2. A drawing of the microscopic structure of the soybean seed coat and underlying structures. Source: Bair (1979).

seed coat generally falls free and can be separated by aspiration. Also if the beans are soaked in water, the hydrated seed coat can be rubbed free easily. Neither of these practices is completely effective, and the small portion of remaining seed coat may serve to identify the presence of soybeans in other food products.

Figure 2.2 shows a cross section of the soybean seed coat. The hourglass cells between the palisade and parenchyma layers are distinctive and readily identifiable under the microscope. The presence of hourglass cells in baked goods or in meat products is good qualitative evidence that soybeans have been used in those products as an added ingredient. Getting good quantitative information on the amount of added soybeans in another food product is an analytical problem that has not been solved satisfactorily.

In the palisade layer a line of highly refractive material can be seen. It is called the "light line," and some have attributed to it a role in limiting imbibition of water although there is no evidence supporting this idea.

If soybeans are soaked in water at 5°C (41°F), some of them (the percentage depends on the variety) will not absorb water and are called "hard" beans. If the

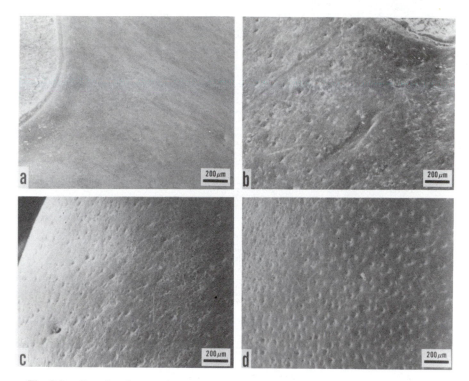

Fig. 2.3.   Scanning electron micrographs of seed coats of several soybean varieties. (a) Ilsoy, (b) Chippewa, (C) Disoy, and (d) Hawkeye, showing varying degrees of pitting. Source: Wolf *et al.* (1981).

soaking is done at high temperatures, 80–100°C (176–212°F) even hard beans will absorb water. The existence of hard beans is important because hard beans will not germinate and because they can affect the quality of soy products. Heating soaked soybeans to inactivate trypsin inhibitor or lipoxygenase is a common practice. But if the soaked soybeans have not imbibed water, the heat treatment will not be sufficient to inactivate proteins that remain dry. Products made from soybeans having active trypsin inhibitor or lipoxygenase may be inferior in nutritive value or in flavor.

The ability of the soybean to imbibe water or to resist imbibition resides in the seed coat. Once the seed coat is removed from hard beans, the cotyledons readily imbibe water. Some studies have implicated the micropyle as the site of water imbibition or lack of it (Saio 1976), but more recent work has shown that soybeans normally imbibe water through the seed coat on the side of the bean opposite the micropyle (Arechavaleta-Medina and Snyder 1981). This study indi-

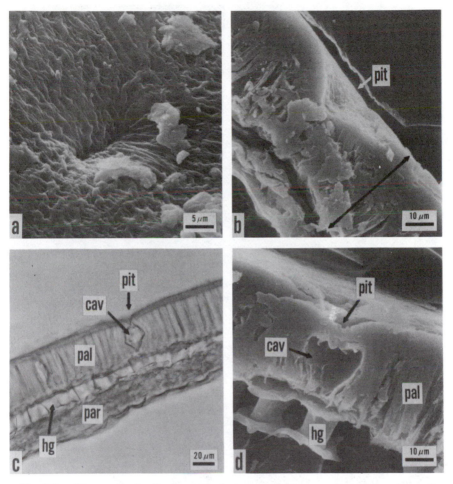

Fig. 2.4. Micrographs showing the continuity between pits in the seed coat and cavities in the palisade layer. Source: Wolf *et al.* (1981).

cated that the cuticle was the likely site of water imbibition blockage. When hard beans were soaked in ethanol or methanol, they became permeable to water, indicating that the outermost cuticular layer had been modified in some way.

Hard beans are frequently small, and the lack of imbibition of water is related to growing conditions at the time of maturation. For example, hot, dry conditions during ripening promote hard beans (Smith *et al.* 1961).

Wolf *et al.* (1981) have studied the surface of the seed coat by scanning electron microscopy. They found that in most soybeans, but not all, the surface is covered with pits. These indentations do not seem to be related to the uptake of

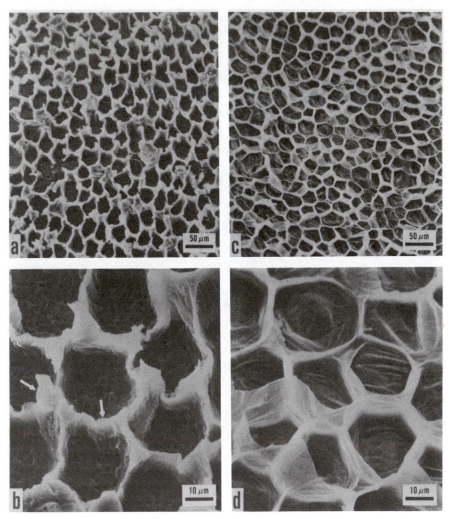

Fig. 2.5.   Seed coat surface (a) and (b) and inner seed pod surface (c) and (d) of PI 339734, a wild-type soybean. Source: Wolf *et al.* (1981).

moisture by the soybean, but their function is unknown (Figs. 2.3 and 2.4). Also, in some soybeans material is deposited in patterns on the seed coat. The patterns correspond with similar patterns found on the inner surface of the seed pod (Fig. 2.5).

Cotyledon

Soybeans are dicotyledonous, meaning that they have two cotyledons. The cotyledons become the first pair of leaves for the young seedling and contain the nutrients required by the seedling before photosynthesis is able to provide the energy and carbon compounds needed for continued growth. The same nutrients that nourish the young seedling are the valuable components of the soybean, because they can also nourish chicks, pigs, dogs, cows, and humans.

The two cotyledons are held together by the seed coat. Once the seed coat is removed, the cotyledons separate readily from each other and from the germ. One of the most common types of damage incurred by soybeans during handling is splits. That is the separation of the cotyledons which in turn leads to deterioration of the oil that is recovered from the split soybeans. The exact cause of the deterioration is not known, but it makes the phospholipid or gum fraction more difficult to remove from the oil.

Figure 2.2 shows diagramatically the structure of the cotyledon epidermis and of the cotyledon cells. These cells are about 15–20 μm in diameter and 70–80 μm long. The predominant features of the cotyledon cells are the protein bodies, the lipid bodies, and cell walls. Of less prominence are the starch granules.

Protein Bodies.   Readily visible with either light or electron microscopy are the protein bodies of soybeans. Figures 2.6 and 2.7 show scanning electron micrographs (SEM) and transmission electron micrographs (TEM) of cotyledon cells. The protein bodies vary in size from 2 to 20 μm but the average size is 8–10 μm. Their shape is spherical. Protein bodies are synonymous with the older term, aleurone grains.

Several isolations of protein bodies have been accomplished (Saio and Watanabe 1966, Tombs 1967, Wolf 1970). The protein bodies contain approximately 60–70% of the total protein in soybeans, and the kinds of proteins found are representative of the proteins generally found in soybeans except that the 2 S fraction is usually missing. The terminology for protein classes is explained in the next section.

As can be seen in Fig. 2.7, lipid bodies tend to adhere to the protein body membrane (visible in the lower right of Fig. 2.7). This adherence may explain why analysis of isolated protein bodies usually includes small amounts (1–5%) of lipid. The protein content of the isolated protein bodies ranges from 60 to 90%, but the best data are for protein body isolates prepared by Tombs (1967), which contained 80–90% protein.

A unique structural feature of protein bodies is that they have globoid inclusions (Lott and Buttrose 1978, Bair 1979). Elemental analyses of the inclusions by energy-dispersive X-ray techniques have shown that they contain considerable phosphorus. This finding has led to the speculation that the inclusions may be phytic acid. The speculation may be true, but the affinity between protein and phytic acid occurs also at the molecular level. When protein bodies are dispersed by breaking the cellular structure of the cotyledon, phytic acid still remains bound to the proteins in solution. Figure 2.8 shows globoid inclusions in soybean protein bodies.

Protein bodies are fragile, and unless appropriate conditions are provided in an

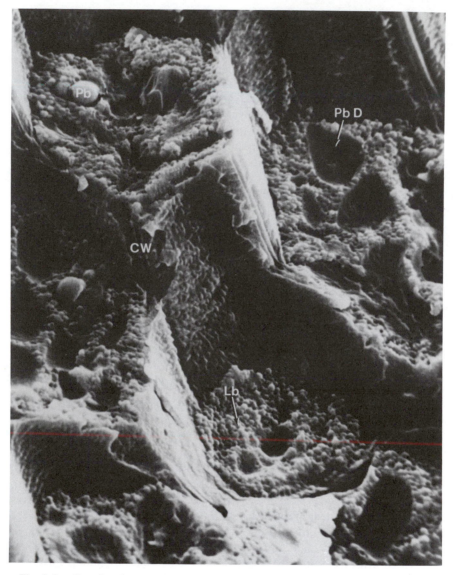

Fig. 2.6.   Scanning electron micrograph of freeze-fractured soybean cotyledon tissue. Cell wall (CW), protein bodies (Pb), and lipid bodies (Lb) are the predominant features. Source: Bair (1979).

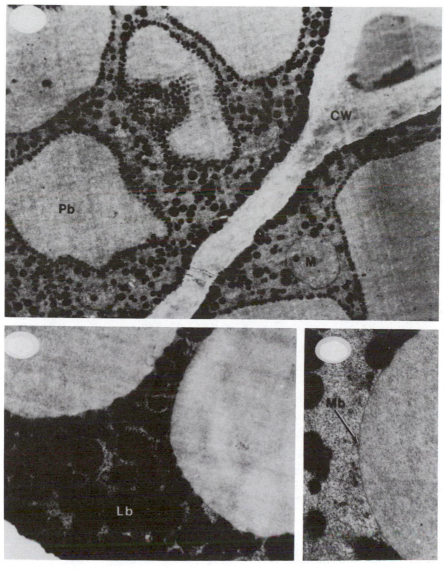

Fig. 2.7. Transmission electron micrographs of subcellular structure in soybean cotyledon cells. Pb, protein body; CW, cell wall; M, mitochondrion; Lb, lipid body; Mb, protein body membrane. Source: Bair (1979).

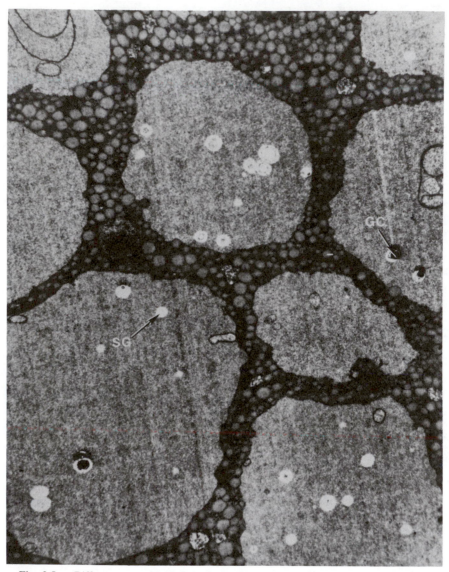

Fig. 2.8.   Different types of inclusions observed in soybean protein bodies. SG, soft globoid; GC, globoid crystal. Source: Bair (1979).

isolation medium, they will lyse as soon as the cell membrane is broken. The lysis will not occur, however, if the intact cells are heated before disruption. Heating hydrated soybeans in boiling water for 3–5 min is sufficient to heat fix protein bodies, and they will survive cell disruption (Johnson and Snyder 1978).

Lipid Bodies. The next most prominent feature of cotyledon cells is the lipid bodies. These structures have been called spherosomes by most investigators and such structures have been found in most plant tissue although they are usually found in lower concentration than in soybean cells. The term "spherosome" has also been used for cytological inclusions other than lipid containing organelles (for example, for glyoxysomes and for lysosomes). Spherosome means literally "round body" and therefore is not very descriptive. Since there is some controversy about spherosomes and since the word does not denote the lipid content of the organelles, we prefer the term "lipid bodies" for these lipid-containing inclusions of oilseeds.

The lipid bodies are a prominent cytological feature of soybeans, but they are not readily seen in the light microscope because of their small size. In the soybean lipid bodies usually fall in a size range of 0.2–0.5 μm. In contrast, the lipid bodies of other plant tissues are often in the 1–2 μm range. Figures 2.6 and 2.7 show the appearance of lipid bodies *in situ* by transmission electron microscopy. A characteristic feature of lipid bodies is that they seem to adhere to other cytoplasmic inclusions. The soybean lipid bodies adhere to the plasmalemma or cell membrane and to protein bodies, but they do not adhere to mitochondria or to nuclei.

To our knowledge, there are only two studies in the literature reporting on the isolation and characterization of soybean lipid bodies. This is surprising in view of the tremendous economic importance of soybean oil. The studies by Kahn *et al.* (1960) and by Bair and Snyder (1980) do not agree on their analysis of the lipid content of soybean lipid bodies. Kahn *et al.* (1960) reported 89% lipid and 10% protein, whereas Bair and Snyder (1980) found 58% lipid and 32% protein. The protein is thought to be due to an integral membrane surrounding the lipid bodies rather than randomly adhering protein. Part of the reason for the discrepancy in lipid content may be due to the size of the lipid bodies isolated. Jacks *et al.* (1967) isolated lipid bodies from peanuts (groundnuts) and found a 98.1% lipid content. The lipid bodies of groundnuts are much larger than those of soybeans; hence the protein membrane would make up a smaller proportion of the weight of the lipid body. Bair and Snyder (1980) found that lipid bodies from soybeans had a wide range of densities depending on their size. Some of the smallest lipid bodies would actually sediment in a centrifugal field, because the high proportion of protein membrane to lipid content would give them a density greater than the suspending aqueous medium.

Starch Grains. Starch grains are prominent in soybean seeds during the development stage but tend to disappear during maturation. Some investigators

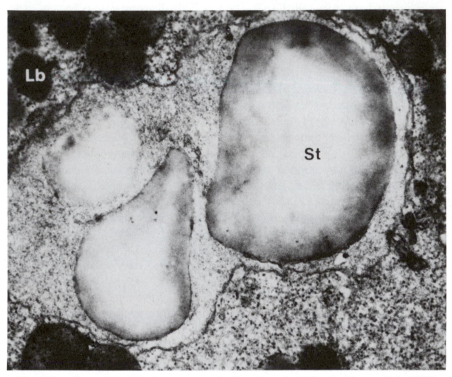

Fig. 2.9. Transmission electron micrograph of soybean starch *in situ*. Source: Bair (1979).

have looked for starch microscopically in mature beans and have not been able to find it, leading to the conclusion that it is not present (MacMasters *et al.* 1941). Wilson *et al.* (1978) found that a small amount of starch (about 0.5%) was always present in the varieties they investigated, but that it was not uniformly distributed. The starch was invariably concentrated near the midline of the bean; hence sampling the tissue near the outer edge would lead to the conclusion that none was present. Starch grains are small in soybeans (1–7 μm) with two to three grains per amyloplast. There is nothing unusual about the starch with respect to amylose to amylopectin ratio or gelatinization temperature. Figure 2.9 shows the appearance of soybean starch by scanning electron microscopy.

Although the amount of starch in soybeans is small, there are situations in which that amount may have significance. The viscosity of soymilk is an important quality criterion, and the viscosity is usually high compared with cow's milk of the same solids content. Denatured protein is thought to be the cause of the high viscosity, but 0.5% starch could contribute considerable viscosity after gelatinization.

Cell Walls.   Another prominent structural feature of the soybean cotyledon cells is the cell wall (Figs. 2.6 and 2.7). Cell walls are composed primarily of cellulose, hemicellulose, and lignin, and the intercellular material is pectin. The insoluble cell wall material is usually removed by filtering during the preparation of soymilk or soy curd.

One of the principal reasons for cooking beans before consumption is to soften them. Most beans have considerable starch content, and the gelatinization of starch during cooking is the main factor causing softening. This is not true of soybeans. Since soybeans do not contain appreciable starch, they do not soften readily during cooking. The factor controlling softening of soybeans is pectin. To speed up softening of soybeans, it is useful to add phosphates or other salts in cooking that will chelate calcium. The binding of calcium causes pectic substances to break down and allows the cellular structure of soybeans to loosen. The result is decreased cooking time to prepare suitably softened soybeans.

Germ

The third part (in addition to the seed coat and cotyledon) of the soybean seed is the germ or hypocotyl, which will become the new soybean plant upon germination. Relatively few studies related to the food or feed value of the germ have been done. In the usual processing methods, the germ may be separated with the cotyledon or with the hull, depending on which structure it adheres to after cracking the soybean.

Generally the germ contains about 10% less fat and 10% more insoluble carbohydrate than the full-fat cotyledon. By weight the germ is about 2.5% of the seed.

There have been reports that the germ is the source of beany off-flavors, and some processors of soymilk have tried to remove the germ to avoid off-flavors in soymilk, but there is little evidence to support this idea.

CHEMICAL COMPOSITION

Proximate chemical composition of soybeans can vary depending on the variety and the growing conditions, but reasonable average figures are 40% protein, 20% lipid, 35% carbohydrate, and 5% ash on a dry weight basis (Fig. 2.10). The water content of stored mature soybeans is usually about 12–14% to insure storage stability.

Water

As the soybean matures, water is lost, and the seed coat shrinks away from the pod. Also, to a lesser extent the cotyledons shrink away from the seed coat,

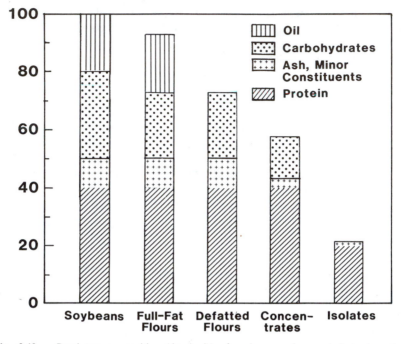

Fig. 2.10.   Proximate composition (dry basis) of soybeans and several derived products. Courtesy of W. J. Wolf.

creating wrinkles. This process may be related to the inability to imbibe water found in hard-seeded soybeans and discussed in the section on the seed coat.

The moisture content at the time of harvest of soybeans is important in their subsequent handling. If the beans are too moist, considerable energy will have to be expended to dry them. If the beans are too dry, they may split during harvest, which leads to deteriorated oil quality and increased refining costs.

Ideally moisture should be about 13% at harvest, but this is difficult to achieve. The moisture content will vary with the time of day, and if beans are too dry, harvesting at night or in early morning when dew is present can prevent some harvest loss. Generally soybeans will have about 15% moisture 5 days after leaf drop if good weather prevails (Tanner and Hume 1978).

Mold will develop in long-term storage if the moisture content is too high. As a result of the metabolism and growth of the mold, moisture and temperature will both increase and thereby further deteriorate the storage conditions.

The ideal moisture level for long-term storage of soybeans (12%) is a little lower than that recommended for other nonoil seeds such as cereal grains. The difference is probably due to the oil content, which should have no or little interaction with water. Hence the water content in relation to the weight of the nonoil constituents would be the important figure. Since soybeans are about 20%

Table 2.1

Typical Composition of Crude and Refined
Soybean Oil[a]

	Crude	Refined
Triglycerides (%)	96	>99
Phospholipids (%)	2	0.03
Free fatty acids (%)	0.5	<0.05
Unsaponifiables (%)	1.6	0.3
Iron (ppm)	2	0.2
Copper (ppm)	0.04	0.04

[a]Source: Pryde (1980a).

oil, the moisture content in equilibrium with the nonoil constituents would be 15% for 12% moisture in whole soybeans. At this moisture content there would be little free water. All the water would be bound to either carbohydrate or protein: bound in the sense that there would be no freezable water in the soybean.

Lipid

One of two valuable components of the soybean is its oil. The crude oil as extracted by hexane has a composition shown in Table 2.1. Most of the crude oil is triglyceride, but minor components are present to some extent, depending on the extraction system. For example, newly developed and experimental techniques for extracting soy oil with carbon dioxide under pressure have shown that the extracted oil has less phospholipid than normally extracted oil. Also the amount of oil extracted from the soybean (usually about 0.5% remains in the defatted flake) will influence the composition. The final portion extracted is rich in phospholipids.

Refined oil composition is also shown in Table 2.1. Note that most minor constituents are decreased as a result of refining, and the triglyceride fraction is increased. However, the refined oil is not 100% triglyceride, and remaining minor components can have an influence on the color and flavor stability of the refined oil.

Fatty Acids.   The triglycerides in any vegetable oil are esters of fatty acids and glycerol. The structures and nomenclature of the predominant fatty acids in soybean oil are shown in Table 2.2. It is useful to become familiar with both common and systematic names of the fatty acids because both are frequently used.

The fatty acid composition of each fat or oil is unique and plant oils are predominantly composed of unsaturated fatty acids. Soybean oil fits this pattern

Table 2.2

Nomenclature and Formulas of the Common Fatty Acids Found in Soybean Oil

Common name	Systematic name	Symbol	Formula
Saturated			
Palmitic	Hexadecanoic	C16	$CH_3(CH_2)_{14}COOH$
Stearic	Octadecanoic	C18	$CH_3(CH_2)_{16}COOH$
Unsaturated			
Oleic	cis-9-Octadecenoic	C18:1	$CH_3(CH_2)_7\overset{H}{C}{=}\overset{H}{C}(CH_2)_7COOH$
Linoleic	cis-9,cis-12-Octadeca-dienoic	C18:2	$CH_3(CH_2)_3\left[CH_2\overset{H}{C}{=}\overset{H}{C}\right]_2(CH_2)_7COOH$
Linolenic	cis-9,cis-12,cis-15-Octadeca-trienoic	C18:3	$CH_3-\left[CH_2\overset{H}{C}{=}\overset{H}{C}\right]_3(CH_2)_7COOH$

with about 80% of its fatty acids being unsaturated. Table 2.3 compares soybean oil fatty acids with those of several other prominent food oils. Note that oleic and linoleic are the predominant fatty acids in all of these plant oils. In all oils there are trace amounts of other fatty acids (even fatty acids with uneven number of carbon atoms), but these have been omitted from Table 2.3.

In all of the oils of Table 2.3, the predominant fatty acids are palmitic, stearic, oleic, and linoleic. In four of the oils (soy, cottonseed, corn, and sunflower) linoleic is the predominant unsaturated fatty acid, and in four of the oils (low-erucic rape, palm, peanut, and olive) oleic is predominant.

Soy and low-erucic rape oils are unique in having about 9% linolenic acid, which oxidizes much faster than oleic acid. The relative rates of oxidation of

Table 2.3

Comparison of Fatty Acid Compositions of Several Edible Oils[a]

Fatty acids	Soy	Cotton seed	Corn	Rape (low erucic)	Sunflower	Palm	Peanut	Olive
Saturated								
Myristic C14	1		1			1		
Palmitic C16	11	29	13	5	11	48	6	14
Stearic C18	4	4	4	2	6	4	5	2
Arachidic C20							2	
Unsaturated								
Hexadecenoic C16:1		2						2
Oleic C18:1	25	24	29	55	29	38	61	64
Linoleic C18:2	51	40	54	20	52	9	22	16
Linolenic C18:3	9			9				

[a]Source: Sonntag (1979b). Copyright by John Wiley & Sons, Inc.

these three fatty acids are roughly in the ratio 1:10:20 for oleic, linoleic, and linolenic, respectively (Sonntag 1979a). The relative rates of oxidation for the corresponding triglycerides are 1:36:180 for triolein, trilinolein, and trilinolenin, respectively (Park *et al.* 1981). The linolenic acid in soybean triglycerides has been blamed for the relative instability of soy oil to oxidation. As a result, a great deal of effort has been expended to find or to breed a soybean variety with low linolenic acid. At the present time (1986), these efforts have resulted in a variety that has about 4% linolenic acid, but there is no indication that the oil from such a variety is superior to the usual soy oil (Hammond and Fehr 1984). The oxidative stability of soy oil will be discussed further in Chapter 4.

In addition to the fatty-acid composition of the oil, the distribution of fatty acids among the three positions of the glycerol molecule may be important in the stability of the fatty acids to oxidation. The positions of the carbon atoms in the glycerol molecule can be stereospecifically numbered (SN) 1, 2, and 3:

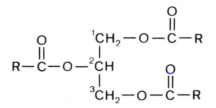

Fatty acids in soy oil are not randomly esterified in the three positions as shown in Table 2.4. The rules for distribution of fatty acids among the triglyceride positions have been stated by Evans *et al.* (1969):

(1) Palmitic and stearic are exclusively assigned to positions 1 and 3.

(2) Oleic and linoleic are treated alike and distributed randomly and equally in all three positions.

(3) All remaining positions are filled by linoleic acid.

Table 2.4

Comparison of Random and Actual Distribution of Saturated and Unsaturated Fatty Acids in Soy Oil[a]

Glyceride class[b]	Random	Analyzed[c]
SSS	0.4	0.07
SUS	2.1	5.2
USS	4.2	0.4
USU	11.2	0.7
UUS	22.4	35.4
UUU	59.7	58.4

[a]Source: List *et al.* (1977).
[b]S, saturated fatty acid; U, unsaturated fatty acid.
[c]Determined by lipase hydrolysis.

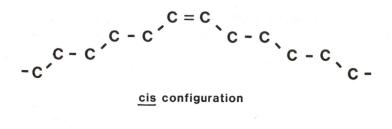

cis configuration

trans configuration

Fig. 2.11.   The effects of cis and trans double bonds on the linearity of fatty-acid hydrocarbon chains.

Whether lipids are liquids (oils) or solids (fats) depends on their fatty-acid composition. Two properties of fatty acids influence the melting point. As chain length increases, it becomes more probable that hydrocarbon chains interact and hence the melting point increases with increasing chain length. As the degree of unsaturation increases, it becomes more difficult for hydrocarbon chains to interact and melting point decreases.

The influence of unsaturation on melting point is modified depending on whether the double bond is cis or trans. As shown in Fig. 2.11, a cis double bond disrupts the linearity of a hydrocarbon chain much more than a trans double bond. Hence cis double bonds make it more difficult for hydrocarbon chains to align as crystals and cause the melting point to decrease more than trans double bonds. Cis double bonds exist almost exclusively in crude soy oil, but during hydrogenation trans double bonds are formed and tend to raise the melting point even more than would be expected from the loss of unsaturation.

The saponification of a vegetable oil refers to treatment of the oil with strong base to cause hydrolysis of the triglyceride to glycerol and fatty acid salts (soaps). If the hydrolysis is done quantitatively, one can calculate a saponification value, defined as the amount (mg) of KOH required to saponify 1 g of oil. The saponification value is useful in judging the chain lengths of the fatty acids in the oil. If the oil has a large number of short-chain fatty acids, the saponification value will be large, and conversely will be smaller if the number of long-chain fatty acids increases. Of course, an increase in the nonsaponifiable content of an oil or fat would also lower the saponification value. For crude soybean oils, saponification values are 190–195.

The unsaponifiable fraction of soybean oil (Table 2.1) includes hydrocarbons such as squalene (a biosynthetic precursor of phytosterols), phytosterols, carotenoids, and tocopherols. These oil-soluble components are probably found in the lipid bodies of the intact soybean cells, but evidence for this is lacking. All we know at present is that these compounds are extracted from the soybean by hexane. The chemistry and significance of the components of the unsaponifiable fraction will be discussed later with other minor components.

The iodine value (IV) is the specification used to judge the degree of unsaturation in oils. IV is defined as the percentage by weight of iodine needed to saturate an oil or fat. Unhydrogenated soy oils have IVs in the range 125–135. After partial hydrogenation for stabilizing soy oil to oxidation, the IVs may be reduced to 110–115. IVs of the hard-fat portion of plastic shortenings may be 0–15.

Other chemical measures of the fatty-acid composition of fats and oils based on volatility and solubility have been used in the past, but with the advent of gas–liquid chromatography (GLC) these chemical tests have become less important. It is now possible to hydrolyze the oils or fats to produce free fatty acids, to convert the free fatty acids into volatile esters, and to analyze completely the fatty acid composition by GLC. This information is far more comprehensive than that obtained from limited chemical tests. In fact, IVs and saponification values may be calculated from fatty acid composition.

As shown in Table 2.1, a portion of the crude soybean oil, about 0.5%, is free fatty acids. Either a small amount of hydrolysis of triglycerides occurs during disruption and extraction of the soybeans, or there may be free fatty acids in the soybean that are extracted with the triglyceride portion. The amount of free fatty acids can be determined by titration of the oil in an alcoholic medium.

Free fatty acids are not desirable components of a food oil, particularly if used for frying, because the smoke point is lowered. The free fatty acids are removed by the alkali-refining step of crude oil refining. Formerly, the free fatty acids were recovered by acidifying and washing to produce soap stock for manufacture of soaps. While this is still being done, because of the large volume of soybean oil being processed, more free fatty acids exist than can be used by the soap industry. Often the free fatty acids are recovered as a feed ingredient to be added to defatted soybean flakes.

Phospholipids. The major nontriglyceride fraction in crude soy oil is the phospholipids. Phospholipids are saponifiable. The phospholipid fraction is heterogeneous and is removed from crude oil by a water washing step referred to as degumming. Soy oil contains 1–3% phospholipid, which is high in relation to other vegetable oils, which usually have less than 1%. The relatively high content

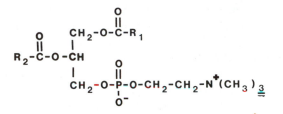

phosphatidyl choline

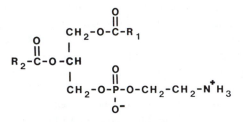

phosphatidyl ethanolamine

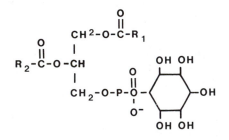

phosphatidyl inositol

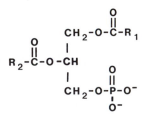

phosphatidic acid

Fig. 2.12.   Structural formulas of the common phospholipids of soybean oil. R_1 and R_2 are C_{15} and C_{17} hydrocarbon chains.

of phospholipid in soybeans could be a consequence of very small lipid body size. The small lipid bodies have more membrane surface than large lipid bodies (for the same total lipid), and phospholipids are located in membranes of various types.

Figure 2.12 shows the structures of the main phospholipids found in soybean oil: phosphatidyl choline, phosphatidyl ethanolamine, phosphatidyl inositol, and phosphatidic acid. Phosphatidic acid is probably a derived product from hydrolysis of one of the other phospholipids.

Total amounts of these four phospholipids in soy oil can be calculated based on the total phosphorus content of soy oil. Of the total phospholipid content, there is about 35% phosphatidyl choline, about 25% phosphatidyl ethanolamine, about 15% phosphatidyl inositol, 5–10% phosphatidic acid, and the remainder is a composite of all the minor phospholipid compounds.

The term lecithin is used in two distinct ways in referring to soybean phospholipids. In one usage, it is the common name for phosphatidyl choline, but "lecithin" is also used as a common name for the entire phospholipid fraction separated from soybean oil by degumming. Usually in this book soybean lecithin will be used in the latter sense.

The fatty-acid composition of soybean phospholipids is roughly the same as the fatty acid composition of the triglycerides.

The degumming step is not 100% effective in removing phospholipids, and much of the remaining phospholipid is removed by the subsequent alkali refining. The phospholipid that is not removed by degumming is sometimes referred to as nonhydratable phospholipid. There has been considerable speculation as to what compounds are not hydrated, and there is evidence that calcium and magnesium salts of phosphatidic acid are the nonhydratable fraction (Hvolby 1971). Nonhydratable phospholipid is increased in soybeans that have been damaged in handling or by early frosts in the field.

Table 2.1 shows that an average figure for phospholipid content of crude soy oil is 2%. Often phospholipid is measured by determining the phosphorus content of an ashed oil sample and then multiplying by a factor (31.7) to convert to phospholipid. Also data on phospholipid content of oils are sometimes reported as ppm phosphorus. A crude oil would contain approximately 700 ppm phosphorus, and a refined oil would contain approximately 10 ppm phosphorus.

The phospholipids in crude soy oil are only about half of the total phospholipid content of the soybean. They probably are extracted primarily from cellular and subcellular membranes. The phospholipid content of the hexane extract increases with time of extraction.

The phospholipids have to be removed from soy oil before it is deodorized. The high heat of deodorization will cause color development in the presence of phospholipids that is undesirable and difficult to remove.

Further discussions of the chemistry of soybean lipids in relation to flavor deterioration due to oxidation and in relation to nutrition appear in later chapters.

Table 2.5
Physical Properties of Soybean Oil[a]

Property	Value
Specific gravity, at 25°C	0.9175
Refractive index n^{25}	1.4728
Viscosity, at 25°C	50.09 cP
Solidification point	-10 to -16°C
Specific heat	0.458 cal/g 19.7°C
Heat of combustion	9478 cal/g
Smoke point	453°F (234°C)
Flash point	623°F (328°C)
Fire point	685°F (363°C)

[a]Source: Pryde (1980b).

Physical Properties. Some of the physical properties of the soybean lipid are useful for judging chemical properties, and others are useful for judging the quality and changes in quality of soybean food oils.

Table 2.5 lists some of the physical properties of soybean oil and their specific values. Refractive index is a useful property because it reflects the fatty-acid composition of an oil. As the chain lengths of fatty acids lengthen, the refractive index increases, and as the degree of unsaturation increases, the refractive index increases. Consequently, refractive index can be used as a distinguishing feature of an oil for identification purposes. For crude oils, free fatty acids, phospholipids, and the unsaponifiable fraction can affect the refractive index, and so it is most reliable as a property of a refined oil.

Viscosity is a property that is used by engineers to design pipes, pumps, and heat exchange equipment for extraction and solvent removal of soy oil.

The solidification point reflects the fatty-acid composition and properties discussed in the section on fatty acids. The solidification point is usually a few degrees lower than the melting point and can vary, depending on the previous history of the oil. For a rapidly cooled oil, the solidification point will be higher than for a slowly cooled oil. Melting points and solidification points are important useful characteristics to know about fats and oils, but they are far from being precise measurements. The difficulty comes from the nature of fats and oils. The triglycerides found in soy oil are a mixture of many different possible combinations of the five main fatty acids esterified at the three hydroxyl positions of glycerol. Consequently, there can be no sharp melting point as would be found for a pure compound. Furthermore, oils are known to crystallize in several different forms, designated by α, β, and β'. The β and β' crystals are most commonly encountered with the β form being most stable and characteristic of soybean, safflower, sunflower, sesame, corn, olive, coconut, and palm kernel oils. Less stable but more desirable because of their fine structure are β' crystals, typical of cottonseed, palm, and rapeseed oils. The β' crystals are thought to be due to a wide variation in chain lengths of the fatty acids that prevent large crystal growth. The largest crystals are α, but they are also the least stable.

The melting point can be determined just as for any organic compound by putting the sample into a glass capillary, sealing the capillary, and chilling the sample for several hours to insure complete crystallization. For reasons noted above, oils can be slow to crystallize. The sample is then heated slowly in a water bath, and the melting point is that temperature at which all crystals have disappeared. This is the method recommended by the Fat Analysis Committee (FAC) of the American Oil Chemists' Society (AOCS method Cc 1-25).

Another widely used method is called the Wiley melting point (AOCS method Cc 2-38), in which the oil is cast in a disk-shaped mold and chilled. The disk of fat is suspended in an alcohol–water bath and heated slowly until it forms a sphere. This is the endpoint, and the temperature is the Wiley melting point. Obviously, considerable judgment is required in this test, and reproducibility between labs may be difficult. Since the fat does not melt in this test, the Wiley melting point will always be lower than the capillary melting point.

Of more interest to manufacturers of shortenings and margarines is the solids contents of these products at several temperatures. Most fats for human consumption are plastic blends of crystalline fat and oil, and the solids content is useful in judging the texture, spreadability, and mouth feel of these products.

The solids fat index (SFI) is one method of judging texture. Solids content is measured at 10, 21.1, 26.7, 33.3, 37.8, and 40°C (50, 70, 80, 92, 100, and 104°F) by dilatometry (AOCS method Cd 10-57). Specific volume is measured by the sample expanding in a carefully controlled volume at a carefully controlled temperature to give a reading on a capillary scale. The specific volume can then be plotted vs. temperature as shown in Fig. 2.13 to give an approximation to SFI. Since the solids line is difficult to establish, a line parallel to the liquid line is drawn exactly 0.1 unit below the liquid line. The SFI at the appropriate temperature can then be calculated as BC/0.1.

A far easier method to measure the solids content in a plastic fat is by nuclear magnetic resonance (NMR). The NMR signal can be tuned so that only the liquid portion of the sample absorbs the signal. Thus the measurement is considerably simplified, but the measurement instrument is considerably more expensive. This modification of the method for measuring solids is AOCS recommended practice Cd 16-81 and is called solid fat content (SFC) rather than SFI.

Smoke, flash, and fire points are important quality criteria of finished oils and reflect largely the free fatty-acid content. Also, the flash point is used as a quality criterion of crude oil to judge the extent to which hexane has been removed.

The smoke point is determined as the temperature at which the sample gives off a thin, bluish stream of smoke continuously under carefully controlled heating and viewing conditions. The flash point is the temperature at which an ignition flame causes the oil to burn with a brief flash. The fire point is the

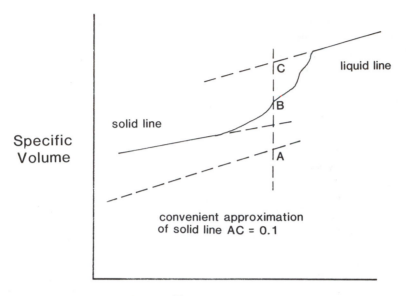

Fig. 2.13.   A convenient method for calculation of SFI. The dashed solids line is constructed parallel to the liquid line and 0.1 unit below. Source: Weiss (1983).

temperature at which an ignition flame causes sustained burning for at least 5 sec. These procedures are described in AOCS method Cc 9a-48. In contrast to the flash point shown for refined soybean oil in Table 2.5 (328°C, 625°F), the minimum flash point acceptable for crude soybean oil is 121°C (250°F).

Much information about oils, the impurities present in them, and the effects of processing steps can be obtained from ultraviolet (UV), visible, and infrared (IR) spectrophotometry.

The pure triglyceride portion of soy oil does not have specific absorption peaks in the UV region of the spectrum but has increasing absorption as the wavelength decreases. The absorbancy of a 1% soy oil solution in hexane (1 cm path length) at 233 nm is about 1.0. Figure 2.14 shows typical UV spectra for pure triglycerides, for conjugated dienes, and for conjugated trienes.

Conjugated dienes and conjugated trienes develop in soybean oil as a result of oxidation of polyunsaturated fatty acids and as a result of the bleaching and hydrogenation steps. Conjugated dienes form when a hydroperoxide is produced as an autooxidation product of polyunsaturated fatty acids still esterified to glycerol or as a result of isomers formed during hydrogenation. The single absorbance peak at about 233 nm is strongly absorbing, with a molar absorptivity of about 25,000. The conjugated triene peaks can be the result of an oxodiene compound or conjugated trienes resulting from bleaching or hydrogenation. Con-

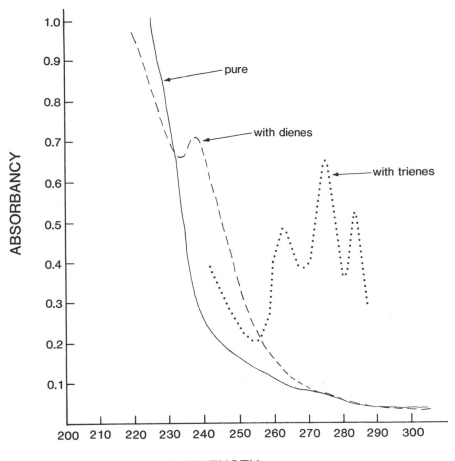

Fig. 2.14. Ultraviolet absorption spectra of pure triglyceride, triglyceride with conjugated dienes, and triglyceride with conjugated trienes.

jugated tetraene spectra also are sometimes visible in soybean oils with peaks in the 300- to 320-nm region.

Tocopherols, which are invariably present in soy oils, have single absorption peaks in the 295-nm region, but their concentration is too low to have much effect on the spectrum of the soy oil.

Light absorption in the visible region of the spectrum causes the color of the oil and is a most important quality criterion. The pigments responsible for the color of soy oil are carotenes, xanthophylls, chlorophyll, and occasionally some nondescript brown pigments developed during mishandling or processing of soybeans or soy oil. Also, a darkening of soy oil can occur as a result of oxidation of colorless tocopherol to chroman-5,6-quinones, red pigments.

The color of soy oil is measured and reported in the Lovibond–Tintometer system. A standardized depth of oil (usually 5.25 in.) is viewed by transmitted light and compared visually with red and yellow glass standards. When a color match is obtained, the color is reported in terms of the units of the red standard. For soybean oil, a ratio of 10 yellow to 1 red is recommended up to 3.9 red, and a ratio of 70 yellow to 1 red for 3.9 red and above. National Soybean Processors Association (NSPA) standards for crude oil call for a maximum of 3.5 red, with a 1.5% price discount from 3.5 to 6 red.

The color specifications of refined oils and fats depend on their end use. For margarines a Lovibond red of 4–5 may be satisfactory, whereas a red of 1.5 maximum may be needed for shortening.

Spectrophotometry is more useful in determining what pigments are responsible for the color than is the Lovibond system. By making absorbance measurements at several specified wavelengths it is possible to calculate a Lovibond red number (AOCS method Cc 13c-50) or to calculate the chlorophyll content (AOCS method Cc 13d-55). Chlorophyll is an unwanted pigment in soy oil and may be present in the oil as a result of an early freeze, causing a cessation of maturation in green soybeans. Chlorophyll would be most objectionable in oils made into shortening, where the green pigment is readily noticed.

The IR spectrum is most useful in identifying specific chemical groupings that may be present such as hydroperoxides, epoxides, hydroxyls, and carbonyls. It is also possible to differentiate between cis and trans configurations in double bonds by IR absorption. Trans double bonds have a characteristic absorption at 970 cm^{-1} (10.3 μm) that can be used to determine the trans double bond content of an oil. Since hydrogenation increases the number of trans double bonds in soy oil, IR spectrophotometry can be used to determine the extent of absorption and incorporation of trans fatty acids into body fat by an animal.

Protein

Protein is the second major chemical component of the soybean that has commercial value. The protein is used mainly as a feedstuff for animals, but a small quantity in the United States, probably less than 5%, is used directly in human foods. Soybean protein is particularly valuable, because it has an amino acid composition that complements that of cereals. Soybeans are limiting in sulfur-containing amino acids for most animal species, including humans, but contain sufficient lysine to help overcome the lysine deficiency of cereals. Table 2.6 shows the essential amino acid composition of soybeans, wheat gluten, milled rice, corn, and broad beans, another legume. A soybean–rice combination is complementary for lysine and methionine and helps to explain the successful use of soy protein products such as soybean curd in the rice-eating cultures of Asia. The nutritional significance of soy protein will be explored further in Chapter 6.

Table 2.6

Essential Amino Acid Composition of Soybeans, Wheat Gluten, Milled Rice, Corn, and Broad Beans (g/16 g N)

	Soybean[a]	Wheat gluten[b]	Rice[c]	Milled corn[d]	Broad beans[a]
Isoleucine	5.1	3.9	4.1	3.7	4.5
Leucine	7.7	6.9	8.2	13.6	7.7
Lysine	6.9	1.0	3.8	2.6	7.0
Methionine	1.6	1.4	3.4	1.8	0.6
Phenylalanine	5.0	3.7	6.0	5.1	4.3
Threonine	4.3	4.7	4.3	3.6	3.7
Tryptophan	1.3	0.7	1.2	0.7	[e]
Valine	5.4	5.3	7.2	5.3	5.2
Histidine	2.6	1.8	[e]	2.8	2.8

[a]Wolf (1977).   [b]Kasarda et al. (1976).   [c]Inglett (1977).
[d]Wall and Paulis (1978).   [e]Value missing.

The amount of protein in soybeans, 38–44%, is larger than the protein content of other legumes, 20–30%, and much larger than that of cereals, 8–15%. This large quantity of protein in soybeans along with excellent quality increases their value as a feedstuff and is one of the reasons for the economic advantage that soybeans have over other oil seeds.

Proteins of soybeans have been studied after extraction from defatted flakes, and in comparison with proteins extracted from full-fat soybeans, no major differences were found (Hill and Breidenbach 1974). In this instance lipid extraction was done by Soxhlet extraction, but even when soybean flakes are extracted commercially with hot hexane (140°F, 60°C) for 30–40 min, there seems to be no major loss of protein solubility, enzymatic or trypsin inhibitor activity.

As a preliminary step to serious study of individual soybean proteins, it is necessary to separate and isolate the proteins. We first discuss the important separation and isolation procedures and follow that discussion with what has been learned about individual soy proteins.

Separation and Isolation.   The proteins of soybeans, as with those of cereals and other legumes, are for the most part devoid of any specific biological activity. Consequently, plant proteins have been separated and named based on the classical solubility pattern

albumins, soluble in water
globulins, soluble in salt solutions
prolamines, soluble in 50–70% ethanol
glutelins, soluble in dilute acid or base.

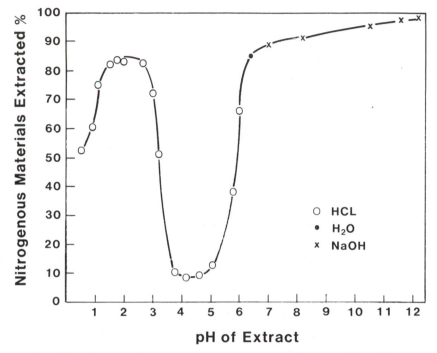

Fig. 2.15.   Extractability of soy protein as a function of pH. Source: Wolf (1978).

Using this oversimplified pattern (limits of solubility are not specified), the major portion of soy proteins are globulins. In contrast, most cereal proteins are prolamines and glutelins.

Although the major fraction of soybean protein is termed globulin, this fraction can be extracted with water. The solubility of soy proteins in water does vary with pH as shown in Fig. 2.15. If no acid or base is added to the extracting water, the pH will usually be about 6.4–6.6, and at this pH range approximately 85% of the soybean protein is extracted. As the pH is raised with the addition of base, more protein is extracted, but the advantage of increased yields of extracted protein is counterbalanced by the disadvantage of protein damage at pH values above 9. The damage to proteins caused by heat and alkali will be discussed in Chapter 6.

As pH is lowered by addition of acid, the solubility of soy protein decreases and reaches a minimum in the region of pH 4.5. This solubility pattern forms the basis for some of the processing steps discussed later for production of soy concentrates, soy isolates, and soy curd.

The globulin or salt-soluble characteristics of soy proteins become apparent at pH 4.5. The solubility increases with increasing concentrations of either calcium chloride or sodium chloride from about 10% at zero ionic strength to 60% at 0.7

ionic strength (Wolf 1977). This increased solubility in the presence of high concentrations of calcium chloride also exists at pH 6–7 and is in direct contrast to the ability of 0.02 M calcium chloride to precipitate soy proteins from water extracts at pH 6.6–7. The precipitation of soy protein by calcium chloride or other divalent cations is the basis for the traditional Oriental soy curd production.

A useful buffer system for extraction of soy proteins was introduced by Wolf and Briggs (1956). It is made up of 0.035 M phosphate (pH 7.6), 0.4 M sodium chloride, and 0.01 M mercaptoethanol with an ionic strength of 0.5. The mercaptoethanol is useful in preventing polymerization of soy proteins through disulfide bond formation during isolation.

The soybean globulins fit into a general pattern for legumes that Danielsson (1949) described for peas consisting of vicilin and legumin. Vicilin and legumin were separated by ultracentrifugation with vicilin being the lighter (7 S) component, and legumin being the heavier (12 S) component. Derbyshire et al. (1976) have reviewed the storage proteins of legumes in general.

Thanh and Shibasaki (1976A) have described a procedure based on solubility for separation of the 7 and 11 S proteins of soybeans. They extract with dilute tris buffer at pH 8, then adjust the pH to 6.4, and centrifuge in the cold. This step causes 11 S protein to precipitate. Next the supernate is adjusted to pH 4.8, which causes the 7 S fraction to precipitate. Currently this is the most effective separation procedure for the major globulin proteins of soybeans.

The cold centrifugation may take advantage of the fact that glycinin does tend to precipitate out of dilute buffers in the cold. This is one of the ways to concentrate the 11 S fraction from a water extract of soybean proteins.

Ultracentrifugation: Soon after Danielsson demonstrated the utility of ultracentrifugation for separating pea proteins, Walter J. Wolf started investigations on ultracentrifugation of soy proteins that in much broader perspective are continuing today. Wolf's extensive use of ultracentrifugation experimentation on soy proteins over a 30 year period at the USDA NRRC in Peoria, Illinois, led to a classification system based on the proteins' relative rates of sedimentation. This system, although imperfect, is probably the most widely known way of designating soy proteins as 2, 7, 11, and 15 S fractions. The S stands for Svedberg units, calculated as the rate of sedimentation per unit field of centrifugal strength: $S = (dx/dt)/\omega^2 x$, where x is the distance from the center of the centrifuge, t is time, and ω is angular velocity.

Values for S range from 1 to 200, with the units being 10^{-13} sec. An ultracentrifuge pattern for a water extract of soy proteins is shown in Fig. 2.16. The terms 2, 7, 11, and 15 S are merely for identification and are not meant to imply that sedimentation constants are always these exact whole numbers.

Analysis of the four fractions separable by ultracentrifugation has shown that the 2 and 7 S fractions are heterogeneous, but the 11 and 15 S fractions probably are each pure proteins. The 2 S fraction includes a 2 S globulin, trypsin inhibitors, and cytochrome C. The 7 S fraction includes α-amylase, lipoxygenase,

2   7   11   15

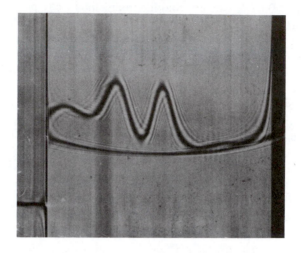

Fig. 2.16. Ultracentrifugal pattern for water-extractable soybean proteins. Numbers are sedimentation coefficients in Svedberg units. Source: Wolf (1970). Copyright by the American Chemical Society

hemagglutinin, and several 7 S globulins. The 11 S fraction is the soybean legumin generally known as glycinin, and the 15 S fraction is thought to be a polymer of glycinin.

Dissatisfaction with the use of the 2, 7, 11, and 15 S designations led to the formation of a Soybean Protein Nomenclature Committee under the Oilseeds Division of the American Association of Cereal Chemists in 1967. One interim report was issued (Wolf 1969), but there was no consensus on a comprehensive system, and the committee is no longer active.

Protein bodies: Another approach to separation of soybean proteins is to isolate those structures known to contain most of the protein, the protein bodies. Several investigators have followed such a separation procedure, and Wolf (1978) has summarized their data. The composition of protein bodies varies widely depending on the method of isolation. Tombs (1967) has isolated a protein body fraction by sucrose density gradient centrifugation that contains 97.5% protein. Furthermore, his data indicated that at least 60–70% of soy protein is located in protein bodies.

Ultracentrifugal analysis of the proteins in protein bodies shows that all four ultracentrifuge fractions are present, but the 2 S fraction is smaller from protein bodies than from water extracts.

Protein bodies are extremely fragile and readily lyse when soy cotyledon cells are ruptured in water. Adjustment to pH5 is effective in minimizing disruption of protein bodies. Saio and Watanabe (1966) solved the lysis of protein bodies by isolating them in a cottonseed oil–carbon tetrachloride mixture, but could only account for 82.7% of the composition of the isolated protein bodies. Oil contamination of the protein bodies was probably a problem with their procedure.

Electrophoresis: Various electrophoretic techniques have been useful in analyzing soy proteins and have contributed to soy protein nomenclature. Most useful have been discontinuous polyacrylamide gel electrophoresis for analyzing subunit composition of various globulins and isoelectric focusing for obtaining isoelectric points.

During the late 1960s, N. Catsimpoolas working at Central Soya laboratories in Chicago did extensive studies of soybean proteins, making use of electrophoresis combined with immunochemical methods. Immunochemical techniques are particularly helpful in identifying and differentiating proteins that do not have specific biological activity. Catsimpoolas and Ekenstam (1969) reported isolation and characterization of three globulin proteins, named α-, β-, and γ-conglycinin based on their rate of migration in gel electrophoresis. This study and others from the laboratories of I. Koshiyama (Koshiyama and Fukushima 1976a,b) and of V. H. Thanh and K. Shibasaki (1978) in Japan have further clarified the identity and structures of the various globulin proteins in soybeans. α-Conglycinin is a globulin of low molecular weight found in the 2 S fraction. The β-conglycinin is the major protein in the 7 S fraction that has the property of dimerizing when ionic strength is lowered from 0.5 to 0.1. The γ-conglycinin fraction is the 7 S globulin that does not dimerize when the ionic strength is lowered.

Hill and Breidenbach (1974) isolated soy proteins by centrifugation in a sucrose gradient under conditions such that a 7 S component (0.5 ionic strength) was isolated, and in a separate experiment, a 9 S component (0.1 ionic strength) was isolated. Both the 7 and 9 S fractions gave identical patterns by discontinuous polyacrylamide electrophoresis showing three components in a nondissociating buffer system. More recent studies have shown seven components for β-conglycinin (see below).

Column chromatography: After isolation of the major soybean protein components, further analysis can be done by ion exchange, gel filtration, and affinity chromatography. By manipulating pH in a pH gradient, charges of soy protein fractions can be changed to achieve separations on diethylaminoethyl (DEAE) or carboxymethyl (CM) cellulose. By combining ion exchange with size separation using agarose or polyacrylamide gels, the separation can be improved. Also, since the 7 S soy proteins contain carbohydrate, affinity chromatography is useful.

Individual Proteins.   Following are brief characterizations of the more widely studied soybean proteins.

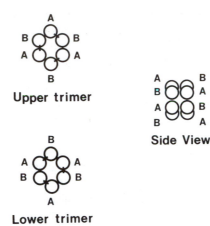

Upper trimer

Side View

Lower trimer

Fig. 2.17.   Quaternary structure of the 6 subunits of glycinin. A and B refer to acidic and basic polypeptide chains.

Glycinin: This is the predominant protein in soybeans and derives its name from the genus name of the plant, Glycine. In the terminology we are using, glycinin is identical to the 11 S fraction.

Peng *et al.* (1984) have reviewed the chemistry and functionality of soybean glycinin. Hughes and Murphy (1983), after studying ten varieties of soybeans, found that glycinin makes up about 35% (31.4–38.3%) of the total protein.

The molecular weight of glycinin is about 350,000, which puts it in the general class of legumins, the higher molecular weight class of the two classes of legume storage proteins.

Glycinin quaternary structure has been studied extensively, and the picture that emerged is of two trimers each consisting of three acidic polypeptide chains with molecular weights of approximately 37,000 and three basic polypeptide components with molecular weights of approximately 18,000. The polypeptide chains are held together with disulfide and hydrogen bonds that can be disrupted by urea, strong acid, strong base, heat, or sodium dodecylsulfate in combination with a disulfide reducing agent. The two layers of trimers with acidic and basic peptides alternating, as shown in Fig. 2.17, make up the glycinin molecule.

The acidic and basic polypeptide chains have been separated and studied. In one such study, Mori *et al.* (1981) isolated seven acidic chains and eight basic chains from 18 different cultivars of soybeans. They classified the 18 cultivars into five groups based on their content of the various glycinin polypeptide chains. With this many different acidic and basic polypeptides, it is obvious that glycinin molecules must differ, but there is relatively little information on glycinin heterogeneity.

Work on the genetics of glycinin biosynthesis has shown that certain specific acidic and basic chains are invariably linked. Thus there is not random association between the acidic and basic chains. This linkage results because both acidic and basic chains are synthesized from a single polypeptide chain precursor (Tumer *et al.* 1981, 1982). After synthesis, the peptide bond linking the acidic and basic chains is cleaved, but they remain associated through disulfide and hydrogen bonds.

An interesting aspect of the genetic control of glycinin quaternary structure is that pairs of acidic and basic chains do not have the same methionine content, differing by as much as eightfold (Staswick *et al.* 1981). Thus by selection of the germplasm with subunits high in sulfur amino acids, it may be possible to increase methionine content of glycinin and hence of soybeans. By contrast, β-conglycinin is very low in or devoid of sulfur-containing amino acids (Thanh and Shibasaki 1978).

As defatted soy meal is stored for months or years, there is a decrease in the amount of protein that can be solubilized (Saio *et al.* 1982). The 11 S fraction, or glycinin, is the soy protein that shows the greatest decrease in solubility during storage, and the decrease is generally attributed to formation of disulfide polymers.

The isoelectric point of glycinin is 4.64. This and other physical properties such as absorbancy at 280 nm, viscosity, diffusion constant, and partial specific volume have been studied and reported by Koshiyama (1972). For a review of other details of the structure and chemistry of soybean globulins see Wolf (1976).

β-*Conglycinin:* This is the name given to the main component of the 7 S fraction, which is the soybean equivalent of vicilin, the lower molecular weight general class of legume globulins. Furthermore, β-conglycinin is the soybean globulin that dimerizes when the ionic strength is lowered from 0.5 to 0.1. This causes the 7 S ultracentrifugation peak to shift to approximately 9 S.

As with glycinin, β-conglycinin has quaternary structure. Some have reported as many as nine subunits (Koshiyama 1968), but it is generally accepted now that there are three subunits with the designations α, α', and β (Thanh and Shibasaki 1978). In contrast to the lack of knowledge about glycinin heterogeneity based on subunit structure, β-conglycinin heterogeneity has been well studied (Thanh and Shibasaki 1976b). Seven different β-conglycinins have been separated by ion exchange chromatography and have been given the designations of B_0 through B_6.

The relationship between the different β-conglycinins and their subunit or quaternary structure is as follows:

β-Conglycinin	Subunits
B_0	3 β
B_1	1 α', 2 β
B_2	1 α, 2 β
B_3	1 α, 1 α', 1 β
B_4	2 α, 1 β
B_5	2 α, 1 α'
B_6	3 α.

Koshiyama (1969) showed that 7 S protein was a glycoprotein with 12 glucosamine and 39 mannose residues per mole of protein. This is another distinguishing feature between 7 and 11 S proteins (11 S does not contain carbohydrate).

Other globulins: The α-conglycinin (2 S globulin) isolated by Catsimpoolas and Ekenstam (1969) migrated in electrophoresis approximately the same as Kunitz trypsin inhibitor, but the authors cited unpublished results showing trypsin inhibitor and α-conglycinin could be distinguished immunochemically. I. A. Vaintraub, a Russian biochemist, has studied 2 S globulins from soybeans, but his results cited by Wolf (1978) on amino acid composition and N terminal amino acids make it difficult to see a clear distinction between trypsin inhibitors and 2 S globulins. Koshiyama *et al.* (1981) studied the 2 S fraction but were unable to find any proteins in the fraction that did not have trypsin inhibitor activity. Obviously they were only looking at major protein fractions (not minor components such as cytochrome C), but the existence of a 2 S globulin is in doubt.

There seems to be agreement that the 7 S globulin fraction contains globulins other than the dimerizing β-conglycinin. Catsimpoolas and Ekenstam (1969) called the other 7 S globulin "γ-conglycinin" and said that it was identical to a 7 S globulin isolated earlier by Koshiyama (1968). Koshiyama and Fukushima (1976a) disputed this, saying their 7 S globulin was identical to β-conglycinin. Koshiyama and Fukushima (1976b) studied the γ-conglycinin and found it to have an isoelectric point of 5.8 and a molecular weight of 104,000. That molecular weight is close to the molecular weight of hemagglutinin and of lipoxygenase, other proteins known to be in the 7 S fraction. However, Koshiyama and Fukushima (1976b) were able to show that γ-conglycinin had neither hemagglutinin nor lipoxygenase activity.

Hemagglutinin: Hemagglutinins or lectins are widely distributed in plants, but their inherent function is not known. These proteins have the ability to cause aggregation of red blood cells, hence the name hemagglutinin. They also can be quite selective in the type of red blood cells aggregated, hence the name lectin (from the Latin *legere,* to choose). Hemagglutinins have specific sites that allow them to react with specific carbohydrates. Since carbohydrates are normally found in cell membranes and since in animal cells the carbohydrates are exposed (no cell walls), hemagglutinins with two binding sites can act as a bridge between cells and cause them to clump.

The hemagglutinin in soybean seeds was first purified and studied by Pallansch and Liener (1953). Part of the growth inhibition caused when raw soybeans are fed to chicks or young rats cannot be explained by trypsin inhibitor activity only, and for a while it was thought that hemagglutinin was the reason for this additional antigrowth effect. But more recently, it has become evident that hemagglutinin does not survive intact the acidity and digestive enzymes of the stomach. Furthermore, experiments by Turner and Liener (1975) have shown that the antigrowth properties of extracts from raw soybeans are not changed when the hemagglutinin is removed from the extracts by affinity chromatography. Hemagglutinins in other plants and seeds can exert toxic effects when eaten, but soy hemagglutinin does not exert much effect on the nutritive value of soybeans.

Hemagglutinin sediments with the 7 S fraction during ultracentrifugation. It has a molecular weight of approximatelyy 120,000 and can be dissociated into four identical subunits (Lotan *et al.* 1974). In addition to reacting with carbohydrates, soy hemagglutinin is a glycoprotein containing five glucosamine and 37 mannose residues per mole.

Trypsin inhibitors: As noted above, raw soybeans when fed to some young animals will cause growth inhibition. This fact has been known since 1917, but it was not until the 1940s that a reasonable explanation arose as a result of the discovery of a trypsin inhibitor in soybeans (Kunitz 1946). Since then the trypsin inhibitors of soybeans have become the most studied and best understood of all the soy proteins. Their effect on nutrition is not the simple inhibition of tryptic proteolysis that people once thought, but that story will be told in Chapter 6. In this section we describe some of the physical and chemical properties of soybean trypsin inhibitors.

Kunitz trypsin inhibitor: The trypsin inhibitor isolated first by Kunitz and all other trypsin inhibitors with approximately the same molecular weight, 21,000, are known as Kunitz trypsin inhibitors. They contrast with the Bowman–Birk inhibitor to be discussed next. With very sensitive isolation and analytical procedures such as gel filtration, isoelectric focusing, and disc gel electrophoresis, it is possible to isolate several Kunitz trypsin inhibitors from the same variety of soybeans. The significance of these several forms of trypsin inhibitors is not known, and the problem is similar to the isolation of isoenzymes with minor differences. There are also different Kunitz trypsin inhibitors associated with varieties of soybeans that are under genetic control. For an excellent review of protease inhibitors in plants see Liener and Kakade (1980).

The complete primary structure of three Kunitz trypsin inhibitors is known (Kim *et al.* 1985), and the primary structure of trypsin inhibitor "a" is shown in Fig. 2.18. The secondary structure shows very little α-helical character. In the 181 amino acid chain there are two disulfide bonds. Kunitz trypsin inhibitor is known to exert its effect by a 1:1 stoichiometric combination with trypsin that is extremely stable. The association constant for the complex is 5×10^9 at pH 7.8 (Green 1953). The active site of the Kunitz trypsin inhibitor is the two amino acid sequence arginine–isoleucine in positions 63 and 64. This site combines with the active site of trypsin, and the peptide bond between arginine and isoleucine is cleaved. But because there are several other combining sites, based on weak noncovalent bonds, the enzyme substrate combination does not break apart, and the trypsin remains bound and inactive.

The trypsins from different species differ, and it is of interest to know how

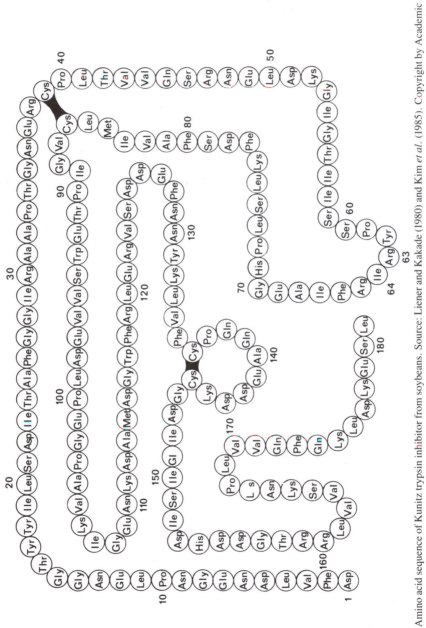

Fig. 2.18. Amino acid sequence of Kunitz trypsin inhibitor from soybeans. Source: Liener and Kakade (1980) and Kim *et al.* (1985). Copyright by Academic Press, Inc.

wide is the range of specificity of Kunitz trypsin inhibitor. It is known to inhibit trypsin from the cow, pig, salmon, stingray, barracuda, and turkey. To some extent proteases other than trypsin may also be inhibited. Of special interest is the interaction between Kunitz inhibitor and human trypsin. There are two forms of human trypsin: a major form with 80–90% of the activity and a minor form with 10–20% of the tryptic activity. The minor form is inactivated by Kunitz trypsin inhibitor, but the major form is only weakly inhibited.

Bowman–Birk trypsin inhibitor: A second group of protease inhibitors found in soybeans is named for Bowman, who first discovered them, and for Birk, who first purified and characterized them. Like the Kunitz inhibitor, there are several differentiable proteins that all have the general characteristics of the Bowman–Birk inhibitor. That is, they have a molecular weight in the 8000 range; they have seven disulfide cross links giving great stability to the molecule; and they inhibit both chymotrypsin and trypsin at different combining sites.

The complete primary structure is known (Fig. 2.19), and the reactive site for trypsin is at residues lysine 16–serine 17 and for chymotrypsin at residues leucine 44–serine 45. The small size of the Bowman–Birk inhibitor gives it some unusual characteristics such as solubility in alcohol and extreme stability to denaturing agents. By the usual assay procedures for trypsin inhibitor activity in soybeans, seldom is all of the activity lost. This residual activity is generally attributed to Bowman–Birk inhibitor that survived the heat of the desolventizer–toaster. The Bowman–Birk inhibitor does have a tendency to polymerize. Inhibitors isolated from soybeans and shown to be different than the Kunitz inhibitor and in the 24,000–30,000 molecular weight range may be polymers of the Bowman–Birk inhibitor.

Lipoxygenase: Soybeans are an excellent source of the enzyme lipoxygenase. It is widely distributed in the plant kingdom and also can be found in fruits, stems, leaves, and seeds of individual plants. This enzyme was first discovered in the 1920s, when its ability to bleach the carotenoids of wheat flour was observed. Raw soy flour is still used to bleach wheat flour in some European countries (see Chapter 9). It was also noted in the 1920s that soybeans had the ability to oxidize lipids. It was not until the 1940s that both activities were found to be due to the same enzyme, then known as lipoxidase.

This enzyme was isolated and characterized in the late 1940s by H. Theorell. Theorell had studied heme proteins that interact with molecular oxygen, and he looked for an iron prosthetic group in lipoxygenase but found none. The finding was surprising because most proteins interacting with molecular oxygen do so through a metal prosthetic group. The idea that lipoxygenase was iron free persisted until the early 1970s, when it was shown that lipoxygenase does contain honheme iron, one atom per molecule of enzyme. Furthermore, the iron is thought to play a role in the oxidation of polyunsaturated fatty acids since it is at the active site and undergoes an oxidation state change during the reaction.

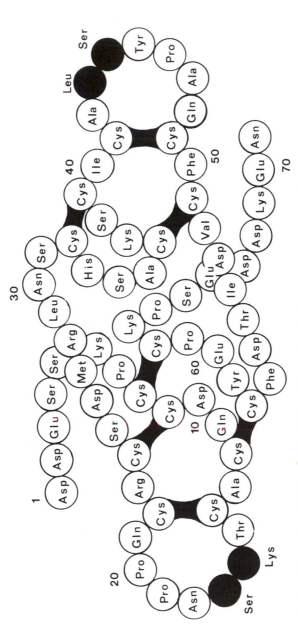

Fig. 2.19. Amino acid sequence of Bowman–Birk inhibitor. The reactive sites for trypsin (Lys 16–Ser 17) and for chymotrypsin (Leu 44–Ser 45) are highlighted. Source: Liener and Kakade (1980). Copyright by Academic Press, Inc.

Lipoxygenase separates with the 7 S fraction in the ultracentrifuge with a molecular weight of about 100,000.

The lipid oxidation reaction catalyzed by lipoxygenase is the combination of molecular oxygen with a polyunsaturated fatty acid to yield a hydroperoxide. Using a methylene-interrupted pentadiene as the segment of fatty acid oxidized, the reaction proceeds as follows:

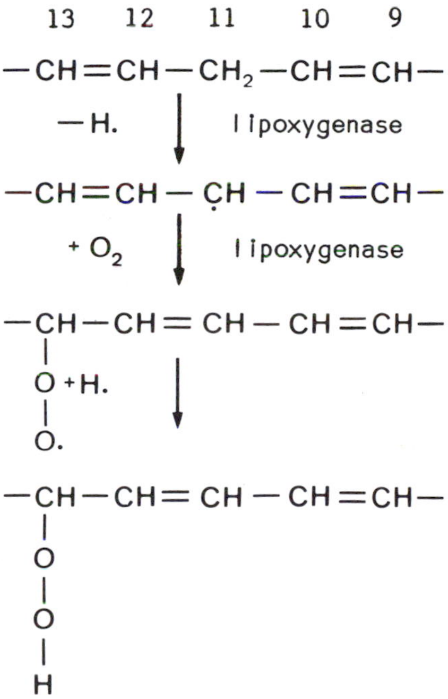

$$
\begin{array}{ccccc}
13 & 12 & 11 & 10 & 9
\end{array}
$$

$$
-CH=CH-CH_2-CH=CH-
$$

$$
-H. \quad \Big\downarrow \quad \text{lipoxygenase}
$$

$$
-CH=CH-\overset{\cdot}{C}H-CH=CH-
$$

$$
+O_2 \quad \Big\downarrow \quad \text{lipoxygenase}
$$

$$
-CH-CH=CH-CH=CH-
$$
$$
\underset{O.}{\overset{|}{\underset{|}{O}}} {}_{+H.} \quad \Big\downarrow
$$

$$
-CH-CH=CH-CH=CH-
$$
$$
\begin{array}{c}
| \\
O \\
| \\
O \\
| \\
H
\end{array}
$$

The hydroperoxide may form at either the 9 or 13 position (for linoleic acid), although some lipoxygenases have specificity for one of the two positions. The newly formed double bond in the 11–12 position is always trans. The double bond system is conjugated in the hydroperoxide product, and a conjugated diene absorbs strongly in the 233-nm region of the spectrum.

Absorbance at 233 nm is one means by which lipoxygenase activity can be assayed. If a clear solution of enzyme and substrate can be achieved, then the increase in absorbance at 233 nm with increase in time is a good assay procedure. For enzyme preparations and substrates that do not give clear solutions, measurement of the disappearance of molecular oxygen with time by means of a Clark oxygen electrode is a good assay procedure. One of the difficulties of assaying activity of lipoxygenase is that the substrate, a fatty acid, fatty acid ester, or triglyceride, is not very soluble in water or buffers. Frequently solvation aids have to be added, and these may affect the activity of lipoxygenase.

Four soybean lipoxygenase isoenzymes (L-1, L-2, L-3a, and L-3b) have been studied (Axelrod *et al.* 1981), the most important being L-1 and L-2. L-1 is the enzyme that Theorell studied. It has a pH optimum of about 9, is most active on free fatty acids as opposed to esters or triglycerides, and is not activated by calcium ions. In contrast, L-2 has a pH optimum of about 7, it is more active on fatty acid esters and triglycerides than on free fatty acids, and its activity is increased by calcium ions.

Lipoxygenases are undoubtedly active when unheated soybeans are soaked and disrupted to prepare soymilk in the traditional Oriental procedure. The activity is evident as a grassy, beany flavor in the soymilk before heating. It is not clear if lipoxygenases are active when soybeans are disrupted by flaking. The low moisture content of the beans (11–12%) minimizes enzymatic activity.

Other enzymes: Undoubtedly in the young maturing seed and in the germinating seed many enzymes associated with metabolic processes are present and active. However, in the mature seed that has not germinated, enzymatic activity is limited. In addition to lipoxygenase, there is probably lipolytic activity producing free fatty acids, but these lipases have not been well studied.

Urease is an important enzyme because it is used as an indicator of sufficiency of heat treatment given to defatted soybean flakes to inactivate trypsin inhibitors. Urease catalyzes the breakdown of urea:

$$(NH_2)_2\!\!-\!\!C\!\!=\!\!O + H_2O \leftrightharpoons 2NH_3 + CO_2$$

and the activity can be followed as a pH rise due to the generation of ammonia.

Both α- and β-amylase have been found in mature soybeans, although starch is present in very low concentration, about 0.5%. As soybeans mature, the amount of starch decreases, and the amount of starch hydrolyzing enzymes increases. See Rackis (1978) for a review of other enzymes that have been found in mature soybeans.

Carbohydrates

The carbohydrate portion of soybean seeds is substantial, about 30% on a dry-weight basis, but economically the carbohydrates are much less important than the lipid and the protein. Consequently, less research has been done and there is relatively less known about soybean carbohydrates compared to the lipids and proteins. Still there are important quality aspects of soybeans and soybean products that hinge on the carbohydrate portion. We have divided our discussion of the carbohydrates into the soluble and insoluble fractions.

Soluble Carbohydrates.   In the immature soybean seed, one can find reducing sugars such as glucose, but in the mature seed the reducing sugars disappear, and predominantly three nonreducing sugars are present. These are sucrose, raffinose, and stachyose. The monosaccharides making up these three oligosaccharides and the linkages between them are shown in Fig. 2.20.

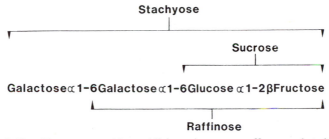

Fig. 2.20. Hexose composition and linkages in sucrose, raffinose, and stachyose.

The quantities of the soluble carbohydrates in soybean seeds are about 10%, with approximately 5% sucrose, 1% raffinose, and 4% stachyose. These amounts vary with variety, maturity, growing conditions, and method of analysis, but the whole numbers given are useful approximations. Smith and Circle (1978) have reviewed data on soluble and insoluble carbohydrates in soybeans.

The soluble carbohydrates, at least raffinose and stachyose, are important in that they are not digested and absorbed as nutrients by humans. However, the intestinal flora in the human digestive tract can digest these oligosaccharides, and the resulting gas production causes flatulence (Steggerda *et al.* 1966). Nutritional aspects of the oligosaccharides are discussed further in Chapter 6. Presumably, raffinose and stachyose are responsible also for the loose and malodorous stools found in infants fed soy flours. This was one of the reasons for development of soy concentrates in which soluble carbohydrates are extracted and for soy isolates in which both soluble and insoluble carbohydrates are removed.

The possibility of breeding soybeans with decreased soluble carbohydrates has been explored. DeMan *et al.* (1975) found varieties that had decreased sucrose, raffinose, and stachyose, and presumably such varieties would be useful in a breeding program aimed at decreasing soluble carbohydrates. The content of soluble carbohydrates can be lowered by allowing soybeans to germinate (Anderson *et al.* 1979). Over a 96-hr period of germination to produce soybean sprouts, 80–100% of raffinose and stachyose is consumed.

Another approach to minimizing the unwanted effects of raffinose and stachyose is to digest them enzymatically (Smiley *et al.* 1976). For a product such as soymilk, it may be feasible to immobilize α-galactosidase and to allow it to react with raffinose and stachyose in soymilk to produce galactose and sucrose, neither of which would cause any digestive problems.

Insoluble Carbohydrates. The insoluble carbohydrates of soybeans are the structural, high-molecular-weight carbohydrates such as cellulose, hemicellulose, and pectins. Also, there is usually less than 1% starch present. The structural components are found in cell walls and in intercellular material. The starch is intracellular.

Kikuchi *et al.* (1971) found 30% pectins, 50% hemicellulose, and 20% cellulose in the insoluble carbohydrate fraction of soybeans. With prolonged cooking, the pectins are solubilized and soybeans are softened. Since the pectins are present between cells, this softening mechanism involves the separation of cells. Soybeans require considerably more cooking time than most other edible beans to produce a palatable texture. The high starch content of most edible beans allows a softening due to gelatinization of starch. But this is not possible with soybeans since the small amount of starch present has negligible influence on texture. The prolonged cooking time (several hours) required to soften soybeans is probably one reason that soybeans are not cooked and eaten regularly as a substantial portion of the diet. Some attempts have been made to overcome the unusually long cooking time for soybeans.

Rockland (1972) patented a process for shortening cooking times from 4–5 hr to less than 1 hr. The process involves a brief blanching of dry soybeans in boiling water followed by soaking in an alkaline solution of sodium chloride plus a chelating agent. Then the soybeans are dried. This process probably influences the pectic substances by removing calcium and substituting sodium. Calcium has a strengthening effect on pectins due to its ability to cross link polygalacturonic acid chains. By removing calcium and substituting sodium, pectins would tend to dissociate and solubilize more readily.

The insoluble carbohydrates contain or are the same as the dietary component, fiber. Processing of soybeans to produce soy isolates removes the fiber portion by solubilizing the protein and centrifuging or filtering out the cell wall and intercellular components. This insoluble fraction is used now as a feed ingredient, but if there is a demand for more fiber in processed foods, this fraction could be used as a source of food fiber.

Minor Components

There are many chemical compounds found in soybeans that do not fit neatly into the categories of lipid, protein, and carbohydrate. These compounds are minor from the quantitative standpoint, but they may be of major importance to the quality or nutritional value of soybeans and of soybean products. We consider first those minor components likely to be associated with the meal or protein products, and subsequently we discuss minor components found in the oil.

Phytic Acid.   This is a phosphorus containing compound that is a derivative of inositol, cyclohexanehexol. Phytic acid is inositol with six phosphate groups esterified to the six hydroxyl groups (Fig. 2.21). The quantities of phosphorus found in soybeans range from 0.5 to 0.7%, and 70–80% of this phosphorus is in the form of phytic acid. The remaining phosphorus is from phospholipids, discussed earlier, and from inorganic phosphorus.

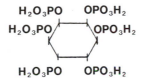

Fig. 2.21.   The structure of phytic acid.

Phytic acid is the term used to describe the free acid. A particular salt of phytic acid having five equivalents of calcium and one equivalent of magnesium per phytic acid is called phytin and is believed to occur in soybeans. O'Dell (1979) has pointed out that there is not enough calcium in soybeans for all of the phytic acid to exist as phytin.

Phytic acid is generally found in the seeds of plants as opposed to stems and leaves, and legumes contain a higher concentration than cereals. The phytic acid content of seeds decreases during germination or fermentation presumably due to inherent or microbial phytases. The role played by phytic acid in seeds is not known.

There is evidence from electron microscopy that phytic acid exists as inclusions in soybean protein bodies (Lott and Buttrose 1978, Bair 1979). Also, when proteins are extracted from soybeans, phytic acid accompanies the protein. Special procedures such as ion exchange chromatography or ultrafiltration are required to remove phytates from soy proteins. The presence of phytates can affect the solubility and the isoelectric point of the proteins.

Phytates are excellent chelators, and their importance in soybean products is linked with their binding of minerals. This binding may change the availability of some minerals as nutrients. Zinc is the mineral that seems to be most affected by the presence of phytic acid from soybeans. This subject of mineral nutrient availability will be pursued further in Chapter 6.

Minerals.   The total mineral content of soybeans is given roughly by total ash. On a dry-weight basis, ash is about 5%. This does not mean that 5% of the weight of soybeans is mineral. During ashing, sulfates, phosphates, and carbonates form, and the oxygen content of the ash accounts for much of the weight.

The major mineral components of soybeans are potassium, sodium, calcium, magnesium, sulfur, and phosphorus. The quantities are shown in Table 2.7. The amounts shown in the table are averages, and the range, for potassium, for example, can be from 0.8 to 2.4%. Both the type of soil and growing conditions can influence the mineral content of soybeans.

Minor mineral constituents are shown in Table 2.8. Again, these values are representative for defatted soybean flakes but are not meant to be invariable. Generally, minerals will follow the protein or meal portion of soybeans rather than the oil. Exceptions are that calcium, magnesium, and phosphorus can be

Table 2.7

Major Mineral Constituents of Soybeans (%)[a]

	K	Na	Ca	Mg	S	P
Whole soybeans	1.7	0.2	0.3	0.3	0.2	0.7

[a]Sources: O'Dell (1979) and Smith and Circle (1978).

extracted with the phospholipids and become part of the oil. Also, the trace minerals iron and copper, which may arise from the original soybean or may arise from metal contact during processing, are important contaminants of oil. Iron and copper have a significant prooxidant effect on soybean oil and are to be avoided.

In addition to the trace minerals noted in Table 2.8 there may be as many as ten others in the 0.5–2 ppm range, and undoubtedly most minerals could be found with sensitive enough analysis techniques.

One problem having to do with heavy metals in soybeans was investigated by Braude *et al.* (1980). With the recommendation of using sewage sludge to fertilize agricultural lands has come the question of build-up of heavy metals in the soil. Those heavy metals could then be assimilated by plants grown in the soil and could represent a potential human health hazard. With respect to cadmium, it was found that sludge-grown soybeans can accumulate 2.3 ppm as opposed to 0.06 ppm normally present. Furthermore, the cadmium tended to stay with the protein or meal portion of soybeans rather than with the oil. This result shows there is reason to monitor and control the heavy-metal content of soils used for food crops and treated with sewage sludge.

Although soybeans contain nutritionally significant quantities of iron and calcium, they are not outstanding nutritional sources of these minerals or of any others.

Saponins.   This is a heterogeneous class of compounds composed of glycosides of triterpeneoid alcohols. Acid hydrolysis of soybean saponins yields sugars plus the triterpeneoid alcohols or sapogenins. Five sapogenins have been isolated from soybeans, and their structures are shown in Fig. 2.22. The sugars found in the glycoside portion of soybean saponins are xylose, arabinose, galac-

Table 2.8

Minor Mineral Constituents of Defatted Soybeans (ppm)[a]

	Fe	Zn	Cu	Mn	Si
Defatted soybeans	137	52	20	38	140

[a]Source: O'Dell (1979). Copyright by Academic Press, Inc.

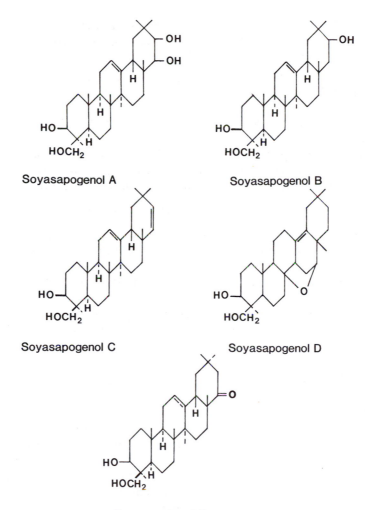

Soyasapogenol A

Soyasapogenol B

Soyasapogenol C

Soyasapogenol D

Soyasapogenol E

Fig. 2.22. Structural formulas of soyasapogenols A, B, C, D, and E. Source: Birk (1969) and Kitagawa *et al.* (1982). Copyright by Academic Press, Inc.

tose, glucose, rhamnose, and glucuronic acid. It has been estimated that an average of three monosaccharides are associated with each sapogenin in soybeans (Birk 1969). The six monosaccharides in different oligosaccharide combinations associated with five different sapogenins explain the heterogeneity of soybean saponins.

The saponins are polar because of the associated carbohydrates and are found in the soybean meal in amounts of approximately 0.5% of the dry weight.

Saponins are strong surfactants and have the ability to lyse red blood cells. For this and other reasons, their toxicity in soybeans has been studied. Birk (1969) concluded that in the amounts present in soy meal, saponins have no detrimental biological activity.

Nonprotein nitrogen.   By extracting soybeans with 0.6–1 N trichloroacetic acid, low-molecular-weight nitrogen compounds can be extracted. These compounds include free amino acids, low-molecular-weight peptides, nucleic acid bases, glutathione, and polyamines.

Polyamines are low-molecular-weight, aliphatic nitrogenous bases. Four have been found in soybeans (Wang 1972): putrescine, cadavarine, spermidine, and spermine. As an example of their structures, putrescine is 1,4-diaminobutane:

$$H_2N—CH_2—CH_2—CH_2—CH_2—NH_2$$

The total amounts of polyamines in soybeans average 130 ppm. Since these kinds of compounds are foul smelling, it was of interest to see if they contribute to soybean flavor. Wang *et al.* (1975) found that the amounts present in soybeans were below flavor thresholds, and so the polyamines probably do not influence flavor.

The total quantity of nonprotein nitrogen ranges around 3–4 mg/g of soybean tissue or about 5% of the protein nitrogen. No specific biological activity has been associated with the nonprotein nitrogen fraction.

Phenolic compounds.   Several phenolic acids often found in plant products have been isolated from soybeans (Arai *et al.* 1966). Figure 2.23 shows names and structures for eight of the compounds isolated. These compounds often are associated with astringent or bitter flavors, but it is not known if they have an influence on soybean flavor. Also, information is lacking on the quantities present.

In addition to phenolic acids, the wide range of compounds under the general heading, flavanoids, are considered phenolic compounds. The pigments found in the seed coat of soybeans are water soluble and are flavanoid in nature but to our knowledge have not been studied. Flavanoids have the general structure of two benzyl rings joined by a three carbon bridge which may or may not be closed in a pyran ring:

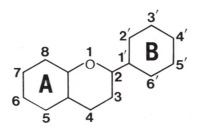

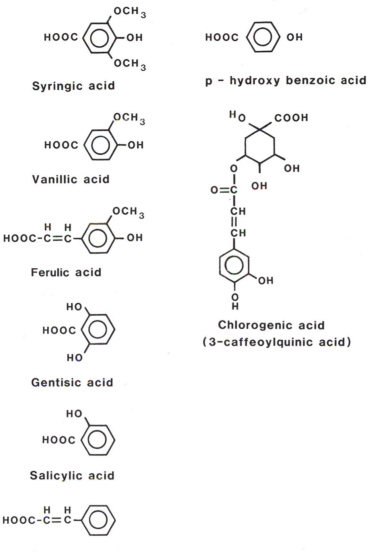

Fig. 2.23. Structures of phenolic acids found in soybeans.

Hydroxy groups are attached to the benzyl rings in various combinations of the 3, 5, 7, 3′, 4′, or 5′ positions. A keto group at position 4 is characteristic of the group of compounds known as flavones.

Three isoflavones have been isolated from soybeans. Isoflavones differ from flavones in that the benzyl ring B is joined at position 3 instead of position 2. The isoflavones are named genestin, daidzin, and glycitin in the glycoside form

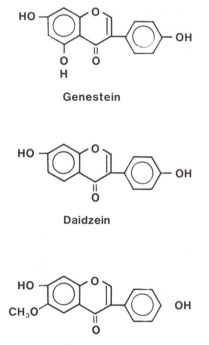

Genestein

Daidzein

Glycitein

Fig. 2.24.   Structures of the three isoflavones found in soybeans.

(glucose or another sugar bound at position 7). The aglycones are named genestein, daidzein, and glycitein, and their structures are shown in Fig. 2.24. Total isoflavone content of soybeans has been estimated to be 0.25% (Wolf 1976).

Extracts of soybeans containing phenolic compounds have been shown to have antioxidant activity in food systems. Hammerschmidt and Pratt (1978) have demonstrated that a pentoside of glycitein was an active antioxidant.

Rackis (1978) reviewed the evidence that genistein and daidzein have estrogenic activity. Although estrogenic activity can be demonstrated for these compounds, the quantities involved would not exert physiological effects in humans.

Vitamins.   Soybeans are not considered to be particularly rich sources of any one vitamin, but they do contain vitamins and can contribute to overall nutritional wellbeing. Table 2.9 lists the water-soluble vitamin content in mature whole soybeans and defatted meal. Generally water-soluble vitamins are not lost during oil extraction and subsequent toasting of the flakes. The data in Table 2.9 show an increase in vitamin concentration, due to oil removal, with the exception of biotin and inositol.

Table 2.9

Water-Soluble Vitamin Content of Mature Whole Soybeans
and of Meal from Defatted Toasted Flakes[a]

	Soybeans	Soybean meal
Thiamin (μg/g)	11–17	12–44
Riboflavin (μg/g)	2.3	2.7–3.3
Niacin (μg/g)	20–26	19–40
Pantothenic acid (μg/g)	12	13–16
Pyridoxine (μg/g)	6.4	8.8
Biotin (μg/g)	0.6	0.2
Folic acid (μg/g)	2.3	4–5
Inositol (mg/g)	1.9–2.6	1.8–2.1
Choline (mg/g)	3.4	3.5–3.8
Ascorbic acid (mg/g)	0.2	—

[a]Source: Liener (1978).

As soybeans germinate, the content of water soluble vitamins increases 50–100%, and so soy sprouts are much better sources of water-soluble vitamins than are mature whole soybeans.

The oil-soluble vitamins present in soybeans are vitamins A and E with essentially no vitamins D and K. Vitamin A exists as the provitamin β-carotene and is present in the green immature bean at higher concentrations than in the mature soybean. The xanthophyll pigment lutein is the predominant pigment in soybeans, but it is not a vitamin A precursor. The bleaching step of oil refining removes most of the pigments, and there is negligible β-carotene remaining in the edible soy oil.

Tocopherols are an important constituent of soy oil due both to the vitamin E supplied for human nutrition and to the antioxidant properties of tocopherol that protect the oil. Soybean oil contains four tocopherol isomers whose structures and corresponding designations are shown in Fig. 2.25. The quantities of three of these isomers in crude and refined oil are shown in Table 2.10. The predominant tocopherol is γ-tocopherol. The amount of β-tocopherol in soybeans is detectable, but it makes up less than 3% of the total. The tocopherols are lost mainly in the deodorization step, and total loss during refining is about 50%. The biggest loss is in γ-tocopherol with less α- and δ-tocopherol being lost. The condensate from the deodorizer is a valuable by-product of soy oil refining, and the more tocopherol it contains, the more valuable it becomes. On the other hand, there is evidence that loss of tocopherol causes refined oil to oxidize more readily. This relationship between tocopherol and oxidative stability of soy oil will be discussed further in Chapter 5.

Table 2.10

Quantities (mg/100 g) of Tocopherol Isomers in Crude and
Refined Soy Oil[a]

	α-Tocopherol	τ-Tocopherol	δ-Tocopherol
Crude soy oil	9–12	74–102	24–30
Refined soy oil	6–9	45–50	19–22

[a]Source: Pryde (1980a).

Sterols and triterpene alcohols.   These compounds are part of the unsaponifiable portion of soy oil. There are three major sterols in soy oil: β-sitosterol,
campestrol, and stigmasterol. Figure 2.26 gives their structures. Several minor
sterols also are present in crude oil, but are nondetectable in refined oils. Sterols
are lost during deodorization and during hydrogenation as shown in Table 2.11.

The sterol present in greatest concentration is β-sitosterol, and it is roughly
three times as much as either campestrol or stigmasterol. In addition to free
sterols and sterol glycosides, some sterols are esterified with the fatty acids
commonly found in soybeans, palmitic, stearic, oleic, linoleic, and linolenic.

Triterpene alcohols have structures similar to sterols (Fig. 2.27) but are present in much lower concentrations (about 20% of the amount of sterols). Seven
different triterpenes in soybean oil have been separated by chromatography, but
not all have been identified (Sonntag 1979b).

There is no indication of either adverse or beneficial effects of the sterols or
triterpene alcohols in soybean oil. Acylated and glycosylated sterols are also
found accompanying the protein fraction of soybeans, but again no specific
effects have been attributed to these compounds.

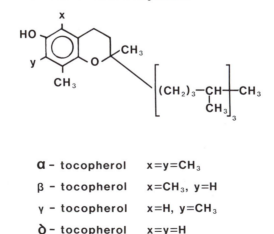

α - tocopherol   x=y=CH₃

β - tocopherol   x=CH₃, y=H

γ - tocopherol   x=H, y=CH₃

δ - tocopherol   x=y=H

Fig. 2.25.   Structures and designations of the four tocopherol isomers found in soybean oil.

Table 2.11
Change in Sterol Content (mg/100 g Oil)
of Soy Oil during Refining[a]

Sterol	Crude	Refined	Refined and hydrogenated
β-Sitosterol	183	123	76
Campestrol	68	47	26
Stigmasterol	64	47	30

[a]Source: Pryde (1980a).

Squalene. Soybean oil contains some hydrocarbons as part of the unsaponifiable fraction. The main hydrocarbon is squalene, a 30-carbon partially unsaturated compound:

$$(CH_3)_2C{=}CH-CH_2-CH_2-\left[\begin{matrix}CH_3\\|\\C{=}CH-CH_2-CH_2\end{matrix}\right]_4$$

The total hydrocarbon fraction is about 15% of the unsaponifiables, and of that 15%, squalene makes up about half. Squalene is a precursor in sterol and triterpene biosynthesis and tends to be volatile enough to concentrate in the condensate fraction of the deodorizer. As with the sterols and triterpenes, no specific effects have been attributed to squalene in soy oil.

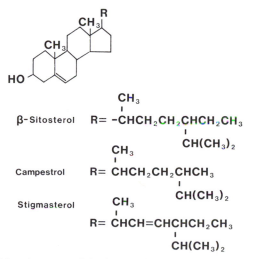

β-Sitosterol $R= -CHCH_2CH_2CHCH_2CH_3$
with CH_3 substituent and CH(CH_3)_2

Campestrol $R= CHCH_2CH_2CHCH_3$
with CH_3 substituent and CH(CH_3)_2

Stigmasterol $R= CHCH{=}CHCHCH_2CH_3$
with CH_3 substituent and CH(CH_3)_2

Fig. 2.26.   Structures of the three predominant sterols in soybean oil.

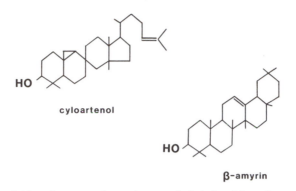

cyloartenol

β-amyrin

Fig. 2.27.   Structures of two triterpene alcohols found in soybean oil.

Carotenoids.   The main pigments of soy oil fall into the carotenoid class although often troublesome amounts of chlorophyll also are present. According to Vogel (1977), crude soy oil contains about 28–30 ppm of total carotenoids and about 97% of this is lutein (3,3′dihydroxy α-carotene). The remaining pigment is a mixture of α- and β-carotenes with β-carotene being predominant. Since lutein would have no provitamin A activity, the act of bleaching oil to improve acceptability has essentially no effect on availability of provitamin A.

There are other known minor chemical compounds found in both oil and meal fractions of soybeans. These are flavor compounds including a wide variety of short chain hydrocarbons, aldehydes, ketones, and acids. They probably arise from lipid oxidation and will be considered further in Chapter 5.

REFERENCES

Anderson, R. L., J. J. Rackis, and W. H. Tallent (1979). Biologically active substances in soy products. *In* "Soy Protein and Human Nutrition" (H. L. Wilcke, D. T. Hopkins, and D. H. Waggle, eds.). Academic Press, New York.

Arai, S., H. Suzuki, M. Fujimaki, and Y. Sakurai (1966). Studies on flavor components in soybean. Part II. Phenolic acids in defatted soybean flour. *Agric. Biol. Chem.* **30:**364.

Arechavaleta-Medina, F., and H. E. Snyder (1981). Water imbibition by normal and hard soybeans. *J. Am. Oil Chem. Soc.* **58:**976.

Axelrod, B., T. M. Cheesbrough, and J. Laakso (1981). Lipoxygenase from soybeans. *In* "Methods in Enzymology," Vol. 71 (J. M. Lowenstein, ed.). Academic Press, New York.

Bair, C. W. (1979). Microscopy of soybean seeds: Cellular and subcellular structure during germination, development and processing with emphasis on lipid bodies. Ph.D. Thesis, Iowa State Univ., Ames.

Bair, C. W., and H. E. Snyder (1980). Electron microscopy of soybean lipid bodies. *J. Am. Oil Chem. Soc.* **57:**279.

Birk, Y. (1969). Saponins. *In* "Toxic Constituents of Plant Foodstuffs," 1st ed. (I. E. Liener, ed.). Academic Press, New York.

Braude, G. L., A. M. Nash, W. J. Wolf, R. L. Carr, and R. L. Chaney (1980). Cadmium and lead content of soybean products. *J. Food Sci.* **45:**1187.

Catsimpoolas, N., and C. Ekenstam (1969). Isolation of alpha, beta, and gamma conglycinins. *Arch. Biochem. Biophys.* **129:**490.

Danielsson, C. E. (1949). Seed globulins of the Graminae and Leguminosae. *Biochem. J.* **44:**387.

Deman, J. M., D. W. Stanley, and V. Rasper (1975). Composition of Ontario soybeans and soymilk. *Can. Inst. Food Sci. Technol. J.* **8:**1.

Derbyshire, E., D. J. Wright, and D. Boulter (1976). Legumin and vicilin storage proteins of legume seeds. *Phytochemistry* **15:**3.

Evans, C. D., D. G. McConnell, G. R. List, and C. R. Scholfield (1969). Structure of unsaturated vegetable oil glycerides: Direct calculation from fatty acid composition. *J. Am. Oil Chem. Soc.* **46:**421.

Green, N. M. (1953). Competition among trypsin inhibitors. *J. Biol. Chem.* **205:**535.

Hammerschmidt, P. A., and D. E. Pratt (1978). Phenolic antioxidants of dried soybeans. *J. Food Sci.* **43:**556.

Hammond, E. G., and W. R. Fehr (1984). Improving the fatty acid composition of soybean oil. *J. Am. Oil Chem. Soc.* **61:**1713.

Hill, J. E., and R. W. Breidenbach (1974). Proteins of soybean seeds 1. Isolation and characterization of the major components. *Plant Physiol.* **53:**742.

Hughes, S. A., and P. A. Murphy (1983). Varietal influence on the quantity of glycinin in soybeans. *J. Agric. Food Chem.* **31:**376.

Hvolby, A. (1971). Removal of nonhydratable phospholipids from soybean oil. *J. Am. Oil Chem. Soc.* **48:**503.

Inglett, G. E. (1977). Cereal proteins. *In* "Food Proteins" (J. R. Whitaker and S. R. Tannenbaum, eds.). AVI Publ. Co., Westport,CT.

Jacks, T. J., L. Y. Yatsu, and A. M. Altschul (1967). Isolation and characterization of peanut spherosomes. *Plant Physiol.* **42:**585.

Johnson, K. W., and H. E. Snyder (1978). Soymilk: A comparison of processing methods on yields and composition. *J. Food Sci.* **43:**349.

Kahn, V., R. W. Howell, and J. B. Hanson (1960). Fat metabolism in germinating soybeans. *Plant Physiol.* **35:**854.

Kasarda, D. D., J. E. Bernardin, and C. C. Nimmo (1976). Wheat proteins. *Adv. Cereal Sci. Technol.* **1:**158.

Kikuchi, T., S. Ishil, D. Fukushima, and T. Yokotsuka (1971). Food chemical studies on soybean polysaccharides. Part I. Chemical and physical properties of soybean cell wall polysaccharides and their changes during cooking. *J. Agric. Chem. Soc.* **45:**228.

Kim, S. H., S. Hara, S. Hase, T. Ikenaka, H. Toda, K. Kitamura, and N. Kaizuma (1985). Comparative study on amino acid sequences of Kunitz-type soybean trypsin inhibitors, Ti[a], Ti[b], and Ti[c]. *J. Biochem.* **98:**435.

Kitagawa, I., M. Yoshikawa, H. K. Wang, Y. Tosirisu, T. Fujiwara, and K. Tomita (1982). Revised structures of soyasapogenol-A, soyasapogenol-B, and soyasapogenol-E oleanene sapogenols from soybean: structures of soyasaponin-I, soyasaponin-II, and soyasaponin-III. *Chem. Pharmacol.* **30:**2294.

Koshiyama, I. (1968). Chemical and physical properties of a 7 S protein in soybean globulins. *Cereal Chem.* **45:**394.

Koshiyama, I. (1969). Isolation of a glycopeptide from a 7 S protein in soybean globulins. *Arch. Biochem. Biophys.* **130:**370.

Koshiyama, I. (1972). Purification and physico-chemical properties of 11 S globulin in soybean seeds. *Int. J. Pept. Protein Res.* **4:**167.

Koshiyama, I., and D. Fukushima (1976a). Identification of the 7 S globulin with beta-conglycinin in soybean seeds. *Phytochemistry* **15:**157.

Koshiyama, I., and D. Fukushima (1976b). Purification and some properties of gamma-conglycinin in soybean seeds. *Phytochemistry* **15:**161.

Koshiyama, I., M. Kikuchi, K. Harada, and D. Fukushima (1981). 2 S Globulins of soybean seeds. 1. Isolation and characterization of protein components. *J. Agric. Food Chem.* **29:**336.

Kunitz, M. (1946). Crystalline soybean trypsin inhibitor. *J. Gen. Physiol.* **29:**149.

Liener, I. E. (1978). Nutritional value of food protein products. *In* "Soybeans: Chemistry and Technology," Vol. 1, Proteins, 2nd ed. (A. K. Smith and S. J. Circle, eds.). AVI Publ. Co., Westport, CT.

Liener, I. E., and M. L. Kakade (1980). Protease inhibitors. *In* "Toxic Constituents of Plant Foodstuffs," 2nd ed. (I. E. Liener, ed.). Academic Press, New York.

List, G. R., E. A. Emken, W. F. Kwolek, T. D. Simpson, and H. J. Dutton (1977). "Zero trans" margarines? Preparation, structure, and properties of interesterified soybean oil–soy trisaturate blends. *J. Am. Oil Chem. Soc.* **54:**408.

Lotan, R., H. W. Siegelman, H. Lis, and N. Sharon (1974). Subunit structure of soybean agglutinin. *J. Biol. Chem.* **249:**1219.

Lott, J. N. A., and M. S. Buttrose (1978). Globoids in protein bodies of legume seed cotyledons. *Aust. J. Plant Physiol.* **5:**89.

MacMasters, M. M., S. Woodruff, and H. Klaus (1941). Studies on soybean carbohydrates. *Ind. Eng. Chem.* **13:**471.

Mori, T., S. Utsumi, H. Inaba, K. Kitamura, and K. Harada (1981). Differences in subunit composition of glycinin among soybean cultivars. *J. Agric. Food Chem.* **29:**20.

O'Dell, B. L. (1979). Effect of soy protein on trace mineral availability. *In* "Soy Protein and Human Nutrition" (H. L. Wilcke, D. R. Hopkins, and D. H. Waggle, eds.). Academic Press, New York.

Pallansch, M. J., and I. E. Liener (1953). Soyin, a toxic protein from the soybean. II Physical characterization. *Arch. Biochem. Biophys.* **45:**366.

Park, K. D., J. Terao, and S. Matsushita (1981). The isomeric composition of hydroperoxides formed by autoxidation of unsaturated triglycerides and vegetable oils. *Agric. Biol. Chem.* **45:**2071.

Peng, I. C., D. W. Quass, W. R. Dayton, and C. E. Allen (1984). The physicochemical and functional properties of soybean 11 S globulin—a review. *Cereal Chem.* **61:**480.

Pryde, E. H. (1980a). Composition of soybean oil. *In* "Handbook of Soy Oil Processing and Utilization" (S. R. Erickson, E. H. Pryde, O. L. Brekke, T. L. Mounts, and R. A. Falb, eds.). American Soybean Association, St. Louis, MO, and American Oil Chemists Society, Champaign, IL.

Pryde, E. H. (1980b). Physical properties of soybean oil. *In* "Handbook of Soy Oil Processing and Utilization. S. R. Erickson, E. H. Pryde, O. L. Brekke, T. L. Mounts, and R. A. Falb, eds.) American Soybean Association, St. Louis, MO, and American Oil Chem. Society, Champaign, IL.

Rackis, J. J. (1978). Biologically active components. *In* "Soybeans: Chemistry and Technology," Vol. 1, Proteins, 2nd ed. (A. K. Smith and S. J. Circle, eds.). AVI Publ. Co., Westport, CT.

Rockland, L. B. (1972). Quick cooking soybean products. U.S. Patent 3,635,728. Jan. 18.

Saio, K. (1976). Soybeans resistant to water absorption. *Cereal Foods World* **21:**168.

Saio, K., and T. Watanabe (1966). Preliminary investigation on protein bodies of soybean seeds. *Agric. Biol. Chem.* **30:**1133.

Saio, K., K. Kobayakawa, and M. Kito (1982). Protein denaturation during model storage studies of soybeans and meals. *Cereal Chem.* **59:**408.

Smiley, K. L., D. E. Hensley, and H. J. Gasdorf (1976). Alpha-galactosidase production and use in a hollow fiber reactor. *Appl. Environ. Microbiol.* **31:**615.

Smith, A. K., and S. J. Circle (1978). Chemical composition of the seed. *In* "Soybeans: Chemistry and Technology," Vol. 1, 2nd ed. (A. K. Smith and C. J. Circle, eds.). AVI Publ. Co., Westport, CT.

Smith, A. K., A. M. Nash, and L. Wilson (1961). Water absorption of soybeans. *J. Am. Oil Chem. Soc.* **38:**120.

Sonntag, N. O. V. (1979a). Reactions in the fatty acid chain. *In* "Bailey's Industrial Oil and Fat Products," Vol. 1, 4th ed. (D. Swern, ed.). Wiley, New York.

Sonntag, N. O. V. (1979b). Composition and characteristics of individual fats and oils. *In* "Bailey's Industrial Oil and Fat Products," Vol. 1, 4th ed. (D. Swern, ed.). Wiley, New York.

Staswick, P. E., M. A. Hermondson, and N. C. Nielsen (1981). Identification of the acidic and basic subunit complexes of glycinin. *J. Biol. Chem.* **256:**8752.

Steggarda, F. R., E. A. Richards, and J. J. Rackis (1966). Effects of various soybean products on flatulence in the adult man. *Proc. Soc. Exp. Biol. Med.* **121:**1235.

Tanner, J. W., and D. J. Hume (1978). Management and production. *In* "Soybean Physiology, Agronomy and Utilization" (A. G. Norman, ed.). Academic Press, New York.

Thanh, V. H., and K. Shibasaki (1976a). Major proteins of soybean seeds: A straightforward fractionation and their characterization. *J. Agric. Food Chem.* **24:**1117.

Thanh, V. H., and K. Shibasaki (1976b). Heterogeneity of beta-conglycinin. *Biochim. Biophys. Acta* **439:**326.

Thanh, V. H., and K. Shibasaki (1978). Major proteins of soybean seeds. Subunit structure of beta-conglycinin. *J. Agric. Food Chem.* **26:**692.

Tombs, M. P. (1967). Protein bodies of the soybean. *Plant Physiol.* **42:**797.

Tumer, N. E., V. H. Thanh, and N. C. Nielsen (1981). Purification and characterization of mRNA from soybean seeds. Identification of glycinin and beta-conglycinin precursors. *J. Biol. Chem.* **256:**8756.

Tumer, N. E., J. D. Ritcher, and N. C. Nielsen (1982). Structural characterization of the glycinin precursors. *J. Biol. Chem.* **257:**4016.

Turner, R. H., and I. E. Liener (1975). The effect of the selective removal of hemagglutinins on the nutritive value of soybeans. *J. Agric. Food Chem.* **23:**484.

Vogel, P. (1977). Determination of xanthophyll content of vegetable oils. *Fette, Seifen, Anstrichm.* **79:**97.

Wall, J. S., and J. W. Paulis (1978). Corn and sorghum grain proteins. *Adv. Cereal Sci. Technol.* **2:**135.

Wang, L. C. (1972). Polyamines in soybeans. *Plant Physiol.* **50:**152.

Wang, L. C., B. W. Thomas, K. Warner, W. J. Wolf, and W. F. Kwolek (1975). Apparent odor thresholds of polyamines in water and 2% soybean flour dispersions. *J. Food Sci.* **40:**274.

Weiss, T. J. (1983). "Food Oils and Their Uses," 2nd ed. AVI Publ. Co., Westport, CT.

Wilson, L. A., V. A. Birmingham, D. P. Moon, and H. E. Snyder (1978). Isolation and characterization of starch from soybeans. *Cereal Chem.* **55:**661.

Wolf, W. J. (1969). Soybean protein nomenclature: A progress report. *Cereal Sci. Today* **14**(3): 75.

Wolf, W. J. (1970). Scanning electron microscopy of soybean protein bodies. *J. Am. Oil. Chem. Soc.* **47:**107.

Wolf, W. J. (1976). Chemistry and technology of soybeans. *Adv. Cereal Sci. Technol.* **1:**325.

Wolf, W. J. (1977). Legumes: Seed composition and structure, processing into protein products and protein properties. *In* "Food Proteins" (J. R. Whitaker and S. R. Tannenbaum, eds.). AVI Publ. Co., Westport CT.

Wolf, W. J. (1978). Purification and properties of the proteins. *In* "Soybeans: Chemistry and Technology," Vol. 1, Proteins, 2nd ed. (A. K. Smith and S. J. Circle, eds.). AVI Publ. Co., Westport, CT.

Wolf, W. J., and D. R. Briggs (1956). Ultracentrifugal investigation of the effect of neutral salts on the extraction of soybean protein. *Arch. Biochem. Biophys.* **63:**40.

Wolf, W. J., and J. C. Cowan (1975). "Soybeans as a Food Source," rev. ed. CRC Press, Cleveland, OH.

Wolf, W. J., F. L. Baker, and R. L. Bernard (1981). Soybean seed-coat structural features: Pits, deposits and cracks. *Scanning Electron Microscopy* **3:**531.

3

Processing of Soybeans

Processing soybeans means all of the steps necessary to convert the intact whole beans to usable final products. One of the simplest ways to use soybeans is to feed them directly to chickens or hogs. Some of this has been done in the past and probably will be done in the future, but even with this direct utilization, it is usual to heat the beans by extrusion or some other means to minimize the growth-inhibiting properties of raw soybeans.

The most common end products resulting from processing are edible oil and a protein feedstuff. We shall concentrate our discussion on the steps necessary to produce these products, but the processing of other products such as full fat flours, soy concentrates, soy isolates, and lecithin will also be described.

PREPARATION

As noted in Chapter 1, the moisture content of soybeans must be 14% or below to prevent deterioration during storage. As beans are moved out of storage for processing, they go through a series of procedures in preparation for extraction (Fig. 3.1). The first step is to clean the soybeans. This is necessary to remove foreign material that might contaminate the final products, to remove non–oil-

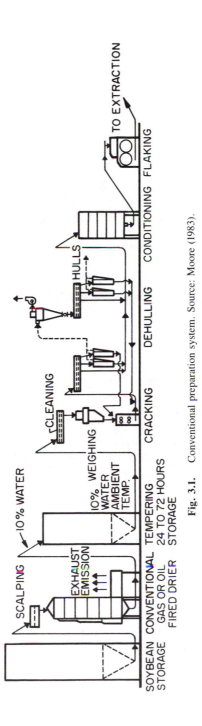

Fig. 3.1. Conventional preparation system. Source: Moore (1983).

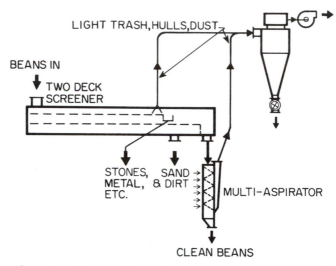

Fig. 3.2. Conventional cleaner with a multiaspirator and cyclone separator. Source: Moore (1983).

containing trash that would reduce the capacity of the extractors, and to remove stones and metal that might damage processing equipment.

Cleaning

Figure 3.2 diagrams a typical cleaner used for soybeans. The soybeans flow onto a two-deck screen sized such that whole soybeans fall through the openings of the first screen. Material larger than soybeans is removed (scalping) as well as light material that can be removed by an airstream. The material smaller than soybeans falls through the holes of the second screen and is discharged as waste. The screens are continuously vibrated to move material across them, and added vibrating rubber balls are useful to prevent clogging. Finally, the beans are put through a multiaspirator to remove fine dust and dirt particles adhering to the soybeans.

Any airstream that cleans by aspiration must be put through a cyclone separator, so that particles are removed before the airstream is discharged to the atmosphere. Cyclone separators have the configuration and dimensions shown in Fig. 3.3. The air to be cleaned enters tangentially at the top, and the centrifugal forces involved allow particles to travel to the periphery and down, while clean air moves to the center and up.

For grade 2 soybeans, 2% foreign material is allowed. In an extraction plant processing 1000 tons daily, this would amount to 20 tons of solid trash removed by cleaning.

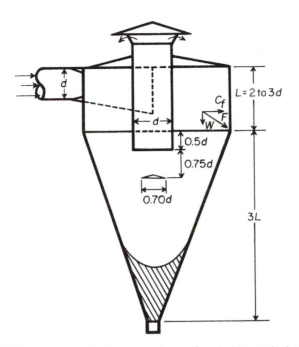

Fig. 3.3. Cyclone separator showing accumulation of particulates at the bottom and center discharge for clean air. Relative dimensions shown are important but overall size depends on quantity to be handled. Source: Henderson and Perry (1955). Copyright by Wiley, Inc.

Drying

Most soybeans processed into meal and oil are first dehulled, because hulls have very little oil content and their presence would decrease the efficiency of the extractors. Also, for some feeds it is desirable to have a high (49%) protein content, which can be achieved only by using dehulled soybeans. For efficient removal of the hulls, it is desirable to have the soybeans at 10% moisture, and so a drying step is required.

For this purpose, it is possible to use the same type of drier that was used originally to bring the moisture content to 14% for storage. Generally, heated outside air is forced through a bed of soybeans, causing some loss of moisture. Then cool outside air is used to remove the warm moist air. The beans are then tempered (held to allow for moisture equilibration within the bean) for 1–5 days.

This type of drier is energy inefficient, because both airstreams are discharged, and all heat is lost. More efficient driers have been designed and built, in which parts of the cooling and drying airstreams are recirculated, saving up to 25% of heating fuel (Moore 1983).

The dried and tempered beans may be put through another screening process and then weighed in preparation for a controlled flow rate through the extraction plant.

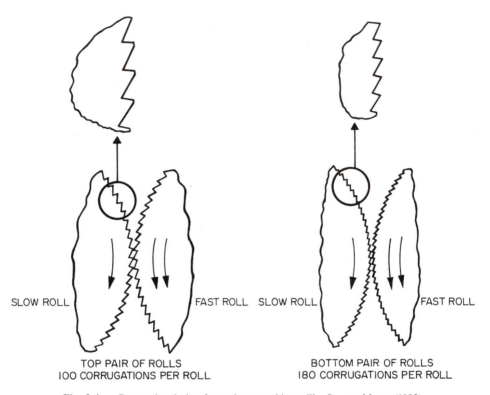

Fig. 3.4.   Corrugation design for soybean-cracking mills. Source: Moore (1983).

Cracking

Conventional cracking mills consist of counterrotating, corrugated, or fluted rolls. There may be a stack of two or three rolls, and soybeans are fed uniformly across the length of the rolls by vibrating feeders. The rolls rotate at different speeds to provide some shearing or nipping action, and the corrugations are fewer and deeper in the first set of rolls compared to the second set (Fig. 3.4).

The size of the cracking rolls is typically 10 in. (25 cm) in diameter with lengths 42 in. (107 cm) or greater. Such cracking rolls can accommodate up to 500–600 tons/day of soybeans. Optimum cracking would give 4–6 pieces or "meats" from each soybean. Of course, this is impossible to achieve practically, since some fines will be produced and some larger pieces.

The hulls are loosened when the beans are cracked and can be separated by aspiration in a multiaspiration process like that shown in Fig. 3.2. The aspirators are set to remove 100% of the hulls—even those hulls that still have meats adhering. These pieces and oversized pieces can be returned to the cracking mills. The meats are sized on vibrating screens, and the fines are separated from the airstream by cyclone separators. It is best to minimize the production of fines because they may cause problems in solvent filtration during extraction and in solvent removal from the oil-bearing miscella. Nevertheless, separated fines are included with the meats to be conditioned for flaking, so that no oil-bearing material is lost.

Conditioning

Conditioning of the meats before flaking is a heating step to give proper plasticity to the soybean particles for optimum flaking. Also moisture may be added during conditioning to achieve 11% moisture in the meats. The heating is done by steam with some direct injection, depending on the amount of added moisture that is needed. Rotary horizontal heat exchangers and vertical stacked types are both used for conditioning.

The heated soy flake is the source of heat for maintaining the solvent extraction system at about 140°F (60°C), and so the temperature achieved during conditioning depends on how much heat loss takes place during flaking and conveyance to the extractor. Generally meats are heated to about 150–160°F (65–70°C).

The next step in preparation is to flake the conditioned meats, but before considering that step, we shall explore the processing needed to produce a full-fat soy flour and some newly proposed alternatives to conventional soybean preparation.

Full-Fat Soy Flour

The meats after separation of the hulls are the raw material for production of full-fat flours. Two types are produced and both are used mainly in the baking industry.

One type is an enzyme-active full-fat flour that is important for its bleaching action on wheat flour. This increases the whiteness due to carotene oxidation associated with lipoxygenase activity. At the same time, one gets some flour oxidation that leads to a dough with better machinability. The limit in the United States is 0.5% of enzyme-active full-fat soy flour in wheat flour. The practice of using soy flour for its bleaching effect on wheat flour is more widespread in Europe than it is in the United States (see Chapter 9).

The other type of full-fat soy flour is made from meats that have been steam treated and dried to inactivate all enzymatic activity.

The grinding of meats to produce flour is done in hammer mills and a fineness is achieved such that 97% passes a 100 U.S. standard screen (Circle and Smith 1978). Full-fat flours are difficult to screen, and so sizing is done by air classification primarily.

A low-cost process for producing full-fat flour for use as a food ingredient in

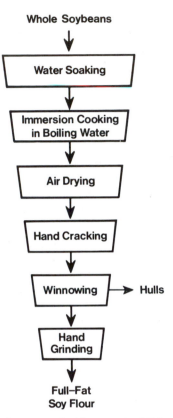

Fig. 3.5.   Flowsheet of a low-cost method for preparing full-fat soybean flour on a village scale. Source: Mustakas *et al.* (1967).

developing countries has been developed by Mustakas *et al.* (1967). Figure 3.5 outlines the steps of the process. A more sophisticated version (Mustakas *et al.* 1970) for producing full-fat soy flour makes use of an extruder for the heating step (Fig. 3.6).

Recent Innovations in Preparation

The abrupt increase in energy costs during the 1970s was an incentive to improve the efficiency of heating steps in soybean processing. One improvement is the use of fluidized beds for the heat exchange steps of drying whole beans and conditioning meats. Fluidized beds are suspensions of solid particles induced by a strong airstream entering from below the particles. The airstream is recirculated to give rapid heat transfer and energy savings.

For preparation of soybeans by a fluid-bed system (Florin and Bartesch 1983), the initial drying step to 10% moisture is done in a fluidized bed. The beans are

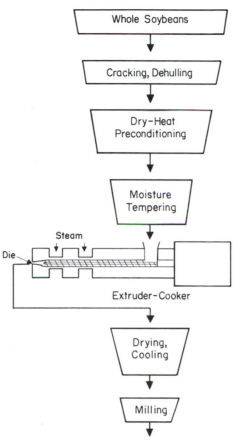

Fig. 3.6. Extrusion cooking process for full-fat flour. Source: Mustakas *et al*. (1970). Copyright by the Institute of Food Technologists.

immediately cracked into halves, and hulls are detached by a combination of cracking mills and hammer mills. Then the warm half-beans are further heated in a fluidized conditioner bed, cracked, and sent to the flaker. As diagrammed in Fig. 3.7, the elimination of cooling and tempering steps of conventional preparation saves time and energy, and the fluidized beds allow finer control and more even heating than conventional heat exchange equipment. This process is named the Escher Wyss hot dehulling system.

A second innovative process that has been designed for soybean preparation is to heat the intact beans with microwave energy under vacuum, crack immediately, and dehull. During dehulling, heat from the magnitron microwave generators is used to heat the airstreams and the product. Thus no conditioning step is required (if proper moisture is maintained), and the meats can be flaked

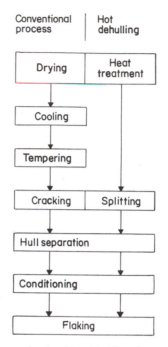

Fig. 3.7.   Comparison of conventional and hot dehulling of soybeans. The initial heat treatment and conditioning are done in fluidized beds. Source: Fetzer (1983).

immediately. Again there is a saving of time and energy. The process outlined in Fig. 3.8 is termed MIVAC (*mi*crowave *vac*uum).

FLAKING

   The conditioned meats are fed directly to flaking mills, which for soybeans are smooth rolls, placed horizontally, with pressure maintained by heavy springs between the two rolls (Fig. 3.9). The size of these rolls is approximately 30 in. (70 cm) in diameter and 48 in. (120 cm) in length. This single flaking step produces soybean flakes about 0.01–0.015 in. (0.025–0.037 cm) in thickness.
   Making thin flakes of the soybean meats in preparation for solvent extraction serves several purposes. These flakes make suitable beds, even when several feet thick, through which solvent can readily flow. The same flow-through capability would not be possible with fine particles. The crushing and shearing action of the flaking rolls tends to disrupt intact cotyledon cells and this disruption may (but this is not certain) facilitate solvent penetration to the lipid bodies.
   A frequently used statement in descriptions of soybean flaking is "flaking disrupts the oil cells." This is difficult to interpret, because it might mean that

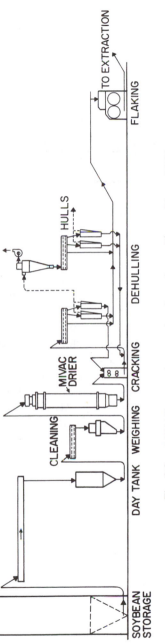

Fig. 3.8. MIVAC preparation procedure for soybeans. Source: Moore (1983).

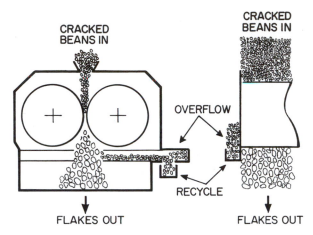

Fig. 3.9.   Flaking rolls and overflow system. Source: Moore (1983).

only some cotyledon cells contain oil, and these oil cells are disrupted, or that there might be subcellular "cells" of oil that are broken by flaking. Neither of these ideas is true. As described in Chapter 2, all the cotyledon cells are rich in oil, which is contained in submicroscopic particles called lipid bodies.

A final advantage of flaking is to reduce greatly the distance the solvent and miscella (solvent–oil mixture) have to travel to complete the extraction process. The influence of flake thickness on extraction of soybean oil is shown in Fig. 3.10. As flake thickness increases, there is a very pronounced increase in the time needed to decrease residual oil to 1%. The role of flake thickness and other factors on solvent extraction of soybean oil will be discussed in more detail when we consider the various theories of solvent extraction.

Flaking rolls demand a lot of attention in preparation of soybeans for solvent extraction. The rolls tend to wear in the center more than at the ends, and this makes the production of uniformly thin flakes difficult. The tension on the springs has to be adjusted frequently, and periodically the rolls have to be reground, so that their surfaces meet. A newly introduced system of overflowing soybean meats at the ends (Fig. 3.9) and recycling helps to give uniform wear to the entire length of the flaking rolls.

A new idea in preparing soybeans for solvent extraction is to feed the flakes into an extruder to produce a material that has better extraction and solvent drainage characteristics than the flakes (Bredeson 1983). The extruder (called an "enhanser" in this instance) is a worm screw inside a solid barrel, which produces high temperatures and pressures. The flakes, after adding moisture to 18%, are extruded. Upon leaving the extruder, the sudden drop in pressure causes some expansion or puffing of the material to produce the desirable extraction and solvent drainage characteristics. Excess moisture has to be removed, and the

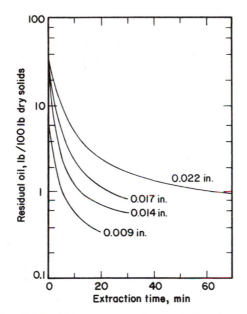

Fig. 3.10. Relation of flake thickness to extraction rate in the solvent extraction of soybean flakes by percolation with hexane. Source: Norris (1982). Copyright by Wiley, Inc.

material is cooled to 140°F (60°C) before solvent extraction. The capacities of the solvent extraction equipment and the desolventizer–toaster are increased with this kind of pretreatment. That advantage has to be weighed against the cost of the extruder and equipment needed for subsequent drying and cooling.

EXPELLERS

The predominant process in Western Europe and the United States for extracting oil from soybeans is by means of solvents. But before discussing that process, we must mention expellers that can also remove oil from oil seeds. For the thousands of small oil extraction plants throughout the world that may need to handle a variety of oil-bearing seeds, expellers are preferred.

Expellers, also called screw presses, consist of a shaft with an attached interrupted worm gear that rotates within a cage (a series of metal bars separated by small openings). As the material to be extracted is introduced at one end of the expeller, it is subjected to high pressures between the edges of the worm gear and the cage. This pressure forces oil out of the material, and the oil flows laterally between the cage bars as the press cake moves parallel to the shaft. Figure 3.11 shows several possible worm arrangements, and Fig. 3.12 shows a two-stage expeller. The first stage is vertical with material being fed by an augur, and the

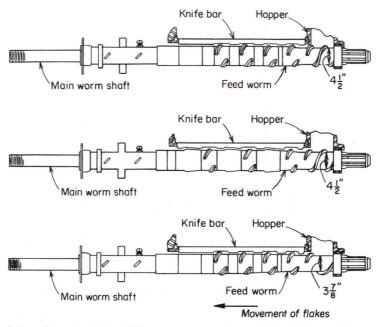

Fig. 3.11. Examples of three different worm arrangements of the Anderson main worm shaft. Source: Norris (1982). Copyright by Wiley, Inc.

second stage is horizontal. Ward (1976) described expeller operation, and Tindale and Hill-Haas (1976) described current expeller equipment.

Material to be extracted by expellers needs to be flaked and cooked before extraction to enhance the oil removal. Modern expellers can handle 50–80 tons/day of material, depending on the type of seed, with a residual oil content in the press cake of 3–4% (Bredeson 1983).

Expellers also are often used as a first extraction step for oil seeds with high oil content such as cottonseed or peanuts (groundnuts). The press cake is solvent extracted as a second step to decrease oil remaining in the cake to less than 1%.

SOLVENT EXTRACTION

After suitable preparation, the soybeans are ready for separation into oil and meal fractions. This is done throughout the world by solvent extraction. This does not mean that only one type of process is involved. Different solvents, different extraction equipment, and different extraction conditions are used. We begin our discussion with the different solvents that have been used and that may be tried.

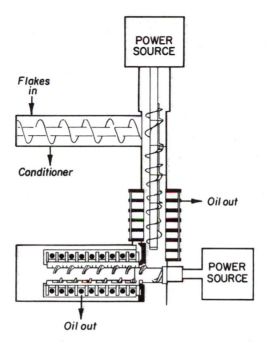

Fig. 3.12.   Arrangement of conditioner and two expellers (vertical and horizontal). Source: Norris (1982). Copyright by Wiley, Inc.

Solvents

To dissolve and remove oil from soybean flakes economically, a solvent must have certain properties. Good solubility for triglycerides is desirable, but also one wants some selectivity so that many unwanted compounds are not dissolved. The solvent or at least the residues of the solvent likely to be found in edible products must be nontoxic. Low specific heat and low heat of vaporization are desirable for low cost of operation. The solvent should not react with the oil seed components or with extraction equipment to form undesirable compounds, nor should it extract undesirable compounds such as pesticides or aflatoxins. Ideally, the solvent would have no inherent safety problems such as explosiveness or flammability and would be cheap and readily available in quantity.

Obviously, no one solvent has all of these desirable attributes, but the solvent that comes closest at the present time is commercial hexane. Some of the properties of commercial hexane are shown in Table 3.1. Other important properties of hexane are heat of vaporization (80 cal/g); specific heat (0.5 cal/g deg); and viscosity at 20°C (0.3 cP). Some undesirable characteristics of hexane are flammability, explosiveness, and high price. Commercial hexane is a product of the petroleum industry, and its price is tied to the price of gasoline. The rapid

Table 3.1

Example of Purchase Specifications for Hexane for Soy Oil Extraction[a]

Property	Value
Specific gravity at 25°C (g/cm^3)	0.6705–0.6805
Distillation range (760 mm)	
Minimum initial boiling point (°C)	65.0
Typical 10% distillation (°C)	67.1
Typical 50% distillation (°C)	67.7
Typical 90% distillation (°C)	68.2
Maximum dry point (°C)	70.0
Maximum nonvolatile residue (g/100 ml)	0.001
Acidity of distillation residue	Neutral
Closed-cup flash point (°C)	−32 to −58
Maximum sulfur	10
Maximum vapor pressure (psia at 35°C)	6.0
Composition (GLC % area)	
n-Hexane	45–70
Methyl cyclopentane	10–25
Total n-hexane and methyl cyclopentane	60–80
Total 2-methyl pentane; 2,3 dimethyl butane; and 3-methyl pentane	18–36
Maximum cyclohexane	2.5
Maximum benzene	0.1
Maximum APHA color	15
General appearance	Free of foreign matter

[a]Source: Johnson and Lusas (1983).

increase in price of petroleum products in the late 1970s has stimulated a search for alternative solvents.

In the 1940s when solvent extraction was just getting started on a large scale, the flammability of commercial hexane led to consideration of trichlorethylene (TCE) (CCl_2=CHCl) as a suitable solvent for oil extraction from soybeans. TCE has the desirable characteristics of being nonflammable, a good solvent for triglycerides, and readily available in quantity. It had a high price, but the other characteristics made it sufficiently useful that commercial extraction plants started using it in 1951. However, its use was short-lived. When soybean meal from these plants was fed to cattle, it proved to be toxic. The problem was the formation of a derivative compound of cysteine (S-*trans*-dichloro-vinyl-L-cysteine) that caused death in cattle when the meal was fed at the rate of 2–3 lb/day. This incident has greatly decreased experimentation with chlorinated hydrocarbons as potential solvents, but other chlorinated hydrocarbons such as dichloromethane may be useful without the toxic effects of TCE.

Alcohols, particularly ethanol and isopropanol, are useful solvents for oil extraction from soybeans, even though at 68°F (20°C) alcohols dissolve very

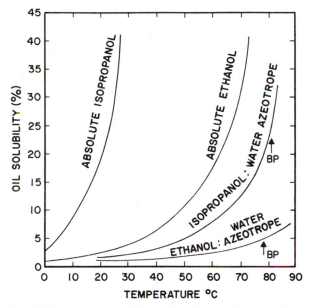

Fig. 3.13.   Solubilities of cottonseed oil in alcohols. Source: Johnson and Lusas (1983).

little triglyceride. Figure 3.13 shows how solubilities of oil change with temperature and with water content of ethanol or isopropanol. At higher temperatures ethanol and isopropanol are good solvents for soybean oil, and therein lies their utility. One can extract at high temperature, cool the miscella, and get a separation into two phases of oil and alcohol at the lower temperature. The oil still contains some alcohol that needs to be stripped out, but the cost for solvent removal from crude oil is greatly reduced.

Karnofsky (1981) has described a four stage process for alcohol extraction in which the solvent and flakes to be extracted move countercurrently (Fig. 3.14). In the first stage, full-fat flakes are contacted with dilute alcohol to remove carbohydrates, phosphatides, and free fatty acids but not oil. The second stage is used to equilibrate the flakes to full-strength alcohol. In the third stage, the extraction of oil takes place using a constant boiling azeotrope of alcohol and water. The miscella is cooled to cause the phase separation, and after removal of oil and reheating, the lean miscella is recirculated through the third-stage extractor. The fourth stage is used to strip remaining oil from the flakes by freshly distilled alcohol coming from desolventizing the flakes and from oil stripping.

The process just described is not currently used for soybeans to our knowledge, but hot alcohol extraction has been used in Darien, Manchuria (Johnson and Lusas 1983).

Baker and Sullivan (1983) have investigated aqueous isopropyl alcohol (IPA) mixtures as an extracting solvent for soy oil. They found an azeotrope of 87.7%

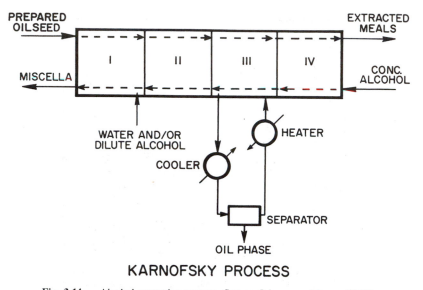

KARNOFSKY PROCESS

Fig. 3.14.   Alcohol extraction process. Source: Johnson and Lusas (1983).

IPA was useful for extraction at 170°F (77°C) and that the extracted oil could be separated by chilling the miscella to 86°F (30°C). The crude soybean oil contained less phospholipids and less free fatty acids than hexane-extracted oil.

In the search for cheap, nontoxic, and nonflammable solvents for soybean oil there has been recent interest in carbon dioxide, particularly at the USDA Northern Regional Research Center in Peoria, Illinois. If a gas is put under sufficient pressure, the gas although not liquefied has some of the solvent and flow characteristics of a liquid. Such gases are called supercritical fluids. Supercritical carbon dioxide has been shown to be an effective extraction solvent for soybean oil (Friedrich and Pryde 1984). Carbon dioxide has the advantages of being cheap, readily available, nontoxic (in the amounts used), nonflammable, and readily removed from the oil by simply reducing the pressure. The oil extracted by supercritical carbon dioxide is equivalent to hexane-extracted oil except that less phospholipid is extracted. The disadvantage of this process is the expensive equipment needed to extract large quantities under pressure. Pressures involved are 1000–10,000 psi.

Although water is not considered a solvent for triglycerides, experimental work has been done at the Food Protein Research and Development Center, Texas A&M University, to separate oil and protein fractions in an aqueous process (Lawhon et al. 1981). The process involves making an aqueous dispersion of finely ground oil seed, centrifuging, and breaking the emulsion (lipid bodies) of the floating layer. Recovery of oil varies widely with this procedure depending on fineness of grind, contact time, and type of oilseed, but the procedure is not as effective as solvent extraction with hexane in which residual oil in flakes is less than 1%.

Theories on Solvent Extraction

The design of solvent extraction equipment and the selection of proper solvents could be helped by a thorough understanding of the extraction process. For this reason considerable study has been devoted to the variables that affect the solvent extraction of soybean oil. While the process has been well studied and much has been learned, we still do not have a complete understanding of solvent extraction and several questions remain unanswered.

The general concept is that hexane dissolves soybean oil from the lipid bodies almost immediately on contact. The time-consuming part of the extraction process is for the solvent to diffuse into the soybean flakes and for the oil-rich miscella to diffuse back out. The intact soybean or soybean cotyledon is almost impenetrable by hexane. Whole soybeans (with or without the seedcoat) can be soaked in hexane for days with virtually no removal of oil. This is the reason that the soybeans are cracked and flaked—to facilitate hexane entry and miscella removal.

The data shown in Fig. 3.10 indicate the dependence of time of extraction on flake thickness. This plot is the usual way of depicting how soybean oil is extracted: log of the residual oil (dry basis) vs. time. One would think that the thinner the flake the better the extraction, but if flakes are too thin, they tend to crumble and produce fines. Fines are bad in that solvent will not easily penetrate and flow through packed fines as it will through a bed of flakes. The usual flake thickness is 0.010–0.015 in. (0.025–0.037 cm). Othmer and Agarwal (1955) calculated that the rate of extraction of oil was related to flake thickness and remaining oil concentration as follows:

$$-dC/dt = kF^{-3.97}C^{3.5}$$

where C is the concentration of oil in the flake, t the time, k the proportionality constant, and F the flake thickness. Since the exponent of F is almost 4, a doubling of flake thickness would decrease the oil extraction rate 16 times.

Flaking of soybeans is presumed to cause rupture of cells as well as decreasing the distance the solvent has to travel. Observation of Fig. 3.10 shows that a large fraction of the oil to be extracted is rapidly removed in the early part of the extraction. This "easily removed oil" may come from ruptured cells; at least this is one explanation that has been used. However, soybeans that have been ground and sized, so that particles have the same dimensions as flake thickness, also have easily removed oil. In this case, one would not expect a large number of cells to be ruptured in the small particles. Hence this is one of the unanswered questions of soy oil extraction: to what extent does cell rupture facilitate extraction?

The effect of diffusion rates in oil extraction has been investigated by Boucher *et al.* (1942). Using porous ceramic platelets impregnated with oil, they found that the extraction rate could be explained by:

$$E = \frac{8}{\pi^2} \sum_{n=0}^{n=\infty} \frac{1}{(2n+1)^2} \exp\left[-(2n+1)^2 \left(\frac{\pi}{2}\right)^2 \left(\frac{D\theta}{R^2}\right)\right]$$

where E is the fraction of total oil remaining, D the diffusion coefficient, θ the time, and R half the plate thickness.

This rather complex formula is simplified greatly as θ increases to give

$$E = \frac{8}{\pi^2} \exp\left(-\frac{\pi^2 D\theta}{4R^2}\right)$$

or $\log_{10}E = -0.091 - 1.07D\theta/R^2$. Thus if the log of the fraction remaining is plotted against time, a straight line should result, with the slope depending on the diffusion coefficient and plate thickness. Figure 3.15 shows such a plot, with line A representing the porous plate. The straight line for porous plates indicates diffusion control of the extraction process. In Fig. 3.15 for comparison purposes, oil extraction rates are shown for peanut slices carefully prepared to be 0.026 in. (0.065 cm) thick by microtome slicing (B), and cottonseed flakes 0.017 in. (0.043 cm) thick (C). A considerable portion (about half) of the peanut oil is extracted rapidly, and the remainder is diffusion controlled as indicated by the straight line. For cottonseed flakes (and soybean flakes behave similarly), none of the extraction process seems to be strictly diffusion controlled. In the production of flakes, there would be volumes within the flake that are easily penetrated by solvent (voids or air spaces) and volumes consisting of relatively undisturbed tissue that would be slowly penetrated by solvent. Perhaps this combination explains the kinetics of extraction shown in Fig. 3.15C.

Diffusion coefficients are influenced by the moisture content of the sample when hexane is the extracting solvent. Fan *et al.* (1948) calculated that the diffusion coefficient decreased about 0.4×10^{-9} cm²/sec for each 1% moisture increase (in the range 10–22% moisture). Absolute values for diffusion coefficients in peanut slices are about 5×10^{-9} to 7×10^{-9} cm²/sec.

Another factor influencing the rate of extraction is that soybean oil is not a pure compound. Soybean oil consists of triglycerides, pigments, free fatty acids, and phospholipids along with other minor components discussed in Chapter 2. Karnofsky (1949) has emphasized that the composition of the extracted oil changes as the extraction proceeds, with the major difference being an increased percentage of phospholipid in the later fractions. This certainly would have an influence on the rate at which remaining traces of oil are extracted, because hexane is not a good solvent for phospholipids. But soybean triglyceride is still

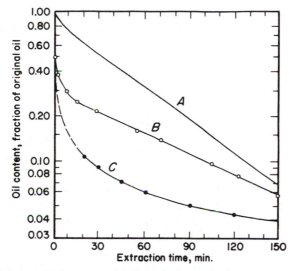

Fig. 3.15. Solvent extraction curves: (A) theoretical curve for homogeneous oil-impregnated platelets of uniform thickness; (B) peanut slices, 0.026 in. (0.66 mm) thick, 14% moisture, extracted at 76°F (25°C) with commercial hexane; (C) cottonseed flakes, 0.017 in. (0.43 mm) thick, 10–25 mesh, 11.6% moisture, extracted at 150°F (67°C) with commercial hexane. Source: Norris (1982). Copyright by Wiley, Inc.

being extracted in the later fractions, and the reason for that hard to extract fraction is another unanswered question.

Coats and Karnofsky (1950) made the interesting observation that soybean oil added back to extracted flakes and reextracted is extracted faster the second time. However, the same general pattern exists for the second extraction: that is, a majority of the oil being extracted very rapidly, but the small amounts of remaining oil being extracted at constantly decreasing rates. One possible explanation for this behavior is that soybean triglyceride is bound to soybean particles of protein and carbohydrate in the same way that soybean triglyceride is bound to the polar surface of silica in chromatography. However, if this binding were happening, making the solvent more polar should readily elute the bound triglyceride.

The effect of increasing the polarity of hexane by incorporating 5% of various alcohols has been investigated by Ayres and Scott (1952). They found that the rates of extraction were increased when 5% methanol was incorporated with hexane, and this increase was particularly noticeable when relatively small amounts of extractable oil remained in the flakes. So there does seem to be some evidence for binding of small amounts of lipids by the flakes.

Temperature is an important variable in the extraction of oil from soybeans

because it influences several other variables. For example, increasing temperature decreases the viscosity of the extracting miscella and thereby speeds up
diffusion. Temperature increases also increase the solubility of soybean triglyceride in hexane. Normally the solvent extraction is done at a temperature close
to the boiling temperature of commercial hexane, which is about 149°F (65°C)
(Table 3.1).

In addition to the factors discussed thus far, it is interesting to know that the
rates of oil extraction differ for different kinds of oilseeds. Coats and Wingard
(1950) used the formula $T = KD^n$, where T is the time needed to extract to 1%
residual oil and D the flake thickness, to compare the constants K and n for
different oil seeds. K is a measure of the ease of extraction of oilseed flakes of
constant thickness, and n is a measure of the influence of changes in flake
thickness on extraction rate. Some approximate values for K were found to be 6–
20 for soybeans, 140–270 for cottonseed 3,600 for flaxseed, and 1.4 for peanuts.
Values for n were 2.3–2.5 for soybeans, 1.5 for cottonseed, 7 for flaxseed, and
3.2 for peanuts. Flaxseed seems to be very difficult to extract, but decreasing
flake thickness has a larger effect on the rate than for any of the other oilseeds
tested.

Equipment

Although solvent extraction equipment for batch processing exists, most of the
equipment is designed for continuous, countercurrent extraction. The basket-
type extractor shown in Figs. 3.16 and 3.17 is one of the first designs for
continuous high-capacity extraction and is still in wide use. The baskets measure
2×4 ft (60×120 cm) and about 18 in. (45 cm) deep and have perforated
bottoms. They travel on an endless chain vertically. The flakes enter the descending leg at the top and concurrently are sprayed with half miscella that percolates
through the beds of flakes formed in each basket. The miscella collecting in the
bottom of the descending leg is full miscella ready for filtering to remove fines
and solvent stripping. As the baskets move into the ascending leg, they are
extracted by a second stream of solvent that starts as pure solvent at the top.
Hence in the ascending leg flakes move countercurrent to the solvent flow. The
miscella collecting in the bottom of the ascending leg is called half miscella and
is the source of the solvent for the descending baskets and flakes. At the top of
the ascending leg the baskets are allowed to drain and then are dumped to
conveyors that take them to the desolventizer–toasters.

The extractor is sealed, and total time of extraction is approximately 1 hr.
Solvent to flake ratio by weight is 1:1.

A more compact arrangement for percolation of solvent through soybean
flakes is provided by the Rotocel (manufactured by the Dravo Engineers and
Constructors) shown in Fig. 3.18. In this extractor the full-fat flakes are charged

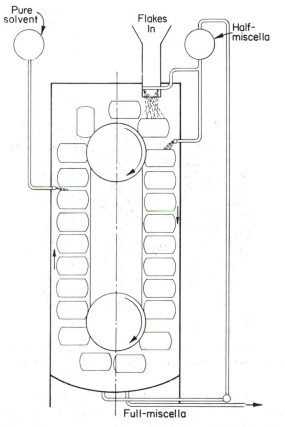

Fig. 3.16.   Basket solvent extractor. Source: Norris (1982). Copyright by Wiley, Inc.

to rotating bins. As the bins rotate, solvent is sprayed over them such that there is a counterflow. Just before dumping for conveyance to the desolventizer–toaster, the extracted flakes are contacted by fresh solvent. The solvent percolates through the flakes and collects in holding tanks in the bottom of the Rotocel. This lean miscella is pumped to the next bed of flakes, and so forth, until the newly entering flakes are used to filter the full-fat miscella before it is stripped of solvent.

A different arrangement for solvent percolation has the baskets of flakes remaining stationary while the solvent sprays move to achieve a countercurrent extraction (Fig. 3.19). This piece of equipment is manufactured by the French Oil Mill Machinery Co.

A fourth arrangement of solvent–soybean flake contact by percolation is manufactured by the Crown Iron Works Co. (Fig. 3.20). A conveyor belt carries

Automatic filling device

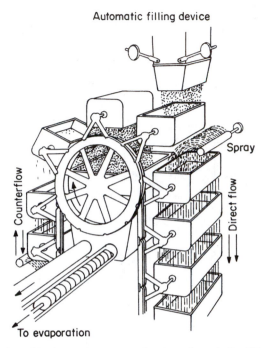

Spray

Counterflow

Direct flow

To evaporation

Fig. 3.17.   Interior of basket-type extractor, showing schematically filling and dumping of baskets and flow of solvent. Source: Norris (1982). Copyright by Wiley, Inc.

horizontal baskets that are loaded with full-fat flakes in the top section. Fresh solvent enters at the lower right and is sprayed over the flakes just before being dumped for conveyance to the desolventizer–toaster. Miscella collects in the several bottom sections of the extractor, where it is picked up by pumps and sprayed over the next section of baskets in a countercurrent manner.

The capacities of solvent extraction plants for soybeans range from a few hundred tons per day to 4000 tons per day. Requirements for new solvent are roughly 1 gal per ton of soybeans processed per day. With the recent increase in price of hexane, this represents a serious operating loss, and with good management solvent losses can be cut approximately in half.

The solvent extraction plants generally have two products: a defatted and desolventized toasted flake and desolventized crude soybean oil. We need to consider how solvent is removed from these two products.

Miscella Stripping

The full-fat miscella contains 25–30% soybean oil, and the 70–75% hexane has to be removed for reuse in solvent extraction. The minimum amount of hexane that needs to be removed is based on the required flash point of the

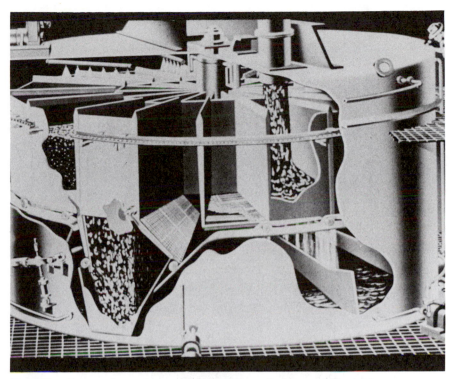

Fig. 3.18.   Rotary solvent extractor. Courtesy of Dravo Engineers and Constructors.

resulting crude oil. A flash point of 250°F (121°C) corresponds to 1000 ppm residual hexane.

A typical system for removing hexane from miscella is shown in Fig. 3.21. The main components are first- and second-stage evaporators and a stripping column. The evaporators are shell-in-tube type, rising film, with miscella entering the tubes at the bottom and collecting in a vapor dome at the top (Fig. 3.22). The heat source for the first-stage evaporator is the excess steam and solvent vapor coming from the desolventizer–toaster. Often there is more heat energy coming from the desolventizer–toaster than can be effectively utilized in the first-stage evaporator, and it is used to heat the solvent condensate in a vapor contactor before the solvent is returned to the solvent–water separator.

The second-stage evaporator makes use of steam as the heat source, and together with the first-stage evaporator concentrates the crude soybean oil to about 90%. As the solvent becomes a smaller part of the miscella, it also becomes harder to remove. The relationship between solvent vapor pressure and temperature (shown in Fig. 3.23) is that solvent vapor pressure decreases as the solvent concentration decreases. For this reason, the final stage of solvent removal is a steam-stripping column such as the one diagrammed in Fig. 3.24. The

Fig. 3.19.   Stationary basket solvent extractor. Courtesy of French Oil Mill Machinery Co.

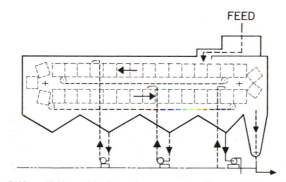

Fig. 3.20. Horizontal-basket solvent extractor. Source: Milligan (1976).

entire system for removing solvent from soybean oil operates under vacuum. The separated solvent is piped to a solvent–water separating tank in which water is allowed to settle out of the mixture before the solvent is removed. This is necessary because in the stripping column and in the desolventizer–toaster steam is mixed with solvent vapor, and condensers subsequently yield solvent–water mixtures.

Solvent Removal from Flakes

The flakes coming from the extractor will contain about 30% by weight of solvent. The most commonly used piece of equipment for producing solvent-free meals for feeding purposes is a desolventizer–toaster (frequently referred to as a DT). The purpose of the desolventizer–toaster is to remove solvent and to heat the meal sufficiently to optimize its nutritional value. This requires inactivation of trypsin inhibitor and denaturation of soy protein to make it susceptible to attack by proteolytic enzymes. Other types of desolventizing equipment are used for soy flakes going into flours, concentrates, or isolates.

Figure 3.25 shows a typical desolventizer–toaster in which the solvent-laden flakes enter at the top and are contacted by steam, so that some of the steam is condensed. This heat of condensation is used to evaporate hexane, and the added moisture is useful in increasing the effectiveness of the subsequent heating. The desolventizer–toaster is fitted with trays that may be heated by steam or that in some cases serve as the source of sparging steam. The upper three trays make up the desolventizing section and the lower four trays the toasting section. Each tray is swept by arms attached to the central rotating shaft to stir up the beds of flakes and to keep them moving down through the unit at a controlled rate through openings in the trays. Flakes are discharged at a moisture content of about 20% and a temperature of 212°F (100°C), and so subsequent drying and cooling units are needed. The arrangement of the desolventizer–toaster and auxiliary equipment can be seen in Fig. 3.26. Recently, drying and cooling capacity have been

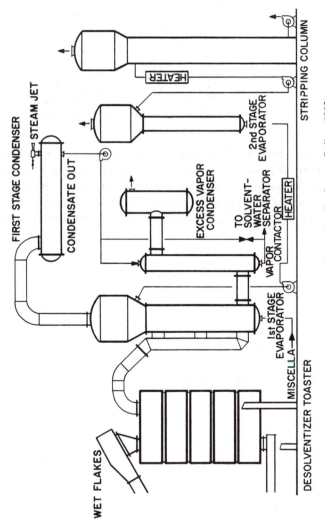

Fig. 3.21. System for removing solvent from miscella. Source: Boling (1982).

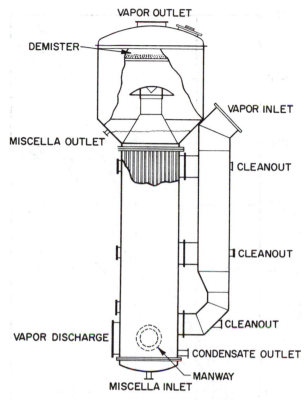

Fig. 3.22.   First-stage evaporator. Source: Boling (1982).

built into the desolventizer–toaster, so that all of these processes can be achieved in one unit.

The protein dispersability index (PDI) is a measure of the amount of heat received by soy protein and will be described further in Chapter 5. PDIs for flakes coming from the desolventizer–toaster range from 10 to 30, indicating that considerable heating occurred. For soy flakes to be used to prepare soy protein isolate, it is essential that the protein remain soluble or dispersable. Hence, different kinds of desolventizers are needed.

A flash desolventizer–deodorizer system is shown in Fig. 3.27. The solvent-laden flakes are introduced into a stream of superheated hexane vapor, which serves to evaporate most of the hexane. The flakes and solvent vapor are separated in a cyclone, with excess solvent being recovered in a condenser, but a portion of the vapor is recirculated to the superheater.

The flakes are fed into a deodorizer in which remaining solvent vapor is stripped from them. With this kind of system, the PDI of the flakes can be 70–90, meaning good solubility of protein for certain food uses (see Chapter 8).

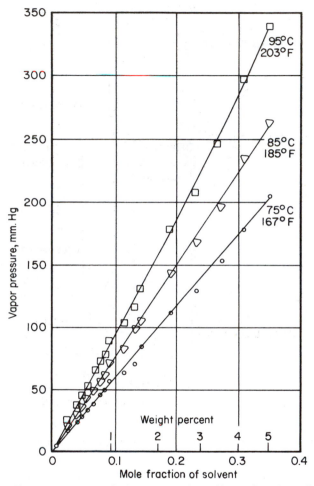

Fig. 3.23.   Vapor pressure curves for low concentrations of hexane in soybean oil. Source: Smith and Wechtar (1950).

A second type of desolventizer is shown in Fig. 3.28. Again the flakes are contacted by hot hexane vapor as a heat source and solvent removal vehicle. The excess hexane is removed by a condenser not shown. Flakes are moved through the desolventizer by a screw conveyor and then through a lock to a mixer, where water can be added. Final removal of hexane is achieved in a deodorizer, where sparge steam is the source of heat. This desolventizer is versatile, because the PDI of the final product can be controlled based on the amount of water added, the temperature of the deodorizer, and the time in the deodorizer. As time, temperature, and moisture increase, the PDI of the final product decreases.

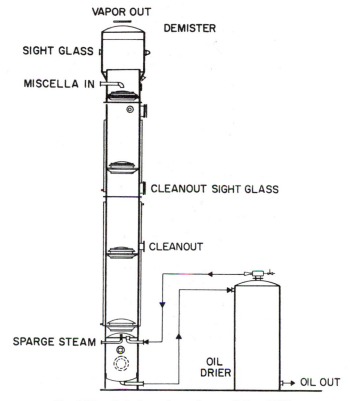

Fig. 3.24. Stripping column. Source: Boling (1982).

Venting

The operation of the solvent extraction plant requires vaportight equipment and seals wherever hexane is used, but the equipment must also be vented. For maintenance of air quality and for recovery of valuable solvent, it is necessary to remove hexane from the vent air before discharging it to the atmosphere.

Figure 3.29 shows a solvent recovery system for vented air consisting of a cold water condenser and a mineral oil adsorption column. After passing through the condenser, the vent air is passed countercurrently through a vertical column of cool mineral oil, which is an efficient way of removing the last traces of hexane from the air. The mineral oil is then heated and passed through a second column with steam moving countercurrently to strip the hexane from the mineral oil.

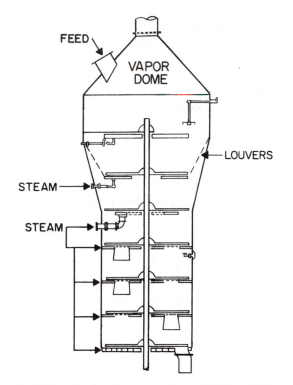

Fig. 3.25. Desolventizer–toaster. Source: Milligan (1976).

OIL REFINING

The general term "refining" refers to all the steps needed to prepare the crude soybean oil for consumption. This type of refining needs to be distinguished from the more specific term "alkali refining" which refers to addition of alkali for the purpose of removing free fatty acids from the crude oil. The refining steps include degumming, alkali refining, bleaching, hydrogenation, winterization, and deodorization. We shall consider each of these processes in the order they normally are used. The refining steps (except for degumming) are usually done at a facility separated from the solvent extraction plant.

Degumming

The purpose of this refining step is to remove the phospholipids that are extracted along with soy oil. Phospholipids can be a problem in that they may cause increased losses of oil during the alkali-refining step due to their emulsification properties. Also, phospholipids can cause excessive browning during deodorization and if they settle out in the hold of an ocean tanker, they are very

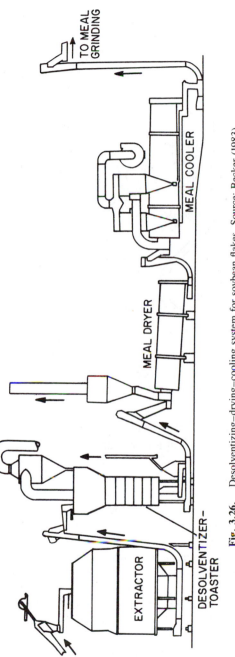

Fig. 3.26. Desolventizing–drying–cooling system for soybean flakes. Source: Becker (1983).

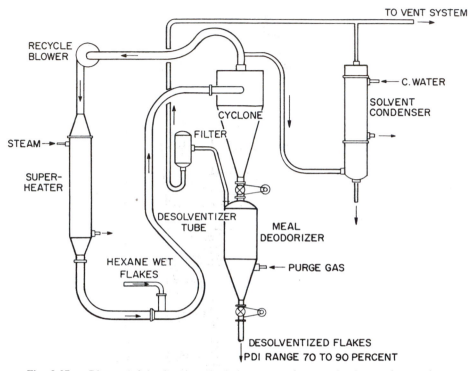

Fig. 3.27. Diagram of desolventizer–deodorizer system for removing hexane from soybean flakes and maintaining protein solubility. Source: Wolf and Cowan (1975).

difficult to remove. Hence, all soybean oil for export must be degummed. It is estimated that about one-third of all soybean oil is degummed (Brekke 1980a).

Phospholipids can be recovered as soybean lecithin (a mixture of the phospholipids in soybean oil and not just phosphatidyl choline) and marketed as an important food additive. The recovered phospholipids also may be added back to soybean flakes to enhance their energy value for feeding purposes. Phospholipids can act as antioxidants in conjunction with tocopherols and in that sense can be a valuable soy oil component.

The amounts of phospholipids in crude soy oil range from 1.5 to 3% on a weight basis or about 500 to 1,000 ppm phosphorous. There is a factor of approximately 30 that is used to convert weight of phosphorous to weight of phospholipid (Wiedermann 1981). As discussed in Chapter 2, the predominant phospholipids are phosphatidyl choline, phosphatidyl ethanolamine, and phosphatidyl inositol.

The procedure for degumming consists of adding water to the oil, mixing well, heating, and centrifuging (Haraldsson 1983). The added water attracts the phospholipids, and they become hydrated, that is, they concentrate in the water phase

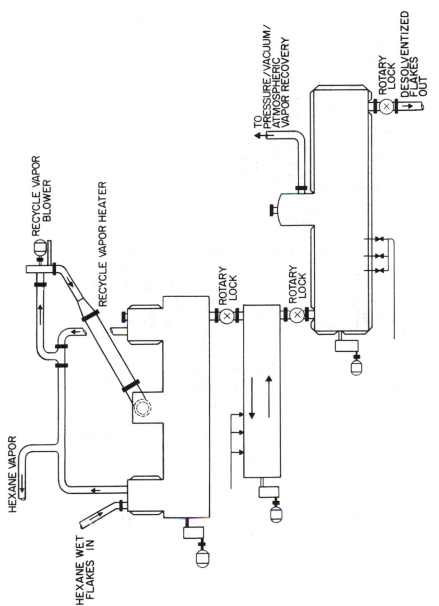

HEXANE VAPOR

RECYCLE VAPOR BLOWER

RECYCLE VAPOR HEATER

HEXANE WET FLAKES IN

ROTARY LOCK

ROTARY LOCK

TO PRESSURE/VACUUM/ ATMOSPHERIC VAPOR RECOVERY

ROTARY LOCK

DESOLVENTIZED FLAKES OUT

Fig. 3.28. Vapor desolventizer–deodorizer system for removing hexane from soybean flakes. Source: Wolf and Cowan (1975).

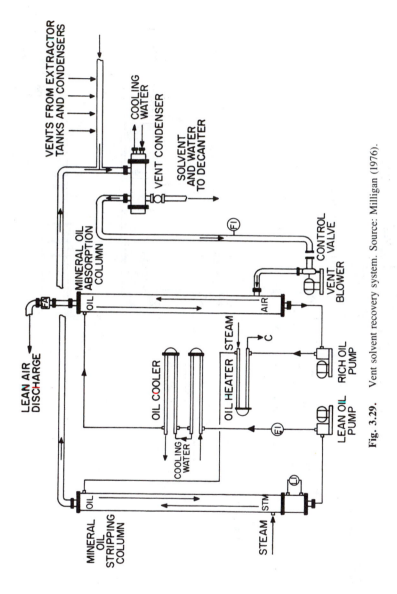

Fig. 3.29. Vent solvent recovery system. Source: Milligan (1976).

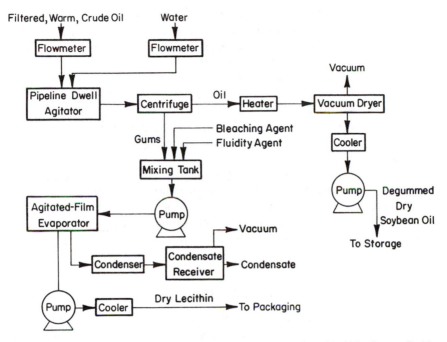

Fig. 3.30.   Flowsheet for degumming soybean oil and production of lecithin. Source: Brekke (1980a).

and are easily removed by centrifuging. The heat helps to break any emulsions that might form. Amounts of water used are about 75% of the weight of the phospholipid present. If too much water is used, a good separation is not achieved during centrifugation, and either too little phosphorous is removed or too much oil is lost with the gums.

Degumming removes about 3.5% of the soy oil. This fraction is about 25% water and 75% oil-soluble material. Of the oil-soluble material, about one-third is oil and the rest is phospholipid (acetone-insoluble fraction). The gums are dried after separation, and if recovered for use as soybean lecithin may be treated to give any one of several different lecithin products. The differences in soybean lecithins are in the fluid consistency (fluid or plastic) and in color (natural, bleached, or double bleached).

A flow diagram for the degumming process and subsequent handling of products is shown in Fig. 3.30. If the degummed soybean oil is going to be stored, it must first be dried under vacuum to avoid hydrolysis, which will produce free fatty acids. If the soybean oil is going immediately to alkali refining, no drying step is needed.

With the best degumming procedures and excellent starting material, the phosphorous content of the soybean oil is decreased to about 10% of the original amount. Trading specifications call for a maximum of 0.02% (0.025% maximum with a price discount) of phosphorous in crude degummed soybean oil. The phosphatide composition of the remaining phosphorous compounds is not greatly different from the composition of the compounds removed.

For various reasons (mainly field or storage damage) some soybeans contain what are known as nonhydratable phosphatides. These phosphorous compounds are not separated by the water washing step and tend to remain with the oil phase. One prevalent idea as to why phospholipids are nonhydratable is that they are calcium and magnesium salts of phospholipids and have greater oil solubility than water solubility (Hvolby 1971).

An aid to removal of phospholipids (and particularly nonhydratable phospholipids) is to acidify the oil with phosphoric acid before adding water. Also, the phosphoric acid addition [0.2% of 85% phosphoric acid at 158–194°F (70–90°C)] can be done just prior to alkali refining. Phosphoric acid does tend to darken the recovered lecithin fraction, and so it is not used if light-colored lecithin is desired.

Alkali Refining

The purpose of this refining step is to remove free fatty acids from soybean oil. The free fatty acids are a problem in finished food oils because they promote foaming and lower the smoke point of heated oil. The normal procedure is to add alkali (NaOH is most commonly used) to convert free fatty acids to soaps and to remove the soap stock from the oil by centrifugation. Although removal of free fatty acids is the main objective in alkali refining, other nontriglyceride material such as phospholipids, pigments, and insoluble materials will also be removed. The soybean oil refiner may receive both degummed and nondegummed oil as starting material (these oils may be blended before alkali refining), but only one procedure is used in refining them.

The free fatty acid content of soybean oil is low in comparison with some other oils. Soybean oil ranges from 0.3 to 0.7% free fatty acids, and the alkali refining process lowers this to <0.05%. For oils coming from wet tissues such as palm oil, hydrolysis of triglycerides to glycerol and free fatty acids is a real problem. Such crude oils may range from 5% upward in free fatty acid content.

The procedure for contacting crude soybean oil with alkali may be in batches or continuous. Most larger refiners make use of a continuous process. The oil to be refined is weighed into a tank sufficient to hold enough oil for an 8-hr run. The oil is well mixed and sampled for analysis of the free fatty acid content. This analysis is important as a basis for determining the amount of alkali needed. Most often NaOH (caustic) is the base used, but other sources of alkali have been used. A formula for calculating the amount of alkali needed is

$$\text{alkali (wt\%)} = \frac{\text{free fatty acids (\%)} \times 0.142 + \text{excess NaOH (\%)} \times 100}{\text{NaOH (\%) in alkali}}$$

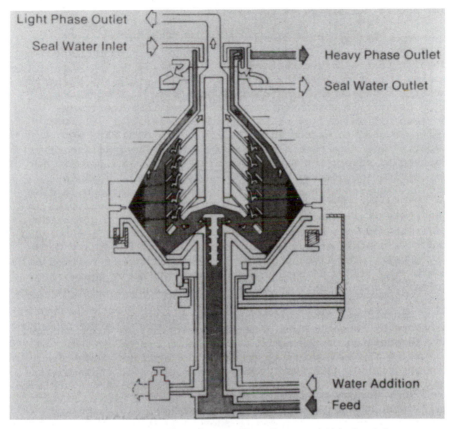

Light Phase Outlet

Seal Water Inlet

Heavy Phase Outlet

Seal Water Outlet

Water Addition

Feed

Fig. 3.31. DeLaval continuous self-cleaning centrifuge. Courtesy of Alpha-Laval, Inc.

Some excess alkali is used based on other impurities or other circumstances. For example, if phosphoric acid has been used as a pretreatment for better removal of phospholipids, excess alkali would be needed to neutralize the phosphoric acid. The strength of NaOH used is generally 12–13%.

In the continuous process, heated oil and alkali are metered continuously into a flow-through mixing tank and then to a centrifuge for separation. The heavy phase is soap stock, and the light phase is refined oil. The centrifuge (Fig. 3.31) is hermetically sealed to minimize contact with air, and the zone of separation into the two phases can be controlled by back pressure on the light phase. By increasing pressure on the light phase, the amount of soap in oil can be reduced, but the amount of oil lost in the heavy phase is increased. Operator experience determines the optimum centrifugal conditions for minimizing oil losses and still achieving a good separation. The centrifuge shown in Fig. 3.31 has the feature of being self-cleaning. That is, the sediment (the dark material at the outermost edge) can be removed while the centrifuge is running by briefly opening the side ports. This alleviates stopping the centrifuge for manual cleaning.

The oil at this stage still contains traces of soap and must be washed once with hot water [about 200°F (94°C)] and centrifuged. After this second centrifuging, the oil is dried under vacuum and stored prior to bleaching or hydrogenating.

The heavy phase contains soap stock, which may be sold to soap manufacturers or incorporated into animal feeds. To convert the soaps back to fatty acids, they are acidulated and separated again by centrifugation.

The process just described is most commonly used, but alternatives have been proposed. It is possible to alkali refine a 40–50% miscella, which saves some of the water washing and may result in less refining loss. The refining loss refers to the amount of soybean oil lost in the soap stock and depends on both the kinds of impurities present in the oil and the skill of the refiner. To measure refining loss, a sample of oil is put through a laboratory-scale alkali refining, and actual losses are measured. This procedure is known as the official cup method (Ca 9a-52, American Oil Chemists' Society 1982).

A more rapid procedure is to measure the neutral oil content by a chromatographic procedure (Ca 9f-57, American Oil Chemists' Society 1982). The neutral oil content when subtracted from the total oil content gives an absolute minimum refining loss. The usual loss for soybean oil is 3 times the free fatty acid content and may be as great as 5–10 times the free fatty acid content. Phospholipids are particularly bad in causing neutral oil to be incorporated into the soap stock fraction. This oil cannot be recovered as high-grade food oil and represents a direct economic loss to the refiner. Refining procedures are often controlled to minimize losses rather than to remove the maximum amount of impurities.

Physical refining is an interesting alternative to alkali refining. If phospholipids in crude oil can be reduced to a low level by phosphoric acid treatment, then it is possible to distill free fatty acids during the deodorization process (to be described later). Phospholipids have to be removed first, because the high heats needed to distill fatty acids cause darkening of oil with phospholipids present. This darkening is very difficult to correct once it takes place. The quality of oil produced by physical refining is no better or worse than that from alkali refining, but economic advantages in operating costs and in initial equipment costs are claimed for physical refining (Mounts and Khym 1980).

Bleaching

In the normal processing of a fully refined soybean oil, bleaching follows alkali refining. Bleaching refers to color removal, and while this is the main purpose of the bleaching step, other impurities such as flavor compounds, phospholipids, traces of soaps remaining from alkali refining, and hydroperoxides are removed as well.

Bleaching sounds like a chemical oxidation process (and some color removal may be chemical), but for the most part bleaching of soybean oil involves

Table 3.2
Typical Characteristics of Adsorbents[a]

Characteristic	Natural earth	Activated earth	Activated carbon
Apparent bulk density, lb/ft³	50	45	30
Oil retention, lb oil/lb adsorbent	0.2–0.3	0.3–0.5	1.0–1.5
Relative bleaching activity	1	1.5–2	—
Soap removal ability	Good	Better	Superior
Filtration rate	—	—	Very poor
Flavor/odor imparted to bleached oil	Musty	Musty	None
Surface area, m²/g	68	165–310	500–900
pH of 10% suspension in distilled water	8	2.8–6.0	6.0–10.0

[a]Source: Brekke (1980b).

adsorption on finely divided solid adsorbents. The solid adsorbents are various kinds of natural earth products and may or may not be acid activated. Also, activated carbon is a useful adsorbent in the oil industry. Table 3.2 shows several properties of different adsorbents. The pigments to be removed are carotenes, xanthophylls, chlorophylls, and browning compounds that may have resulted from mishandling of the oil at some stage. Normally there is little chlorophyll present, but if it is present due to an early freeze and immature beans being harvested, it is a particularly difficult pigment to remove. Activated carbon is useful for chlorophyll removal and may be added to an acid earth in 5–20% amounts. Activated carbon is also useful in removing browning pigments.

As described in Chapter 2, the Lovibond system is traditionally used in the soybean oil industry to specify color. Even when determined by a spectrophotometer, the color is described in terms of the Lovibond system, because Lovibond red standards are used in trading. For example, the maximum red for a refined oil is 3.5, although the value may go up to 6, with a corresponding reduction in price. The amount of bleaching needed varies with the intended use of the oil. If the oil is intended for shortening, it should have a bleached red value of approximately 1 with no chlorophyll. If the oil is to be used for margarine, a value of 4 may be satisfactory, and for other products red values of 2–2.5 are usual.

Theory. There is a great deal of art remaining in the process of bleaching soybean oil, and experience in judging the right kind and amounts of adsorbents for soybean oils having different colors is invaluable. Still a formula developed by Freundlich (1922) relates the quantity of a material adsorbed, the amount remaining unadsorbed, and the amount of adsorbent:

$$x/m = KC^n$$

where x is the amount adsorbed, m the amount of adsorbent, C the amount unadsorbed, and K and n constants. The form of this equation is such that by taking the logarithm of both sides:

$$\log x/m = \log K + n \log C$$

and plotting $\log x/m$ vs. $\log C$, the constants K and n can be evaluated. While the Freundlich equation fits the adsorption of pigments in some instances, there are problems with its application. For example, Gutfinger and Letan (1978) found widely varying values of K and n calculated for different adsorbents. Also, the Freundlich equation was developed to describe reversible adsorption processes, but the binding of pigments to bleaching clays is irreversible.

The adsorption of pigments is typically better in regular plant runs than one would expect from bleaching a sample of oil under laboratory conditions. This is regularly observed and is called the "press bleaching" effect. This implies that due to concentration effects achieved during filtering to remove the activated earth, additional bleaching is achieved. This would not be expected based on Freundlich's equation, but it does happen, and this again raises a question about the applicability of Freundlich's equation to bleaching.

Procedure.   As with alkali refining, both batch and continuous procedures are possible and are being used. Also, it is possible to bleach in air or under vacuum. The procedure that we shall describe is diagrammed in Fig. 3.32 and is for a continuous vacuum bleaching, but other procedures are also used.

The oil to be bleached is mixed with adsorbent in a slurry tank. About 1% by weight of adsorbent is a reasonable amount, but this varies depending on the final color desired, adsorption capacity of the adsorbent, and pigments in the oil that need to be removed. The oil–adsorbent mixture is sprayed into a tank under vacuum to remove air and then is heated to 220–230°F (104–110°C). The mixture again is sprayed into the bleaching tank to maximize chances for air removal. The liquid level in the bleacher is controlled, and the mixture is continually agitated. About 20 min contact time is sufficient, and the adsorbent–oil mixture is then cooled and filtered. The bleached oil is removed to storage, and the adsorbent is generally steam treated or water washed to recover as much oil as possible.

The bleaching clay after use still contains oil and poses a waste disposal problem. The only solution at the present time is to haul this spent earth to a landfill for disposal. If an adsorption process could be worked out in which the adsorbed pigment and other impurities would be eluted from the adsorbent, and it could be reused, that would be a definite advantage. Also, with such a procedure the oil loss might be minimized.

Effect of Bleaching Variables.   Important variables affecting the bleaching process are the type of adsorbent, adsorbent dosage, contact time, temperature,

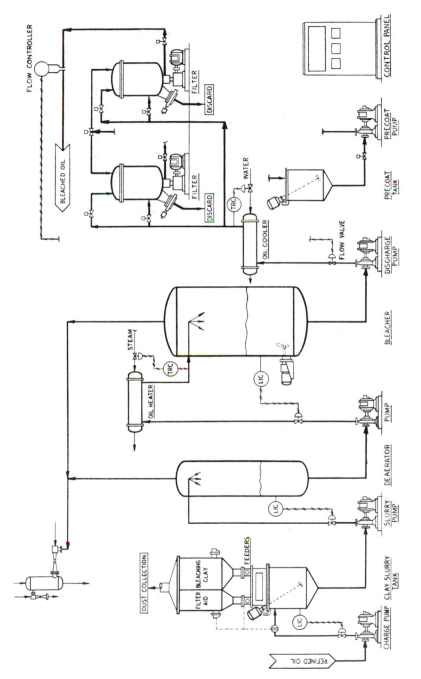

Fig. 3.32. Flowsheet for Votator continuous vacuum bleaching system. Courtesy of Votator Division, Chemetron Corp.

Table 3.3

Decolorization of Soybean Oil with Various Percentages of
Several Commercial Bleaching Earths[a]

Earth	Earth dosage	Lovibond		Chlorophyll density
		Yellow	Red	
None		70	7.7	0.80
A	1.0	35	3.0	0.18
	1.5	35	2.1	0.90
	2.0	20	1.0	0.041
B	1.0	35	3.9	0.40
	2.0	35	2.8	0.16
	3.0	25	1.7	0.071
	4.0	20	1.6	0.037
C	2.0	35	2.3	0.53
D	2.0	35	3.9	0.75
E	1.0	70	7.2	0.58
	2.0	35	4.8	0.43
	4.0	35	2.3	0.19
	6.0	35	1.7	0.093

[a]Source: Brekke (1980b).

oil loss, and final oil quality. The adsorbent should have some moisture present for optimum activity. Optimum moisture varies with different types of adsorbents, but may be in the 20% range.

Treating natural earths with acid (activation) increases their adsorption capacity, and so less can be used, and less oil is lost with disposal of the spent earth. This advantage has to be weighed against the increased cost of activated earths. A finer state of subdivision of the activated earth increases surface area and increases adsorption capacity. However, finely divided earths are difficult to filter. Generally adsorbents are fine enough so that 100% will pass a 100-mesh screen.

As dosage of adsorbent increases, the amount of pigment adsorbed also increases. This is shown in Table 3.3 for five different types of adsorbents. The refiner has to consider adsorbent cost, dosage, and oil lost during adsorption to decide the most economical course. There are nomographs and equations that are useful in reconciling the contrasting factors involved.

Time and temperature for contact between the oil and adsorbent have the general effect shown in Fig. 3.33. As time increases at any given temperature, a minimum amount of Lovibond red is reached, and as the temperature increases, the time necessary to reach the minimum red value decreases. This is the reverse of the temperature effect for most adsorption processes, in which the amount adsorbed decreases as temperature increases.

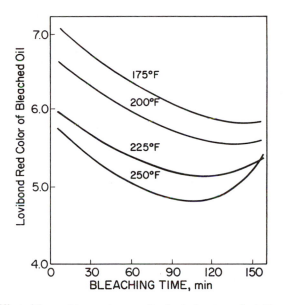

Fig. 3.33. Effect of time and temperature on oil color during atmospheric bleaching of an alkali-refined, washed soybean oil. Source: Brekke (1980b).

Final oil quality is improved by bleaching since one can produce a color of oil to fit the end usage, with better flavor and flavor stability of the finished oil. Bleaching does cause oxidized products (hydroperoxides) to be converted into conjugated trienes. There is some evidence that this change can make the oil less stable to oxidation, and so bleaching is not a cure-all for a poor starting material, and in some instances, it may cause oxidative instability.

Hydrogenation

Hydrogenation is the addition of gaseous H_2 to double bonds of unsaturated fatty acids. It has been used industrially in the United States since 1911 and has two main purposes when applied to edible oils. First, hydrogenation changes the texture by increasing the solidification point. Thus a liquid vegetable oil can be changed to a plastic shortening or margarine. This is useful because the plastic products are in demand for baking and for table use, but there is insufficient naturally plastic fat from animal sources to meet the demand. Second, the poly-unsaturated fatty acids present in vegetable oils are highly subject to oxidation, and addition of hydrogen to those double bonds can increase the oxidative stability of the oils, thus maintaining desirable flavors for a longer time.

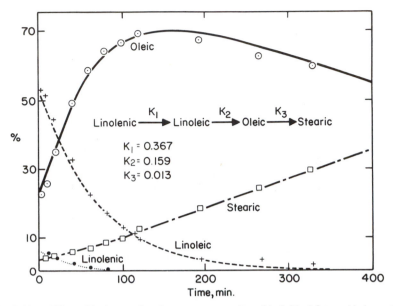

Fig. 3.34. Effect of hydrogenation time on concentration of individual fatty acids in partially hydrogenated soybean oil. Source: Mounts (1980).

Selectivity. The process of hydrogenation does not proceed at equal rates for all double bonds. In general the rate is faster as the degree of unsaturation increases, and therein is the idea of selectivity. The time course for hydrogenation of the four main fatty acids in soybean oil is shown in Fig. 3.34. The K values are first-order rate constants for hydrogenation of each of the three unsaturated fatty acids. Although the rates do not always follow first-order kinetics, using first-order rate constants is a reasonable approximation. Under the particular conditions stated for Fig. 3.34, linolenic ester hydrogenates 2.3 times faster than linoleic ester (0.367/0.159), and linoleic ester hydrogenates 12.3 times faster than oleic ester (0.159/0.013). The numbers generated by comparing first-order rate constants are referred to as selectivity ratios and should be designated for linolenic or linoleic acid. Thus hydrogenation can be said to be selective for the more unsaturated fatty acids, and the larger the ratios the more selective is the hydrogenation. The concept of selectivity is important because oil refiners are interested in eliminating linolenic acid while retaining a considerable amount of linoleic and oleic fatty acids. Linolenic acid is thought to be the cause of soybean oil reversion (development of off-flavors at very low oxidation levels), and hence the interest in a selective hydrogenation that would effectively eliminate linolenic acid.

The selectivity of hydrogenation can be changed by changing conditions such as temperature and pressure and by changing the catalyst. In general as the temperature of hydrogenation increases, the nickel catalyst becomes more selective for linolenic acid, but as the pressure of hydrogen increases, the selectivity for linolenic acid decreases.

Research on catalysts for hydrogenation that would be more selective than nickel has been going on at the USDA Northern Regional Research Center in Peoria, Illinois, for about 20 years. Copper catalysts have been developed that have superior selectivity, but they have the problem of low activity and are difficult to remove completely. Copper is a potent prooxidant for lipid oxidation.

Positional Isomers. An important aspect of hydrogenation that has not yet been mentioned is that the compounds resulting from hydrogenation may differ from the natural compounds found in soybean oil. For example, linoleic acid has double bonds in the 9 and 12 positions. If during hydrogenation the double bond in the 9 position is saturated with hydrogen, that would leave an 18-carbon fatty acid (still esterified to glycerol, of course) with one double bond in the 12 position. Since the naturally occurring 18-carbon fatty acid with one double bond (oleic acid) has that double bond in the 9 position, a new isooleic acid has been produced by hydrogenation. Similarly, new fatty acid isomers can be produced by hydrogenation of linolenic acid. The problem with fatty acid isomer production is that it is very difficult to predict rates and final compositions of hydrogenated oils, since isomers do not always hydrogenate at the same rate as the native compounds.

Another important aspect of hydrogenation is that often new double bonds are generated by shifting to new positions, and the new double bonds are in the trans configuration. As explained in Chapter 2, trans fatty acids have higher melting points than cis fatty acids. Hydrogenation thus can raise the melting point of vegetable oils both by saturating fatty acids and by generating new trans double bonds.

The Hydrogenation Reaction. For hydrogenation to take place, it is necessary to achieve close contact with solid nickel catalyst, gaseous hydrogen, and fluid soybean oil. The catalyst is most often finely divided and reduced nickel (Raney nickel) that is protected by a hardened oil or tallow and diluted with a carrier such as kieselguhr. Final nickel concentration in the catalyst preparation is 25%. The hydrogen is usually generated at the oil refinery and is at least 98% pure. It also must be dry. The liquid oil has already been alkali refined and bleached. Soap content should be less than 25 ppm to avoid inactivating (poisoning) the catalyst system.

The course of the reaction is followed by refractive index changes for lightly hydrogenated oil done for increased stability (above an IV of 95). The refractive index can be used as a measure of unsaturation in fluid oils, but it is not useful to predict when hydrogenation has proceeded sufficiently to produce a satisfactory shortening or margarine. For this, congeal points or SFI is a better control measure (see Chapter 2).

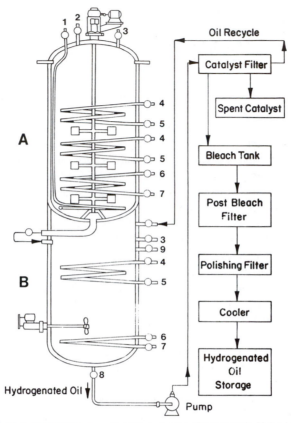

Fig. 3.35. Batch hydrogenation apparatus (A) converter, (B) drop tank. Valve identification: (1) hydrogen inlet, (2) vent, (3) vacuum, (4) heating coil inlet, (5) heating coil outlet, (6) cooling water inlet, (7) cooling water outlet, (8) oil transfer, (9) nitrogen gas inlet. Source: Mounts (1980).

For hydrogenation, batch reactions are most commonly in use, but equipment for continuous hydrogenation does exist and is in commercial use at the present time.

Equipment. An apparatus for batch hydrogenation is shown in Fig. 3.35, and a system for continuous hydrogenation is shown in Fig. 3.36. In the batch process, the oil and catalyst are fed into the converter section (A) and heated to the desired temperature [250–390°F (120–200°C)]. Since the hydrogenation reaction is exothermic, the converter is fitted with cooling coils for temperature control. Newer designs with energy-saving concepts make use of the heat generated during hydrogenation to heat the next batch of oil to be hydrogenated. Hydrogen is introduced as small bubbles in the bottom of the converter, and the hydrogen gas–oil–catalyst mixture is agitated by the centrally mounted agitation

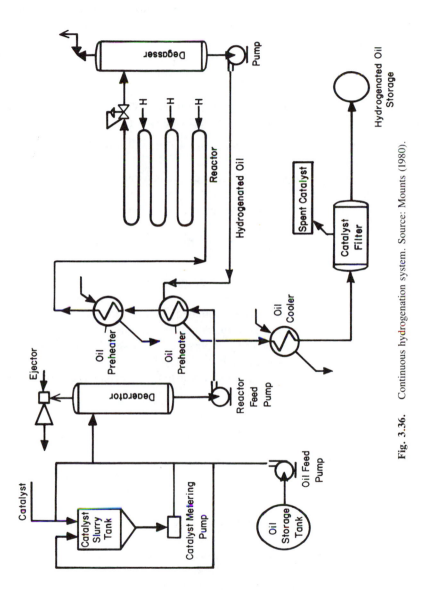

Fig. 3.36. Continuous hydrogenation system. Source: Mounts (1980).

system. Frequently, agitation is a limiting factor in hydrogenation. It is necessary for maintaining a high enough level of dissolved hydrogen in the oil and for maintaining good contact between the oil to be hydrogenated and the catalyst. Hydrogen pressures in batch converters range from 15 to 90 psig, and vessels have capacities from 5 to 20 tons.

After hydrogenation, the batch can be dropped to vessel B for cooling and subsequent filtration. This frees the converter for accepting a new batch for hydrogenation.

The effects of the main process variables on selectivity, trans formation, and rates of hydrogenation may be summed up as follows (Allen 1978):

(1) temperature increase: selectivity increases, trans formation increases, hydrogenation rate increases;

(2) pressure increase: selectivity decreases, trans formation decreases, hydrogenation rate increases;

(3) agitation increase: selectivity decreases, trans formation decreases, hydrogenation rate increases;

(4) catalyst concentration increase: selectivity increases; trans formation increases, hydrogenation rate increases.

The catalyst is filtered out of the hydrogenated oil and can be recycled, although catalytic activity is slowly lost. Filtering of the catalyst does not remove all metal contamination from the hydrogenated oil. For this a light bleaching is done with 0.1–0.2% bleaching earth (usually activated) in the presence of phosphoric or citric acids. It is particularly important to remove all copper catalyst (at least below 0.03 ppm) to insure stability of the oil.

The hydrogenated oil may go through a polishing filter (paper) to remove any bleaching earth and is then stored.

Winterizing

When liquid oils are kept at refrigerator temperatures [about 40°F (5°C)], the more saturated triglycerides crystallize, forming an opaque suspension that may confuse the consumer. In the early days of marketing liquid oils, it was observed that outdoor tanks of oil during the winter would separate into crystals of saturated triglycerides and a clear oil fraction. The clear oil was marketed as "winter salad oil" and remained clear when stored cold. This practice of producing an oil that is stable to cold temperatures continues today and is called winterizing.

The procedure today, as used for soybean oil, generally follows a light hydrogenation to produce a salad and cooking oil that does not crystallize in the refrigerator. The separated crystalline fraction is known as stearine and is recovered for use in shortening. The yield of salad oil is 75–85% of the starting weight, and this recovery depends on the degree of hydrogenation. As hydrogenation proceeds, less salad oil would be recovered. The 75–85% yield is reasonable for oil with an IV of 110 after hydrogenation.

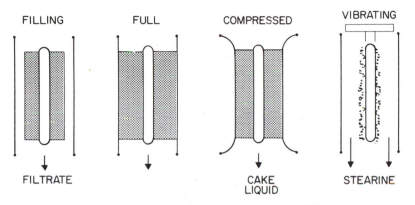

Fig. 3.37. Diaphragm filter. Source: Latondress (1983).

Although the concept is simple, the actual practice of winterizing is difficult, because large crystals must be formed to allow any possibility of separation of the liquid and solid phases. With small crystals there is a tendency to knit the liquid and solid phases together in a plastic mass such as exists in shortening. The key to producing large crystals is slow cooling.

The bleached and partially hydrogenated soybean oil is pumped into large horizontal tanks fitted with cooling coils and called chillers or graining tanks. The initial temperature is 85–95°F (30–35°C), and the substance used in the cooling coils is brine or propylene glycol slurry. The temperature differential between cooling coils and liquid is kept at about 23°F (13°C) so that cooling is slow, and a system of agitation is used to speed up heat transfer. The agitation must be slow to avoid formation of small crystals and may be done by mechanical paddles or by bubbling cold, dry air. As the temperature drops, the cooling rate is slowed further, so that the temperature differential between coolant and oil is 9°F (5°C) at the final temperature of 40°F (5°C). When the oil is depositing a lot of crystalline material near the end of the process, the temperature rises a degree or two due to heat of crystallization, and this can be used as an endpoint indicator. The total cooling and crystallization time for soybean oil may be 24 hr.

Once crystals have formed, the even more difficult job of separating them must be done. This is usually accomplished by some kind of filtration, but centrifugation is also used. The original filtration equipment was the plate and frame filter press. The same equipment was originally employed for removing bleaching earth and for removing hydrogenation catalyst from oil, but cleaning plate and frame filters is time consuming and involves a lot of labor. Also, filtration rates are slow, about 31 lb oil/ft²/hr.

Improved filtration equipment includes the pressure leaf filter, the continuous vacuum filter, and the self-cleaning filter. The pressure leaf filter has a diaphragm between filter leaves that can exert pressure on the filter cake (Fig. 3.37). The pressure on the filter cake forces remaining oil to be released, and the

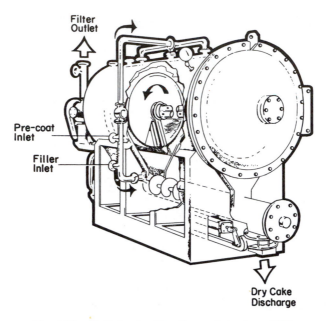

Filter
Outlet

Pre-coat
Inlet

Filler
Inlet

Dry Cake
Discharge

Fig. 3.38.   Self-cleaning filter. Source: Latondress (1983).

cake can then be freed easily from the filter leaves by vibration (Latondress 1983).

The continuous vacuum filter is a rotating filter drum half immersed in the cold oil–stearine mixture. A vacuum on the immersed filter section causes the pressure differential to force cold oil through the filter. As the filter rotates above the layer of the oil, air forces remaining oil through the filter and dries the stearine layer. A doctor blade removes the stearine before the filter is again immersed.

The self-cleaning filter (Fig. 3.38) is enclosed with rotating filter leaves mounted vertically. The tank is filled with oil–crystalline stearine mixture and filtration continues until a preset pressure drop across the filter signals that the filter cake needs to be cleaned from the leaves. Then the tank is drained, and filter brushes automatically scrape the cake from the rotating leaves. The crystalline sludge falls to a screw conveyor in the bottom, which removes it. The filter can then be recharged for another filtering cycle.

With these improved filter systems, filtration rates may be as high as 50 lb/ft²/hr.

Deodorization

The purpose of the deodorization step is to remove unwanted flavor and other compounds from soybean oil just before it is utilized or packaged. The method used is to steam distill at high temperatures under vacuum. This process, along with hydrogenation, was largely responsible for making suitable food products out of vegetable oils.

The main reason for deodorization is to remove flavor compounds such as aldehydes, ketones, and hydrocarbons that result from hydroperoxide breakdown. But deodorization also removes free fatty acids, sterols, tocopherols; causes hydroperoxides to breakdown; may cause breakup of some prooxidants; and bleaches carotenoids. Under the conditions used, even small amounts of triglyceride are lost due to distillation and entrainment.

The deodorization process may be batch, semicontinuous, or continuous. Each type will be described briefly.

Batch Deodorization. Batch deodorization is useful for small refiners who want to minimize equipment costs. The deodorizer consists of a single vertical tank with a capacity ranging from 5000 to 60,000 lb (2273 to 27,273 kg) of oil. The tank is filled approximately half full leaving a large headspace for minimizing carryover of oil droplets into the vapor system. The tank is supplied with heating and cooling coils, and stripping steam is introduced through small holes in piping at the bottom. A vacuum of 6 mm Hg can be obtained by a three-stage steam ejector. The oil is first deaerated, then heated, deodorized, and cooled. The entire process including filling and emptying may take 8 hr, but the actual deodorization would be completed in 1–3 hr at 420–480°F (215–249°C) and 6 mm Hg. For maximum protection of the hot oil, the surfaces of the tank contacting the oil should be constructed of type 304 or 316 stainless steel or nickel. However, this increases the price, and so it is more common to use carbon steel in construction.

Semicontinuous Deodorization. Semicontinuous deodorization is achieved by doing deaeration, heating, deodorization, and cooling in successive sections in one vessel as shown in Fig. 3.39. A batch of oil is measured into the top section, where it is heated and sparged to remove air. After about 15 min, which is the approximate holding time in each section, a valve opens, and the oil drains by gravity into the heating section.

The temperatures for deodorization, 400–525°F (204–275°C), are far beyond what can be achieved by steam at usual pressures from commercial boilers. What is needed is a compound that will condense at the temperatures wanted so that the large heat of condensation can be captured. A mixture of diphenyl and diphenyl oxide with the trade name Dowtherm A serves this purpose. The high-temperature heat source shown in Fig. 3.39 achieves vaporization of Dowtherm A by direct heating. The vapor then gives up its heat of vaporization to the oil in the second section of the deodorizer, and the Dowtherm A condensate returns to the heat source for vaporization. The Japanese have questioned the safety of Dowtherm A for heating edible oils because of an incident with contaminated rapeseed oil in which people died. Investigators of the incident could find no

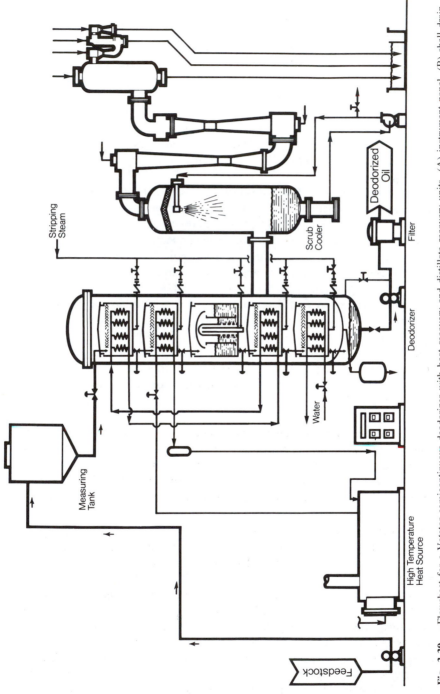

Fig. 3.39. Flowsheet for a Votator semicontinuous deodorizer with heat recovery and distillate recovery units. (A) instrument panel, (B) shell drain condensate collection tank, (C) deodorizer distillate, (D) steam jet ejector system. Courtesy of Votator Division, Chemetron Corp.

Stripping Steam

Scrub Cooler

Deodorized Oil

Filter

Deodorizer

Water

Measuring Tank

High Temperature Heat Source

Feedstock

evidence that Dowtherm A was the cause of the deaths, but the Japanese government has stopped the use of Dowtherm in deodorizers (Norris 1985).

After heating to the desired temperature, the oil again flows by gravity to the deodorizing section. This section has a column of oil surrounding a steam distribution system from which jets of steam penetrate the oil, causing violent boiling action. The oil is splashed against the curved dome from which thin layers drain to the outside of the oil column. The next two sections achieve cooling of the oil, with the heat recovered in the first of these sections used to warm the incoming oil.

Vapors of compounds removed in each section are carried out with the sparging steam through the annular gap between the wall of the tray and the overhead deflector. Some condensation of these vapors will occur inside the deodorizer and drain down the walls and off the deflectors to form a shell drain condensate. Most of the vapor removed will leave the deodorizer through the vacuum line and enter the scrub cooler. Here the vapor is contacted by a liquid spray of cooled condensate. The condensation process promotes the vacuum, but temperatures are not low enough to condense stripping steam. Further condensation occurs after the second-stage ejector in which steam is condensed in the barometric condenser. Condenser cooling water is usually kept free of any condensate contamination by using a separate closed system. This is necessary to prevent obnoxious odors from escaping into the air. The deodorizer condensate is valuable mainly for its tocopherol content.

Continuous Deodorization. Continuous deodorization is commonly used by the largest of the refiners and employs either a single- (Fig. 3.40) or a double-shell deodorizer (Fig. 3.41). For capacities of 15,000–60,000 lb/hr (6818–27,273 kg/hr) the double-shell deodorizer is used, but for capacities less than 15,000 lb/hr (6818 kg/hr) the initial cost of the single shell is more economical (Brekke 1980c).

The continuous systems have deaeration sections external to the deodorizers and make provision for direct heat recovery by passing incoming oil through the first oil-cooling stage. Dowtherm heating, vacuum production, and condensate recovery are similar in the continuous system to the semicontinuous system already described.

With oil being heated to such high temperatures, it is necessary to avoid any contact with air. This is why oil is held inside the deodorizer until cooled sufficiently, so that air contact is not a problem. The double-shell deodorizer has the advantage that any air leaking into the deodorizer system is removed by the vacuum system and thus kept from contacting the oil. Deterioration of oil quality is thereby minimized.

The ability of deodorizers to produce a high-quality oil is usually gauged in terms of the compounds removed. If free fatty acid levels can be reduced from 0.1 to 0.03%, and if total losses are held to 0.5–0.6%, the deodorizer is operat-

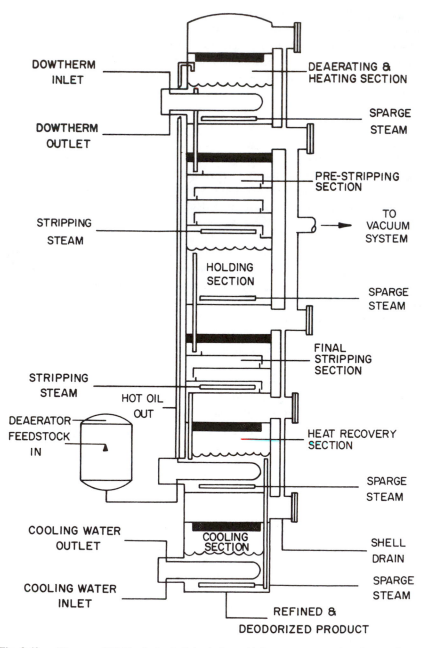

Fig. 3.40. Diagram of EMI's single-shell deodorizer with heat recovery section. Source: Gavin (1978).

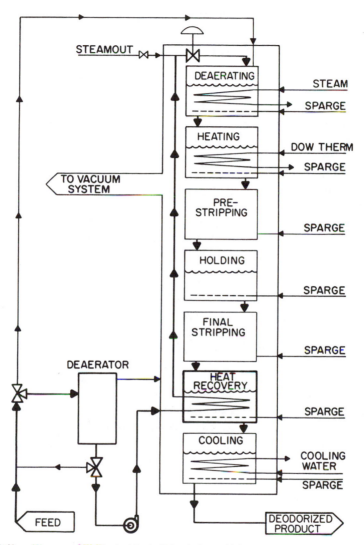

Fig. 3.41. Diagram of EMI's double-shell deodorizer with heat recovery section. Source: Gavin (1977).

ing well. Some triglyceride is hydrolyzed during deodorization, some is distilled off, and some is lost by entrainment of small droplets in the vacuum system.

To maintain the high quality of deodorized soybean oil, it is necessary to add 0.005–0.01% citric acid. This is added as an aqueous solution to the final cooling section of the deodorizer. The citric acid is thought to act by chelating prooxidant metals.

PROTEIN PRODUCTS

Meals, Flours, and Grits

The principal protein product coming from defatted soybean flakes is soybean meal for feeding purposes. The meal may contain a minimum of 44% protein if hulls have been added back or 47.5–49% protein if free from hulls. Trading rules require that the type of process used for removing the oil (solvent extraction or expeller) be included as part of the name of the defatted meal. The soy meals are not fed directly but are feed components valued mainly for their high-protein quantity and quality.

Grinding of defatted flakes to produce meal is done with hammer mills. The specification used by the industry is that all meal should pass a 10-mesh screen with a maximum of 50% passing a 24-mesh screen and a maximum of 1% passing an 80-mesh screen (Thomas 1981). This means that the grinding should be done without excessive production of fines.

To minimize fines, the flakes should move through the hammer mill rapidly, and this means there should be ample screen area in the mill. Another factor in moving meal through the mill is good air flow created by the fanlike action of the hammers rotating at 1800 rpm. Well-operated hammer mills can handle 350 lb/hp/hr (159 kg/hp/hr) and motors of 100 hp are used. Figure 3.42 shows a mill with associated dust-collecting system used for grinding defatted soy flakes into meal.

Products intended for human use are called soybean flour or soybean grits, depending on the state of subdivision. Soybean flour is fine enough that 97% will pass a 100-mesh screen. Soybean grits are produced in a range of sizes with coarse passing 10- to 20-mesh screens, medium passing 20- to 40-mesh screens, and fine passing 40- to 80-mesh screens.

The processing of soybean flakes for human consumption is essentially the same as for animal feeds, except that more attention is given to cleanliness and sanitation of the raw materials and equipment. There are also full-fat products made for human consumption, as mentioned earlier in this chapter, and some products are made with intermediate amounts of fat. Low-fat flour has 5–6% soy oil added and high-fat flour has about 15% soy oil added (still less than a full-fat flour at 20%). Both low- and high-fat flours may have lecithin added to a specified level up to 15%.

To enhance the protein level in soy protein products above 50%, it is necessary to remove some of the soy constituents other than oil. This is done in the processing of soy protein concentrates and of soy protein isolates.

Soy Protein Concentrates

The protein concentration of defatted soy flakes can be increased by removal of the soluble carbohydrate portion. Since soluble carbohydrates (at least

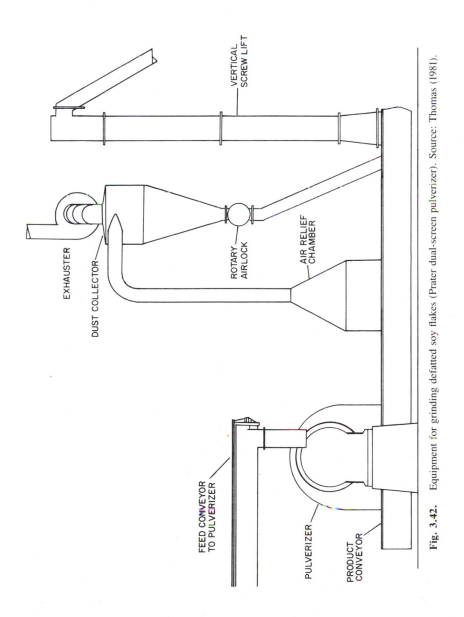

Fig. 3.42. Equipment for grinding defatted soy flakes (Prater dual-screen pulverizer). Source: Thomas (1981).

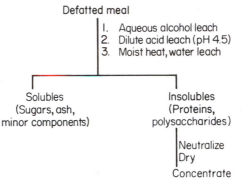

Fig. 3.43.   Processes for preparing soy protein concentrates. Source: Wolf and Cowan (1975).

raffinose and stachyose) are responsible for adverse flatus production, their re-moval is a distinct improvement in defatted soy flakes for human use. The product made by removing soluble carbohydrates is called soy protein concen-trate or simply soy concentrate and should contain a minimum of 70% protein on a dry-weight basis. The term "protein concentrate" has also been used to describe other vegetable protein products containing at least 70% protein.

The soluble carbohydrates can be extracted from defatted soybean flakes in three principal ways and with many minor variations. All processes require that the protein be insolubilized in some way, so that it is not extracted. The three methods (Fig. 3.43) for insolubilizing the protein are by heating, by acid, and by ethanol. Details of these processes are commercial secrets and not readily avail-able, but the general processes are known.

The starting material for concentrates prepared by extraction with hot water is toasted, defatted flakes with low nitrogen solubility index (NSI) and low PDI. The temperature of the water is in the range of 150–200°F (66–93°C), and the solids are usually separated by centrifugation. The pH is in the range of 5.5–7.5, and processes may be either batch or continuous.

For acid extraction, the starting material is defatted flakes that have been desolventized under vacuum or in a flash desolventizer. Any food grade acid can be used to extract the flakes at pH 4.5. Since soy protein has a minimum solubility at this pH (Chapter 2), mainly soluble carbohydrates are extracted. The unextracted protein concentrate may be dried directly or first neutralized and then dried, depending on the need for soluble protein in the final concentrate product. In this process lipoxygenase and trypsin inhibitors would still be active, but they are not precipitated at pH 4.5, and so these unwanted proteins are probably extracted by the dilute acid. There is evidence that soy concentrates and isolates contain about 25% of the original trypsin inhibitor activity (Rackis *et al.* 1979).

The third means of producing a concentrate is by extraction with aqueous ethanol. Using 60–80% ethanol in water, most of the protein is insoluble, but soluble oligosaccharides are extracted. This process has the added expense of

Table 3.4
Proximate Compositions of Soy Protein Concentrates[a]

	Alcohol	Acid	Hot water
Protein (N × 6.25, %)	66	67	70
Moisture (%)	6.7	5.2	3.1
Fat (petroleum ether extract)	0.3	0.3	1.2
Crude fiber (%)	3.5	3.4	4.4
Ash (%)	5.6	4.8	3.7
NSI	5	69	3
pH of 1:10 water dispersion	6.9	6.6	6.9

[a]Source: Wolf and Cowan (1975).

recovery of the valuable ethanol. To compensate for that cost, the soluble oligosaccharides are concentrated during ethanol recovery, making them more valuable as feed ingredients. Normally, the oligosaccharides are present at concentrations too low to justify recovery as a feed ingredient.

In addition to the soluble oligosaccharides, some proteins and minerals are also extracted during the processing of soy protein concentrates. Table 3.4 shows proximate composition and NSI for soy concentrates made by the three processes. The unaccounted for 20% dry weight in Table 3.4 is insoluble carbohydrate coming predominantly from cell walls and intercellular material. Yields of soy concentrates are 65–70% of the starting material.

Soy Protein Isolates

Products that contain more than 90% protein are named soy protein isolates, and as with concentrates, the term "isolate" is also used for other vegetable proteins in essentially pure form. The process for producing soy isolates is outlined in Fig. 3.44.

The starting material for soy protein isolate is finely ground, defatted flakes that have been desolventized by a vacuum process for maintaining protein solubility. The protein is extracted with dilute alkali, pH 9, at 121–130°F (50–55°C). Stronger alkali would extract more protein but would also cause more damage particularly to the sulfur-containing amino acids. The solids are separated by screening and centrifuging and contain primarily insoluble carbohydrates and proteins.

The soluble protein portion is adjusted to pH 4.5 with food grade acid. This results in a protein precipitate and a whey fraction (analogous to casein precipitation by acid from fluid milk). The precipitate is washed and then may be spray dried to yield an isoelectric protein or may be neutralized first and then dried to yield a soy proteinate. For end uses in which good solubility is important, the proteinate form is needed.

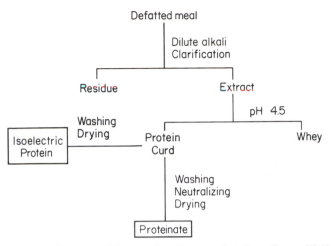

Fig. 3.44. Diagram for commercial production of soy protein isolates. Source: Wolf and Cowan (1975).

Approximately one-third of the original starting material is retrieved as soy protein isolate, with one-third remaining as insoluble residue after alkali extraction, and one-third remaining in the whey fraction. The insoluble residue is normally recovered as a feed ingredient by drum drying. The whey fraction would contain the trypsin inhibitors of soy, and would need to be heat inactivated before being used as a feedstuff, but a more serious problem is that it contains only 1–3% solids and is too dilute for economic recovery.

The protein isolate has less lysine and less sulfur containing amino acids than the starting soy flakes, and so protein quality is somewhat diminished by the isolation process.

Minor constituents accompanying protein isolates include saponins, phosphatides, and phytin. Washing a protein isolate with alcohol is known to produce an isolate that forms a stable foam. Presumably foam-inhibiting lipids are removed by the alcohol washing step. Phytate can be removed from protein isolates by dialysis or ultrafiltration.

The properties of soy protein isolates and soy protein concentrates can differ widely even though proximate composition is about the same. Solubility of protein obviously varies depending on whether the protein was neutralized before drying. Other functional properties (see Chapter 5) will also be modified, depending on processing conditions.

The processing of soy protein concentrates and soy protein isolates makes use of stainless steel equipment, cleaning-in-place arrangements, and emphasis on sanitation. Bacteriological standards of 10,000/g total count and essentially free of *E. coli, Salmonella,* and coagulase-positive *Staphylococci* are not hard to meet. The biggest bacteriological problem is from thermophilic spore formers, but it still is possible to meet standards required for ingredients going into canned foods.

Texturizing Protein Products

Meals, flours, grits, concentrates, and isolates do not make palatable foods because of their consistency. Therefore, there has been considerable research on processes for converting flourlike textures to more desirable meatlike textures. Three processes have been sufficiently successful to be used commercially: extrusion, spinning, and steam texturizing.

Extrusion. Extruders are widely used in the food industry as low-cost cookers that also have the capability of texturizing and shaping products. The basic configuration of an extruder (Fig. 3.45) is a screw rotating within a barrel to cause friction and thus heat for cooking products. The heated mixture is forced through a restricted opening at the end of the barrel. High temperatures and pressures are reached for short times, and then the product is extruded through an opening that can be used to shape the foodstuff. Simultaneously the extruded ribbon of material is cut by a revolving knife that controls the particle size. When the product suddenly is exposed to atmospheric pressure, the steam tends to flash off, giving some puffing or product expansion. The final steps are cooling and drying the products.

Harper and Jansen (1981) have conveniently grouped the many types of extruders available into three categories (Table 3.5): those capable of handling low- (<20%), intermediate- (20–28%), and high-moisture foods (>28%). For processing soybeans, all three types of extruders are useful. If the primary aim is to cook soy mixtures such as corn–soy–milk food supplements, a low-cost extruder that heats mainly by friction (uses low- or intermediate-moisture raw materials but has little capability for texturizing or shaping) is suitable.

For texturizing, the extruders that handle intermediate- and high-moisture mixtures are most suitable. These extruders are more costly initially than extruders handling low-moisture products and are more costly to operate. The difference is that less friction is developed with high moisture, and the barrels of the extruders have to be jacketed to allow heating and cooling.

The raw materials for producing an extruded, texturized soy product are soy flour (usually defatted), flavors, colors, oil, and water. An outline of a process for production of textured soy protein by extrusion is shown in Fig. 3.46. The important variables for controlling texture during extrusion are the moisture content, temperatures and times at different temperatures (heating and cooling), and pressures.

Soy protein products texturized by extrusion are frequently used as part of a meat–soy mixture rather than as meat analogs. Meat analogs from soy protein are produced by a process known as spinning.

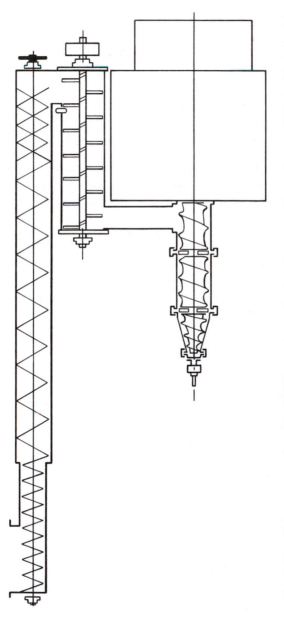

Fig. 3.45. High-temperature/short-time (HTST) extrusion cooker including preconditioner. Courtesy of Wenger International, Inc.

Table 3.5
Changing Characteristics of Extruders Based on the Initial Moisture Content of Feed Ingredients[a]

	Low moisture	Intermediate moisture	High moisture
Initial moisture	$\leqslant 20\%$	20–28%	$\geqslant 28\%$
Energy source	Most from mechanical input to extruder	Half from mechanical; half from added steam	Most from added steam
Mechanical energy needed (kW-hr/kg)	0.1	0.04	<0.02
Product drying	None required	Some needed to bring product to <12%	Extensive drying needed
Product shape	Limited available	Many shapes available	Many shapes
Product density	Low	Moderate	Range of densities
Ingredients	Should have >7% fat	Few limitations	Few limitations
Capital cost	Low to high	Low to moderate	Moderate to high
Maintenance cost (per MT)	$1.20	$0.50–0.60	$0.40–0.50
Manufacturers	Brady, Meals for Millions, Insta-Pro, Collet, Manley Dorsey-McComb, Adams	Anderson, Wenger	Anderson, Bonnot, Sprout-Waldron, Wenger

[a]Source: Harper and Jansen (1981).

Spinning. The spinning process has been borrowed from the synthetic textile industry and has similarities to processes for producing rayon and nylon fibers. Proteins from many sources such as casein, animal proteins from organ meats, and single-cell proteins can be spun into fibers, but the process only has been developed to a commercial production scale with soy protein. The starting material for spun soy fibers is soy protein isolate with good solubility.

One process for spinning soy fibers (several have been developed) is shown in Fig. 3.47. In this process the soy protein isolate is being produced as the first step of the spinning technique. A concentrated solution of soy isolate, referred to as spinning "dope," is made in alkali. The concentration of protein is about 20%, which along with a pH of 12–13 gives a high viscosity of 50,000–100,000 cP. The viscous dope is pumped through a filter and into the coagulating bath.

To form fibers the spinning dope is forced through a spinneret, which is a platinum plate [about 1 ft (0.3 m) in diameter] with thousands of holes approx-

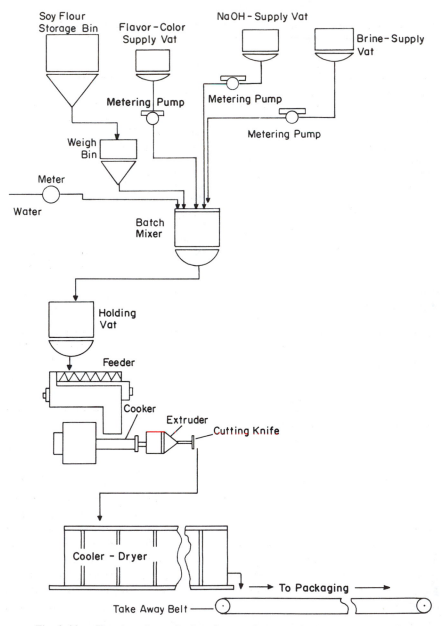

Fig. 3.46. Flowsheet for production of textured soy protein. Source: Horan (1974).

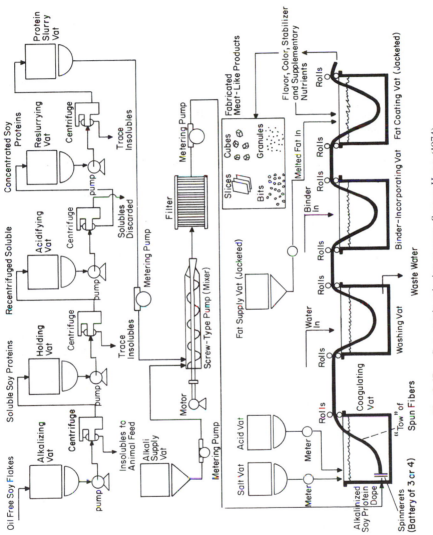

Fig. 3.47. Soy protein spinning process. Source: Horan (1974).

imately 0.003 in. (0.075 mm) in diameter. As the alkaline dope leaves the small holes of the spinneret, it contacts the acid and salt of the coagulating bath and forms a thin fiber. The pH of the coagulating bath is about 2.5, and phosphoric acid is used along with 8% NaCl.

It is essential to texture formation that the newly formed fibers be picked up as a tow and stretched. The fibers from a single spinneret would make a tow of about 0.25 in (0.62 cm) in diameter and several of these tows may be combined into a larger size. As a result of stretching, the original fiber size of 0.003 in (0.075 mm) may be decreased to one-fifth that size. If fibers are not stretched, they tend to kink and readily fall apart, leaving very little texture or chewiness.

After forming the stretched fiber, the tow goes through a washing vat to remove acid and salt. The tow may be heated at this point as a hardening process. Subsequent processing of the spun fibers includes the addition of binders to help make cohesive pieces out of the many individual fibers. Egg albumen is a frequently used binder.

For production of the final product, a meat analog, it is necessary to incorporate fat, flavors, colors, supplementary nutrients, and any other needed ingredients. Finally the tow is cut or further processed into the shape and size of pieces wanted. The concentration of soy protein in a particular meat analog may be only 25%, with fat and moisture making up most of the remaining weight. Thus soy meat analogs are not 100% soy protein, even though all of the protein is from soy.

Steam Texturizing. The third commercially proven process for introducing texture into a soy protein product makes use of a process patented by Strommer and Beck (1973) and illustrated in Fig. 3.48. This texturizing process uses soy flour as the starting ingredient with a moisture content of 18–24%. Soy concentrates and soy isolates may also be used as starting materials, but for reasons of economy, soy flour is preferred.

The apparatus consists of a high-pressure steam source, a rotating valve that is alternately charged and fired, and a barrel through which the textured protein is discharged. Figure 3.48a shows the entire apparatus. In Fig. 3.48b the rotating valve shows one section in position for discharge through the barrel. The discharge is forced by high-pressure steam. At the same time, another section of the rotating valve is being charged with new starting material. In the intermediate position of the rotary valve (Fig. 3.48c), the high-pressure steam is released through an exhaust pipe, so that a new charge of soy flour can enter.

The steam pressures for satisfactory operation of this type of texturization are at least 25 psig and more commonly 80–110 psig. The temperature of saturated steam at 100 psig would be about 338°F (170°C).

In both the steam texturization just described and in extrusion, the soy proteins are probably unfolded by the temperatures involved as a result of breaking hydrogen bonds. The high temperatures along with shear forces cause the pro-

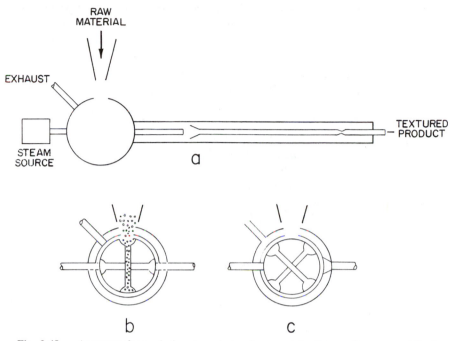

Fig. 3.48. Apparatus for producing steam textured soy protein. Source: Strommer and Beck (1973).

teins to realign in an elongated form, and as the proteins cool new hydrogen bonds tend to form that hold the protein molecules in the new arrangement. Also, undoubtedly new cross linkages form between protein molecules that stabilize the new textured structure.

Steam-textured soy protein products are used mainly as meat extenders. The process was developed by General Mills and was sold to Central Soya in the late 1970s, but it is no longer being used. The soy product produced by steam texturization is generally less dense than that produced by extrusion, and the less dense product has a desirable brightening effect on fresh red meat color.

There are other methods of introducing texture into soybean protein products. If soy protein at a concentration of 8% or more is heated at 159°F (70°C) for 10–30 min, a gel forms. Anson (1958) suggested that such gels could be manipulated to simulate meat texture. The gelling of soy protein will be discussed further in Chapter 5.

The basis for another soy protein texturizing method is in the observations made by Tombs (1972) and Van Megen (1974). If soy protein solutions or suspensions are exposed to a pH of 4.5 and an ionic strength of 0.7 or higher (sodium salts) solutions form. But if the ionic strength is lowered below 0.7 and the solution is centrifuged (50,000 g for 60 min), two phases form. The upper phase is a normal protein solution, but the lower phase is high in protein concentration, viscous, and dense. This lower layer (mesophase) from centrifugation can be texturized easily by throwing droplets against a heated surface or by extruding into water at 176°F (80°C) or higher. The same mesophase can be achieved with calcium salts at pH 4.5, but at lower ionic strengths. Also, mesophases can be achieved at pH 7, but not until a certain ionic strength is exceeded.

Protein isolates can be texturized by a process known as "jet cooking" (Wilcke *et al.* 1979). The protein isolate is heated by pumping under high pressure through a heat exchanger. The pumping and heat cause the isolate to form fibers. The textured protein is cooled and frozen for later use to add texture to mechanically deboned meat products.

Two books exist that summarize the extensive patent literature on processing soybean protein products. Hansen (1974) covers the general field of vegetable protein processing, and Gutcho (1977) covers textured protein products.

REFERENCES

Allen, R. R. (1978). Principles and catalysts for hydrogenation of fats and oils. *J. Am. Oil Chem. Soc.* **55**:792.

American Oil Chemists Society (1982). "Official and Tentative Methods." American Oil Chemists Society, Champaign, IL.

Anson, M. L. (1958). Potential uses of isolated oil seed protein in foodstuffs. *In* "Processed Plant Protein Foodstuffs" (A. M. Altschul, ed.). Academic Press, New York.

Ayers, A. L., and C. R. Scott (1952). A study of extraction rates for cottonseed and soybean flakes using n-hexane and various alcohol–hexane mixtures. *J. Am. Oil Chem. Soc.* **29**:213.

Baker, E. C., and D. A. Sullivan (1983). Development of a pilot-plant process for the extraction of soy flakes with aqueous isopropylalcohol. *J. Am. Oil Chem. Soc.* **60**:1271.

Becker, K. W. (1983). Current trends in meal desolventizing. *J. Am. Oil Chem. Soc.* **60**:216.

Boling, F. (1982). Energy and efficiency in the solvent plant. *Oil Mill Gaz.* **86**(9):42.

Boucher, D. F., J. C. Brier, and J. O. Osburn (1942). Extraction of oil from a porous solid. *Trans. Am. Inst. Chem. Eng.* **38**:967.

Bredeson, D. K. (1983). Mechanical oil extraction. *J. Am. Oil Chem. Soc.* **60**:163A.

Brekke, O. L. (1980a). Oil degumming and soybean lecithin. *In* "Handbook of Soy Oil Processing and Utilization" (D. R. Erickson, E. H. Pryde, O. L. Brekke, T. L. Mounts, and R. A. Falb, eds.). American Soybean Association, St. Louis, MO, and American Oil Chemists Society, Champaign, IL.

Brekke, K. W. (1980b). Bleaching. *In* "Handbook of Soy Oil Processing and Utilization" (D. R. Erickson, E. H. Pryde, O. L. Brekke, T. L. Mounts, and R. A. Falb, eds.). American Soybean Association, St. Louis, MO, and American Oil Chemists Society, Champaign, IL.

Brekke, K. W. (1980c). Deodorization. *In* "Handbook of Soy Oil Processing and Utilization" (D. R. Erickson, E. H. Pryde, O. L. Brekke, T. L. Mounts, and R. A. Falb, eds.). American Soybean Association, St. Louis, MO, and American Oil Chemists Society, Champaign, IL.

Circle, S. J., and A. K. Smith (1978). Processing soy flours, protein concentrates, and protein isolates. *In* "Soybeans: Chemistry and Technology," Vol. 1, Proteins (A. K. Smith and S. J. Circle, eds.). AVI Publ. Co., Westport, CT.

Coats, H. B., and G. Karnofsky (1950). Solvent extraction II. The soaking theory of extraction. *J. Am. Oil Chem. Soc.* **27**:51.

Fan, H. P., J. C. Morris, and H. Wakeham (1948). Diffusion phenomena in solvent extraction of peanut oil. Effect of cellular structure. *Ind. Eng. Chem.* **40:**195.

Fetzer, W. (1983). Head- and tail-end dehulling of soybeans. *J. Am. Oil Chem. Soc.* **60:**203.

Florin, G., and H. R. Bartesch (1983). Processing of oilseeds using fluidbed technology. *J. Am. Oil Chem. Soc.* **60:**145A.

Freundlich, H. (1922). "Colloid and Capillary Chemistry" (transl. by H. S. Hatfield, from the 3rd German ed.). Dutton, New York.

Friedrich, J. P., and E. H. Pride (1984). Supercritical CO_2 extraction of lipid-bearing materials and characterization of the products. *J. Am. Oil Chem. Soc.* **61:**223.

Gavin, A. M. (1977). Edible oil deodorizing systems. *J. Am. Oil Chem. Soc.* **54:**528.

Gavin, A. M. (1978). Edible oil deodorization *J. Am. Oil Chem. Soc.* **55:**783.

Gutcho, M. H. (1977). "Textured Protein Products." Noyes Data Corp., Pea Ridge, NJ.

Gutfinger, T., and A. Letan (1978). Pretreatment of soybean oil for physical refining: Evaluation of efficiency of various adsorbents in removing phospholipids and pigments. *J. Am. Oil Chem. Soc.* **55:**856.

Hansen, L. P. (1974). "Vegetable Protein Processing." Noyes Data Corp., Pca Ridge, NJ.

Haraldsson, G. (1983). Degumming, dewaxing and refining. *J. Am. Oil Chem. Soc.* **60:**203A.

Harper, J. M., and G. R. Jansen (1981). "Nutritious Foods Produced by Low-Cost Technology." Departments of Agricultural and Chemical Engineering, and Food and Nutrition, Colorado State University, Fort Collins.

Henderson, S. M., and R. L. Perry (1955). "Agricultural Processing Engineering." Wiley, New York.

Horan, F. E. (1974). Soy protein products and their production. *J. Am. Oil Chem. Soc.* **51:**67A.

Hvolby, A. (1971). Removal of nonhydratable phospholipids from soybean oil. *J. Am. Oil Chem. Soc.* **48:**503.

Johnson, L. A., and E. W. Lusas (1983). Comparison of alternative solvents for oils extraction. *J. Am. Oil Chem. Soc.* **60:**181a.

Karnofsky, G. (1949). The theory of solvent extraction *J. Am. Oil Chem. Soc.* **26:**564.

Karnofsky, G. (1981). Ethanol and isopropanol as solvents for full-fat cottonseed extraction. *Oil Mill Gaz.* **85**(10):34.

Latondress, E. G. (1983). Oil–solids separation in edible oil processing. *J. Am. Oil Chem. Soc.* **60:**209A.

Lawhon, J. T., L. J. Manak, K. C. Rhee, and E. W. Lusas (1981). Production of oil and protein food products from raw peanuts by aqueous extraction and ultrafiltration. *J. Food Sci.* **46:**391.

Milligan, E. D. (1976). Survey of current solvent extraction equipment. *J. Am. Oil Chem. Soc.* **53:**286.

Moore, N. H. (1983). Oilseed handling and preparation prior to solvent extraction. *J. Am. Oil Chem. Soc.* **60:**141A.

Mounts, T. L. (1980). Hydrogenation practices. *In* "Handbook of Soy Oil Processing and Utilization" (D. R. Erickson, E. H. Pryde, O. L. Brekke, T. L. Mounts, and R. A. Falb, eds.). American Soybean Association, St. Louis MO, and American Oil Chemists Society, Champaign, IL.

Mounts, T. L., and F. P. Khym (1980). Refining. *In* "Handbook of Soy Oil Processing and Utilization" (D. R. Erickson, E. H. Pryde, O. L. Brekke, T. L. Mounts, and R. A. Falb, eds.). American Soybean Association, St. Louis MO, and American Oil Chemists Society, Champaign, IL.

Mustakas, G. C., W. J. Albrecht, G. N. Bookwalter, and E. L. Griffen, Jr. (1967). Full-fat soy flour by a simple process for villagers. ARS 71-34, Agricultural Research Service, USDA, Washington, D.C.

Mustakas, G. C., W. J. Albrecht, G. N. Bookwalter, J. E. McGhee, W. F. Kwolek, and E. L. Griffen, Jr. (1970). Extruder processing to improve nutritional quality, flavor, and keeping quality of full-fat soy flour. *Food Technol.* **24:**1290.

Norris, F. A. (1982). Extraction of fats and oils. *In* "Bailey's Industrial Oil and Fat Products," Vol. 2, 4th ed. (D. Swern, ed.). Wiley, New York.

Norris, F. A. (1985). Deodorization. *In* "Bailey's Industrial Oil and Fat Products," Vol. 3, 4th ed. (T. H. Applewhite, ed.). Wiley, New York.

Othmer, D. F., and J. C. Agarwal (1955). Extraction of soybeans: Theory and mechanism. *Chem. Eng. Prog.* **51:**372.

Rackis, J. J., J. E. McGee, M. R. Gumbman, and A. N. Booth (1979). Effects of soy proteins containing trypsin inhibitors in long term feeding studies in rats. *J. Am. Oil. Chem. Soc.* **56:**162.

Smith, A. S., and F. J. Wechter (1950). Vapor pressure of hexane-soybean oil solutions at low solvent concentrations. *J. Am. Oil Chem. Soc.* **27:**381.

Strommer, P. K., and C. I. Beck (1973). Texturization by passage through elongated pipe in presence of steam at elevated pressure and temperature. U.S. Patent 3,754,926, Aug. 28.

Thomas, G. R. (1981). The art of soybean meal and hull grinding. *J. Am. Oil Chem. Soc.* **58:**194.

Tindale, L. H., and S. R. Hill-Haas (1976). Current equipment for mechanical oil extraction. *J. Am. Oil Chem. Soc.* **53:**265.

Tombs, M. P. (1972). Protein products. British Patent 1,265,661, March 14.

Van Megen, W. H. (1974). Solubility behavior of soybean globulins as a function of pH and ionic strength. *J. Agric. Food Chem.* **22:**126.

Ward, J. A. (1976). Processing high oil content seeds in continuous screw presses. *J. Am. Oil Chem. Soc.* **53:**261.

Wiedermann, L. H. (1981). Degumming, refining, and bleaching soybean oil. *J. Am. Oil Chem. Soc.* **58:**159.

Wilcke, H. L., D. H. Waggle, and C. K. Kolar (1978). Textural contribution of vegetable protein products. *J. Am. Oil. Chem. Soc.* **56:**259.

Wolf, W. J., and J. C. Cowan (1975). "Soybeans as a Food Source," rev. ed. CRC Press, Cleveland, OH.

4

Quality Criteria for Soy Products

To judge whether a soy product is suitable for a particular food or feed use, several different tests are available. In this chapter we consider those quality criteria that give information on some aspects of functionality and flavor of protein products, and on oxidative stability for oils. Chapter 5 will cover functionality properties of soy protein in more detail, and Chapter 6 will consider nutritional quality.

PROTEIN PRODUCTS

NSI and PDI

Probably the most frequently used quality criterion for soy protein products is solubility as determined by NSI or PDI. We discussed solubility of individual soy proteins and soy protein classes in Chapter 2 but not methods. NSI and PDI are based on procedures that need to be followed carefully to achieve reproducibility. These two procedures are useful in judging the kind of processing involved in producing a defatted soy flour, suitability of the flour for further processing to a soy concentrate or soy isolate, suitability for feeding (as a measure of how much heating was used), and suitability for specific food uses.

The NSI procedure is the AOCS Official Method Ba 11-65 revised in 1969 and reapproved in 1975. As contrasted with PDI, NSI is referred to as the slow-stir method. The procedure calls for stirring 5 g of sample in 200 ml of distilled water for 120 min at 86°F (30°C). The stirring is done mechanically at 120 rpm.

Next the solids are separated by diluting to 250 ml, decanting 40 ml into a centrifuge tube, and centrifuging at 1500 rpm for 10 min. The clear supernatant liquid is decanted through glass wool in a funnel, and the nitrogen content is determined by Kjeldahl analysis.

Total nitrogen content of the sample also needs to be determined by Kjeldahl analysis.

The NSI is calculated by first finding the percentage of water-soluble nitrogen:

weight of nitrogen in supernatant/weight of sample × 100

Then NSI is calculated:

water-soluble nitrogen (%)/total nitrogen (%) × 100

The size of particles being stirred will influence the NSI. The method suggests a grinding procedure such that 95% of the sample passes a 100-mesh screen, but makes the grinding optional. That is NSI can be reported on an "as is" basis or on a "ground sample" basis.

The PDI or fast stir method is detailed as AOCS Official Method Ba 10-65 revised 1982. The fast stirring is achieved by an automatic blender at 8500 rpm. Depending on the size of the blender jar, either 10 g in 150 ml water or 20 g in 300 ml is blended for 10 min. The slurry is decanted, allowed to settle, and a portion decanted into a 50-ml centrifuge tube. The suspension is centrifuged for 10 min at 2700 rpm, and the clear supernatant liquid is analyzed for nitrogen by the Kjeldahl procedure.

The PDI is calculated by first determining the percentage of water-dispersible protein:

weight of nitrogen in supernatant × 6.25/weight of sample × 100

Then PDI is calculated:

water-dispersible nitrogen (%)/total protein (%) × 100

The fact that one method (NSI) is based on nitrogen and the other (PDI) is based on protein has no significance. PDI would be exactly the same if calculated as nitrogen since both numerator and denominator are multiplied by 6.25 to convert to protein. PDI is always higher than NSI because the rapid shearing action of the blender blades disperses more protein than slow stirring. However, there is no simple relationship between the two measures. Figure 4.1 shows some data in which PDIs and NSIs are compared for the same samples (Horan 1974).

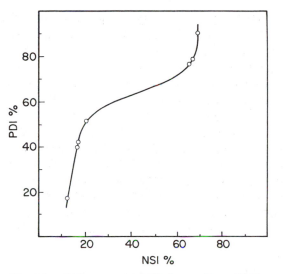

Fig. 4.1. PDI compared to NSI. Source: Horan (1974).

Trypsin Inhibitor Assays

The early assays of trypsin inhibitor in soy products were done by measuring trypsin activity on a protein such as denatured hemoglobin or casein. The assays were done in the presence and in the absence of trypsin inhibitor (usually a water extract of the soy product) and the difference was a measure of the inhibition.

Kakade *et al.* (1969) introduced a method in which the protein substrate was replaced by benzoyl–arginine–*para*-nitroanilide (BAPA). When trypsin attacks this substrate, *p*-nitroaniline is released as a colored product, and the rate of enzymatic hydrolysis can be followed easily by spectrophotometry. Again, assays are made in the presence and absence of trypsin inhibitor, and the difference is a measure of trypsin inhibitor activity. The procedure works well, but there were difficulties with reproducibility in different laboratories. A collaborative study (Kadade *et al.* 1974) helped with this problem by improving the trypsin inhibitor extraction procedure. The procedure of Kakade *et al.* (1974) has been adopted as the official method by the American Oil Chemists' Society and by the American Association of Cereal Chemists.

As the amount of trypsin inhibitor in the assay system increases, the trypsin inhibitor activity per milliliter of extract decreases. Consequently, the method suggests extrapolation of a plot of trypsin inhibitor units/milliliter vs. milliliters of extract for a true value of trypsin inhibitor units/milliliter. Hamerstrand *et al.* (1981) studied this situation and suggested a single assay in the region of 40–60% trypsin inhibition would give the best measure of trypsin inhibitor activity.

Values obtained by this assay method range from 90–100 trypsin inhibitor units/mg soy flour for raw, unheated flour, to 7–10 units/mg fully toasted soy flour. A trypsin inhibitor unit is based on a change in absorbancy at 410 nm (the absorbancy peak for the yellow colored p-nitroaniline) of 0.01 under the conditions specified for the assay. The fact that trypsin inhibitor units never reach zero with heating is generally attributed to the presence of the Bowman–Birk inhibitor, which is much more resistant to heat denaturation than the Kunitz inhibitor.

A different assay procedure was proposed by Stinson and Snyder (1980) and by Hill *et al.* (1982). In this procedure an alkaline extract of the soy product is made as a source of both the trypsin inhibitor and protein substrate for trypsin. Trypsin is added to this substrate–inhibitor mixture in a pH stat at pH 9, and the amount of base added per unit time to keep the pH at 9 is used as a measure of tryptic hydrolysis. As trypsin attacks peptide bonds, newly formed carboxyl groups and amino groups tend to dissociate, thus causing the pH to drop:

$$R-\overset{\overset{\displaystyle O}{\|}}{C}-\overset{\overset{\displaystyle H}{|}}{N}-R'+H_2O \xrightarrow[\text{pH 9}]{\text{trypsin}} R-\overset{\overset{\displaystyle O}{\|}}{C}-O^- + NH_2-R'+H^+$$

If several of these kinds of assays are made with different amounts of trypsin, a plot can be made of the rate of hydrolysis vs. amount of trypsin used (Fig. 4.2). By extrapolation of the straight line to zero rate of hydrolysis, a measure can be obtained of the amount of trypsin needed to overcome the trypsin inhibitor. In effect trypsin inhibitor is being titrated with trypsin, and this gives an assay of the amount of trypsin inhibitor in the sample.

In addition, the slope of the plot shown in Fig. 4.2 gives the rate of protein hydrolysis per milligram of trypsin. This rate increases as a result of heat treatment of soy protein. Thus a measure of susceptibility to tryptic hydrolysis and a trypsin inhibitor assay can both be obtained by the pH stat method.

Urease Activity

The simplest test to gauge how much heat was used in processing a soy meal or flour is the urease assay. It is the test most frequently used in oil extraction mills and is the AOCS Official Method Ba 9-58 reapproved in 1973.

The urease activity determination makes use of added urea, which serves as a substrate for the inherent urease in soybeans:

$$\begin{matrix} H_2N \\ \searrow \\ C{=}O+H_2O \\ \nearrow \\ H_2N \end{matrix} \xrightarrow{\text{urease}} 2NH_3+CO_2$$

As a result of urease action the pH rises. The reaction is run in 0.05 M phosphate at pH 7, 86°F (30°C), for 30 min. After this time the pH of the reaction mixture

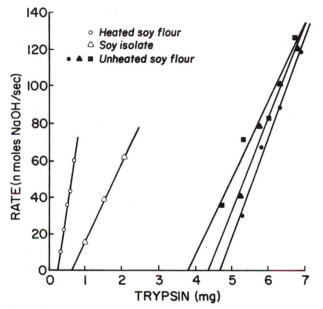

Fig. 4.2.   Assays for trypsin inhibitor (intercept on trypsin axis) and for proteolysis susceptibility (slope) by the pH stat. Source: Hill *et al.* (1982). Copyright by the Institute of Food Technologists.

and the pH of a blank (without urea added) are measured. If meal has been properly heated for trypsin inhibitor inactivation, the urease activity (pH difference between blank and sample) should be less than 0.25. It is worth emphasizing that urease is not a factor in soybeans that needs to be inactivated but is used only as an indicator of sufficient heat treatment much as phosphatase activity is used to judge pasteurization of milk.

Flavor

Along with the texture of soy protein products, the flavor of these products is a major quality concern in their utilization. Off-flavors of soy protein products are variously described as green, grassy, beany, bitter, and astringent. The off-flavors are apparent in soy flours, soy concentrates, and soy isolates. Also, these off-flavors can be found in some soy milks, but seldom are they a factor in soy curd acceptance.

If no heating is involved in the processing of soy protein products, lipoxygenase can play a major role in off-flavor development. Soybeans that have imbibed water are extreme in their grassy, beany off-flavors, but if the soaked soybeans are blanched by immersion in boiling water for 2 min, the off-flavors disappear. The development of off-flavors is generally attributed to lipoxygenase activity during chewing of unblanched beans. The heat denaturation of lipoxygenase prevents off-flavor development in blanched soybeans just as blanching prevents off-flavor development in other peas and beans. In addition to heat, the treatment of soybeans or soybean protein products with aqueous ethanol can minimize off-flavor development by inactivating lipoxygenase.

It is less certain what role lipoxygenase plays in off-flavor development in defatted products such as flours, concentrates, and isolates that have not been toasted. Since even defatted products have 0.5–1% lipid remaining, it is conceivable that lipoxygenase would be active during storage. Most dry products are rehydrated before or after incorporation into a food product, thus making it possible for lipoxygenase to act.

OIL PRODUCTS

Many quality criteria are used to judge the suitability of oils for food use. Textural quality is influenced by melting point, SFI, and crystal size. Nutritional quality depends on degree of unsaturation and fat-soluble vitamin content among other things. Frying and baking qualities depend upon free fatty acid content, emulsification properties, smoke points, and color as examples. These quality aspects of soybean oil are treated in Chapters 2, 6, and 10. In this chapter we consider the oxidative deterioration of oil and resulting off-flavors. Also, we describe some of the more important tests used to judge oxidative stability of soy oil.

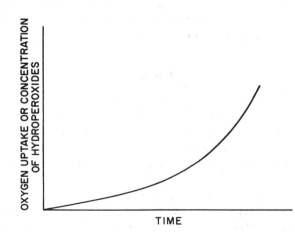

Fig. 4.3.   Typical curve of hydroperoxide development in fatty materials. Source: Lundberg (1962).

Autoxidation

Highly unsaturated oils, such as soybean oil, react readily with molecular oxygen to form hydroperoxides, which in turn can break down to secondary oxidation products or can polymerize and cause film formation. Film formation is a desirable reaction in a drying oil used as a base for paints, but in a food oil the off-flavors are very objectionable. The time sequence for reaction with oxygen is shown in Fig. 4.3. A long induction period is followed by a rapid uptake of oxygen.

Conventionally, autoxidation of oils is divided into three phases: initiation, propagation, and termination. The initiation phase corresponds to the long time period shown in Fig. 4.3 before rapid oxidation begins. It is a time during which antioxidative properties of the oil are being overcome, and the concentration of free radicals is slowly building. In terms of what is happening to long-chain polyunsaturated fatty acids, a hydrogen is abstracted from the methylene group to form a free radical:

$$\begin{array}{ccccc} 9 & 10 & 11 & 12 & 13 \end{array}$$

$$-CH=CH-CH_2-CH=CH-$$

$$\downarrow$$

$$-CH=CH-\overset{\cdot}{C}H-CH=CH-$$
$$+\ H\cdot$$

which rapidly reacts with molecular oxygen to form a hydroperoxy free radical:

$$-\overset{|}{\underset{\underset{\overset{|}{O}}{O}}{C}H}-CH=CH-CH=CH-$$

The propagation reaction continues to generate free radicals and hydroperoxides:

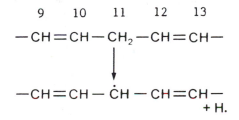

Finally the reaction terminates as a result of depletion of the groups susceptible to hydrogen abstraction and free radical formation.

Autoxidation of linoleic acid will produce hydroperoxides in either the 9 or 13 position with equal probability. A new double bond is generated in either the 10 or 11 position giving a conjugated double-bond system. The original double bonds are cis in most naturally occurring fatty acids, but the newly formed double bond due to autoxidation is trans . In autoxidizing soybean oil the fatty acid being oxidized is still part of a triglyceride molecule. The unsaturated fatty acids in soybean oil are oleic, linoleic, and linolenic, and they autoxidize (as fatty acids not triglycerides) at relative rates of 100, 1200, 2500, respectively.

Because of the very high rates of oxidation of linolenic acid and because of some experiments by Dutton *et al.* (1951) in which linolenic acid was interesterified into cottonseed oil, linolenic acid is generally assumed to be the culprit responsible for the flavor instability of soybean oil. The interesterification of linolenic acid into cottonseed oil produced an off-flavor in the cottonseed oil that resembled the off-flavor normally found in oxidized soy oil. There is some doubt that the 8–10% linolenic acid would be more important than the approximately 50% linoleic acid present in soy oil, even if linolenic acid oxidizes twice as fast as linoleic acid, unless there is a unique oxidation product coming from the linolenic acid that could explain the off-flavors found in soybean oil. Good evidence for such a unique flavor compound is lacking.

If autoxidizing lipids are exposed to ultraviolet radiation, the rate of autoxidation can be greatly accelerated. Ultraviolet radiation can catalyze the formation of free radicals in autoxidizing lipids. Normally this type of photocatalyzed oxidation is not a problem, because the containers for oil, either metal or glass, are not permeable to ultraviolet radiation.

More of a problem may be the presence of photosensitizers such as chlorophyll in soy oil. Such photosensitizers can activate oxygen to the singlet state in the presence of visible light. Singlet oxygen can cause lipid oxidation at approximately 1500 times the rate of normal free radical oxidation (Frankel 1980). Metals can also generate activated or singlet oxygen and thus catalyze lipid oxidation. Carotenoids can quench the singlet oxygen and thereby protect the lipid from oxidation, but the tendency today is to bleach the soy oil sufficiently so that it has a light appearance. This bleaching removes the carotenoids as effective quenchers. The problem could also be eliminated by packaging, but the trend seems to be toward packages that allow the customer to see the product and consequently allow light passage.

Soybean Oil Reversion

Unique to soybean oil is the appearance of off-flavors at very low peroxide values. These off-flavors are grassy or beany in nature and do not resemble the original flavor of crude soy oil. Hence, all agree that the term "reversion" is a misnomer, but it has been used so extensively to describe the flavor change that its use will undoubtedly continue.

There was doubt at one time about the autoxidative nature of soy oil reversion,

Table 4.1

Partial List of Volatile Compounds Identified in Oxidized Soybean Oil[a]

Class	Compounds
Aldehydes	Alkanals: C1, C2, C3, C4, C5, C6, C7, C8, C9
	Alkenals Δ^2: C4, C5, C6, C7, C8, C9, C10, C11
	Alkenals Δ^3: C6
	Dienals $\Delta^{2,4}$: C6, C7. C8. C9. C10. C12
	Dienals $\Delta^{2,5}$: C8
	Dienals $\Delta^{2,6}$: C9
	Trienals $\Delta^{2,4,7}$: C10
	Dialdehydes: malonaldehyde, maleic dialdehyde, hexene-1,6-dial
Ketones	2-Alkanones: C4, C5, C6. C7. C8
	3-Alkanones: C8
	Unsaturated: 1-pentene-3-one, 4-octene-3-one, 2-methyl-5-octene-4-one,
	7-methyl-2-octene-4-one, 3,6-nonadiene-5-one
	Alkadiones, 2,3-: C4, C5
	Alkadiones, 4,5-: C8
Alcohols	Saturated: C2, C3, C4, C5, isoC5, C6, C7
	Unsaturated: 1-pentene-3-ol, 1-octene-3-ol
Ethyl esters	C1,C2
Hydrocarbons	Saturated: C2, C3, C5, C6, C8, C9, C10, C11
	Unsaturated: 2-pentene, 1-hexene, 2-octene, 1-decene, 1-decyne
Other compounds	2-Pentyl furan, lactones, benzene, benzaldehyde, acetophenone, water

[a]Source: Frankel (1980).

because it can occur at peroxide values of only 2 to 3. Research at the Northern Regional Research Center of the USDA at Peoria, Illinois has shown that autoxidation is involved, but it is autoxidation that is catalyzed by very low levels of iron and copper. Proper use of citric acid as a chelator of metals can control the flavor deterioration known as reversion. Soy oil seems to be more susceptible to metal-catalyzed autoxidation than are other vegetable oils.

It is generally accepted that the hydroperoxides generated by autoxidation are not the off-flavor compounds, but breakdown products of the hydroperoxides, known as secondary products, cause the off-flavors. The chemical nature of these secondary products has been the object of much research, and Table 4.1 lists the many aldehydes, ketones, and other compounds that have been identified. It is not correct to assume that all of these compounds are present in reverted soy oil, because often soy oil is highly oxidized to generate sufficient secondary products for identification. Furthermore, for simplification of the study of autoxidation, frequently soy oil researchers will study pure fatty acids such as linoleic or linolenic acids or their methyl or ethyl esters as starting materials for the generation of secondary oxidation products.

One generalization about the formation of secondary products from hydroperoxides is that the carbon chain of the fatty acid is most susceptible to scission immediately adjacent to the hydroperoxide. Hence from linoleic acid hydro-peroxide (13 position) still esterified with glycerol, one would expect to obtain pentane and hexanal as secondary oxidation products:

$$
\begin{array}{l}
CH_2-O-R \\
\mid \\
R'-O-CH \qquad O \\
\mid \qquad\qquad \parallel \\
CH_2-O-C-(CH_2)_7-CH=CH-CH=CH+C+(CH_2)_4-CH_3 \\
\qquad\qquad\qquad\qquad CH_3(CH_2)_4\,CHO \qquad CH_3(CH_2)_3-CH_3
\end{array}
$$

Analysis of the headspace of autoxidizing soy oil has shown that pentane and hexanal are major products.

Since the production of hydroperoxides is going on concurrently with their breakdown to secondary products, the unique feature of soy oil reversion (off-flavor at low peroxide values) may be due to an accelerated breakdown of hydroperoxides.

Smouse (1979) has reviewed the various ideas about the cause of soy oil reversion. The predominant idea is that linolenic acid in soybean oil is responsible for the reversion flavor, but a satisfactory explanation of how linolenic acid causes soy oil reversion does not exist. So far the many attempts to improve soy oil by breeding a soybean with low linolenic acid content have not been successful. A second idea is that isolinoleic acid, produced during hydrogenation, is the source of the off-flavor. Frankel (1980) has described how isolinoleic acid might be the cause of off-flavors generated in heated oils. A hydrogenation catalyst that would minimize isolinoleic acid by hydrogenating selectively the 15 double bond of linolenic acid is thought to be a possible solution.

Other compounds that have been suggested as causes for soy oil reversion are the phosphatides and the unsaponifiable content of the oil, although the evidence in support of these compounds being responsible is not strong. Oxidative polymers of linolenic acid can form during autoxidation, and these polymers can break down to generate off-flavor compounds under anaerobic conditions.

Although several reasonable ideas exist to explain soy oil reversion, none of the explanations fits the facts well enough to be accepted by those most knowledgable about soy oil reversion. One of the real difficulties in explaining reversion flavor is the detection of the off-flavor. This requires the availability of a highly trained sensory panel, which is expensive and time consuming and not generally available in most laboratories.

Control of Autoxidation.   The most obvious ways of controlling autoxidation are to minimize exposure to prooxidants or to add antioxidants.

Prooxidants.   Molecular oxygen, being a reactant in autoxidation, is an important prooxidant and is generally controlled by removal. Storing partially or completely refined soy oil in nitrogen atmospheres is effective in minimizing autoxidation. Furthermore, retail-sized packages of soy oil are sometimes protected by replacing air with nitrogen in the headspace.

Metals are known to be potent catalysts of autoxidation of lipids and can act at the initiation or propagation stages. Most common metals are involved in the breakdown of hydroperoxides, particularly those that can cycle between two oxidation states (Ingold 1962):

$$M^+ + ROOH \rightarrow RO^{\cdot} + OH^- + M^{2+}$$
$$M^{2+} + ROOH \rightarrow ROO^{\cdot} + H^+ + M^+$$

$$2ROOH \rightarrow RO^{\cdot} + ROO^{\cdot} + H_2O$$

Citric acid is used as an antioxidant agent in soy oil, and its effectiveness has been attributed to its ability to chelate metals. There are some puzzling aspects to this, however. Citric acid would only effectively chelate metals when the protons of its carboxyl groups are dissociated. But dissociated citric acid would not be oil soluble. Also there is no general rule that chelation of a metal would decrease its catalysis of autoxidation. In fact, iron chelated very firmly in the heme or hematin molecules remains a potent catalyst of autoxidation. In soy oil autoxidation, iron and copper are particularly potent prooxidants. Control (in addition to the use of citric acid) is achieved by avoiding contact of the oil with metals that would dissolve in the oil. This means that equipment for processing of soy oil should be made of stainless steel. Even as little as 0.3 ppm iron can catalyze off-flavor development in soy oil, while cottonseed oil is not affected by this amount of iron.

Ultraviolet radiation and visible light act as prooxidants in soy oil. Ultraviolet radiation is not much of a problem because all containers used for soy oil are impermeable to ultraviolet radiation. Formerly it was a common practice to protect soy oil from the possible harmful effects of visible light by packaging in brown bottles, but this is no longer done. Minimizing the chlorophyll content of oil to avoid a photosensitizer is effective in controlling the effect of visible light.

While not strictly classified as a prooxidant, the temperature of soy oil is an important factor in its stability to autoxidation. Studies on autoxidation rates at different temperatures have shown that a 10°C increase in temperature will bring about a doubling of the autoxidation rate. Thus temperature affects lipid autoxidation in the same way it does other chemical and biological reactions.

Antioxidants. In addition to minimizing contact with prooxidants, one can control autoxidation of soy oil by use of antioxidants that may be inherent in the oil or added. The most effective antioxidants are phenolic compounds that interact with free radicals to break the chain reaction

$$R^{\cdot} + AH \rightarrow RH + A^{\cdot}$$

or

$$ROO^{\cdot} + AH \rightarrow ROOH + A^{\cdot}$$

The antioxidant free radical does not generate more lipid free radicals, but autoxidizes to a quinone. Reducing agents such as ascorbic acid are not effective antioxidants in soy oil.

The inherent antioxidant in soy oil and in most other vegetable oils is tocopherol. As discussed in Chapter 2, four isomers of tocopherol are found in soy oil with the most prominent one being γ-tocopherol:

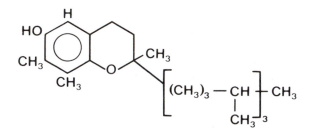

The tocopherol isomers differ in their antioxidant capacity as well as in their vitamin E potency, with γ-tocopherol having the greatest vitamin E activity. Antioxidant activity varies with the type of oil being oxidized and the other antioxidants that may be present. The reactive site of the tocopherol molecule is the —OH group.

Added antioxidants that are useful in some instances in protecting oils from autoxidation are butylated hydroxyanisole (BHA), butylated hydroxytoluene (BHT), propyl gallate (PG), and tertiary-butyl hydroquinone (TBHQ), whose structures are shown in Fig. 4.4. The tertiary butyl group is effective in slowing reactions at the active —OH group by steric hindrance and thereby extending the active life of the antioxidant.

Amounts that can be used in food products are controlled by the U.S. Food and Drug Administration, and the maxima are 0.01% for any one antioxidant or 0.02% for a mixture of antioxidants. The antioxidants do differ in some of their properties. For example, BHT is more volatile than BHA and therefore has less "carry through" activity after extensive heating of a food such as in baking or frying. TBHQ does seem to have some protective effect when used to stabilize crude soy oil during storage, whereas BHA and BHT are less effective. Also, BHA and BHT do not afford protection from off-flavor development during accelerated autoxidation of refined soy oil samples.

An important aspect of the use of antioxidants to control autoxidation of lipids is the phenomenon of synergism. Synergism refers to any interaction between two components in which the mixture of the two gives more than an additive effect. For example, if component A has an activity of 10 when acting alone and component B has an activity by itself of 15, then a mixture showing no synergism would have an activity of 25. But if the mixture is synergistic, it would have an activity of more than 25. Often in mixing phospholipids with the naturally

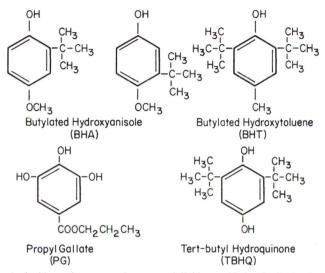

Fig. 4.4. Antioxidants that are used to control lipid autoxidation in foods. Source: Dugan (1976). Copyright by Marcel Dekker, Inc.

occurring tocopherols in soy oil, synergistic effects on antioxidant activity are noted. Phospholipids alone do not exhibit antioxidant activity in soy oil, but they enhance the activity of the inherent tocopherols. Citric acid and phosphates are also characterized as synergists. The mode of action of synergists is not known for certain, but is often attributed to metal chelation.

Measures of Autoxidation

To understand the subject of autoxidation properly, it is important to know the methods that are used to measure autoxidation and their shortcomings. Since autoxidation is a complicated process with an induction period whose length depends upon the interaction of prooxidants and antioxidants and since the accumulation of hydroperoxides depends upon the induction period and upon the rate of hydroperoxide breakdown to secondary products, it is no wonder that single measures of autoxidation do not always give the expected answers.

Peroxide Value (PV). Perhaps the oldest, best known, and most widely used measure of autoxidation of lipids is a measure of the first stable product of autoxidation, the hydroperoxide. Since hydroperoxides are strong oxidants, they can be measured by incubating with an iodide solution in chloroform–acetic acid. The hydroperoxide oxidizes the iodide to iodine, which in turn can be measured by titrating with thiosulfate. For each equivalent of hydroperoxide present an equivalent of iodine is formed. The PV is calculated as the milliequivalents of hydroperoxide present per kilogram of oil. The determination of PV, AOCS Official Method Cd 8-53, requires 5 g of oil, and it is a destructive method. Consequently, one needs a fairly large quantity of oil to complete a storage stability test.

Oxygen Uptake.   To produce a hydroperoxide it is necessary for molecular oxygen to be taken up by the unsaturated triglyceride. Thus another measure of autoxidation is the amount of oxygen absorbed. Early studies on autoxidation made use of respirometers in which oxygen uptake was measured manometrically with the sample in a closed container of known volume under careful temperature control. This can still be a useful technique, but if the sample is liquid, the shaking rate to renew the surface is a large factor in the rates obtained.

The sample can be incubated in a closed container with oxygen under pressure and a pressure gauge (oxygen bomb). The induction period of the sample can be measured as the time needed for the pressure to start dropping rapidly.

Perhaps the simplest way of measuring oxygen absorption is to incubate the sample at a known temperature under controlled conditions of light and surface exposure and to measure the weight increase. This procedure was used by Olcott and Einset (1958) and is still useful, but one must keep some precautions in mind. The weight increase is due to oxygen uptake but as soon as an appreciable amount of hydroperoxide is formed, the hydroperoxide breaks down to volatile compounds, which cause a weight loss. Thus the rate of weight increase is a combination of oxygen uptake and loss of volatiles. If the sample is incubated long enough, the oxygen uptake slows and loss of volatiles becomes dominant, so that the sample is losing weight. Also, it has been noted by Olcott and Einset (1958) and by others that the weight increase per unit weight of sample depends on the surface to weight ratio of the sample. In other words, the amount of oxygen taken up is a function of the surface of sample exposed. The larger the surface per unit weight of sample the more oxygen is absorbed. This means that oil at the surface is oxidizing faster than oil beneath the surface and has important implications for all measures of autoxidation.

Ultraviolet Absorption.   When the hydroperoxide forms, the double-bond system of the unsaturated fatty acid shifts from methylene interrupted to conjugated, and the conjugated double-bond system absorbs at 234 nm in the ultraviolet region of the spectrum. The absorbancy is strong, with an absorbancy coefficient of about 25,000 mol liter^{-1}cm^{-1}. This means that changes in absorbancy at 234 nm are sensitive to small changes in hydroperoxide concentration. Freshly extracted soy oil always has some hydroperoxide present, and absorbancy values for 1% solutions of crude oil in highly purified hexane are about 1.6–2.0. The crude soy oil completely devoid of hydroperoxide has an absorbancy at 234 nm of 0.9–1.0 for a 1% solution.   This is inherent end absorption due to the triglyceride itself. Ultraviolet absorbancy measurement is a rapid and easy procedure, but can be difficult to interpret due to changes that take place during bleaching and hydrogenation that produce other compounds that absorb in the ultraviolet region.

Thiobarbituric Acid (TBA) Reaction Products. The hydroperoxides are primary reaction products of autoxidation but are not the flavor compounds that are actually responsible for off-flavors in oils or oil-containing foods. The off-flavor compounds are secondary reaction products from the breakdown of hydroperoxides, and one way of measuring them is by the TBA reagent. The structure of TBA is

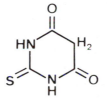

and the principal secondary oxidation product with which it reacts is malondialdehyde

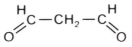

The analysis of oxidizing lipids with TBA leads to a TBA number defined as malondialdehyde (mg) per 1000 g of sample. TBA numbers are most frequently used to analyze foods containing fats rather than pure oils or fats.

The analysis is done by acidifying the sample, distilling, and reacting the distillate with TBA. The reaction product absorbs at 532 nm and by means of a standard curve the absorbancy can be related to milligrams of malondialdehyde. Malondialdehyde is not available as a pure standard, but tetraethoxy propane can be used for this purpose.

Gas Chromatographic Analysis. Since flavor compounds are volatile, a direct analysis of the flavor compounds by gas chromatography seems reasonable, and several research programs are moving in this direction (Legendre *et al.* 1979, Min 1981, Jackson and Giacherio 1977). The best correlations with gas chromatographic analysis and sensory evaluation have been achieved by special sampling techniques. The oil to be sampled is placed into the carrier gas stream ahead of the column in a special holder. The carrier gas sweeps volatiles from the sample onto the column. After an appropiate sampling time of perhaps 30 min, the sample is taken out of the gas chromatographic system, and the volatiles that have collected on the column are separated and recorded. As can be seen in Table

4.1, many compounds can be isolated from oxidizing lipid, and good correlations with sensory evaluation can be obtained with several of these flavor compounds. The off-flavor is not uniquely associated with any one of the compounds. Good correlations have been achieved between sensory scores and the content of pentane, butenal, pentanal, octane, 2-heptenal, or 2,4-decadienal as determined from gas chromatographic analysis.

Stability Tests.   In the testing of soy oil, it is necessary to have information on how the stability of the oil changes with time as well as the accumulation of autoxidation products at any one time. The most frequently used stability tests are the active oxygen method (AOM) and the Schaal oven test. The autoxidation of soy oil to generate off-flavors takes considerable time at room or refrigerator temperatures. To speed up the testing procedure both of the methods mentioned above make use of elevated temperatures.

The AOM consists of bubbling air through the oil sample held at 208°F (98°C) under carefully controlled conditions (AOCS Official Method Cd 12-57). Peroxide values are determined as the sample oxidizes, and the endpoint is chosen as the time needed to develop a PV of 100. With experience one can smell the air coming from the samples and judge when off-flavor compounds have been produced. This also can be used as an endpoint for the method.

The Schaal oven test uses less drastic conditions than the AOM. The oil sample is incubated at 140°F (60°C) in the presence of air but without bubbling air through the sample. Generally PV is measured as the sample oxidizes but there is no accepted endpoint as with the AOM. As already described for measuring weight increases during autoxidation, it is necessary to control the surface area of the oil exposed to air in relation to the weight of oil being incubated.

Both stability tests are most useful in direct comparisons of the stabilities of oils to evaluate antioxidants or some similar variable. The methods are not particularly useful in predicting actual shelf life of an oil under normal storage conditions.

Sensory Evaluation.   The measures of autoxidation discussed thus far are based on physical and chemical procedures and are considered objective measures. It is generally agreed, however, that objective measures of autoxidation are only useful when they correlate with sensory evaluation. That is, the ultimate criterion of the autoxidative quality of soy oil is: How does it smell and how does it taste?

The field of sensory evaluation encompasses a wide range of procedures. The procedures pertinent to judging oil quality are analytical and are done by an expert panel. Mounts and Warner (1980) have described the procedures for sensory evaluation of soy oil that have been found most useful. Details of these procedures are given in Table 4.2.

Although it is generally agreed that sensory evaluation provides the ultimate

Table 4.2

USDA Northern Regional Research Center Sensory Evaluation Procedures[a]

Oil Sample Preparation and Presentation
1. Glass beaker (50 ml) containing 10 ml of oil is covered with a watch glass and is placed in a recess (or hole) in an aluminum block.
2. Aluminum blocks containing the beakers are heated on an electrically controlled aluminum hot plate. The blocks have the dimensions: 49.5 × 20 in. × 1 in. thick; or 131 × 51 cm × 2.6 cm thick.
3. Hot plate temperature is set at 140°F (60°C); aluminum blocks are heated to 131°F (55°C); temperature of the oil is 122°F (50°C).
4. Oils are heated for 30 min prior to serving.
5. Sample presentation order is randomized; e.g., if there are two samples, one-half of the panel receives the oil in 1–2 order and the other half receives the oil in 2–1 order.

Order- and Flavor-Testing Procedures
1. Odor
    a. Oils are evaluated for odor in order which samples are presented left to right.
    b. Covered beaker is swirled and lifted to nose.
    c. Cover is removed as tester sniffs the volatiles.
    d. Beaker is swirled again if necessary.
    e. Panelist records type and intensity of odor, then overall intensity of odor.
2. Flavor
    a. Sample with the weakest odor is evaluated first.
    b. Panelist takes 10 ml of the oil into mouth and thoroughly swishes throughout the mouth.
    c. Oil is expectorated into paper waste cups.
    d. Carbon-filtered tap water heated to 100°F (38°C) is used to rinse the mouth before testing and between testing each sample.
    e. Panelist records type and intensity of flavor, then overall intensity of flavor.
    f. Panelist allows sufficient time between samples for saliva to bathe the taste buds again.

---

[a]Source: Mounts and Warner (1980).

authority for judging flavor of oil samples, it is not generally appreciated how time consuming and difficult it is to generate reliable sensory data. Not the least of the difficulties is the selection of panel members. They must demonstrate the ability to detect and describe flavor changes of the magnitude being investigated and the ability to replicate their judgments over weeks of tasting.

The procedure used at the Northern Regional Research Center of the USDA in Peoria, Illinois (Mounts and Warner 1980) is one of the best documented procedures and makes use of recommendations of the Flavor Nomenclature Committee of the American Oil Chemists' Society. It calls for tasting of warm samples, 122°F (50°C) kept at that temperature by holding the 10-ml samples in a heated aluminum block. The number of samples presented to a panel member at any one time is limited to four or less. The panel members make judgments on the odor of the samples, followed by flavor analysis starting with the sample with least odor. For both odor and flavor, individual descriptions are made for bland, buttery, beany, grassy, rancid, or painty sensations on a three-point intensity scale. Then the odor and flavor or the sample overall is judged on a scale of 1 (extreme) to 10 (bland). From the descriptions of individual off-flavors, a flavor intensity value (FIV) can be generated by weighting the responses of individual panel members.

The entire sensory evaluation procedure needs to be treated statistically to determine the reliability of individual panel members and to judge the reliability of results. Because of the time and money needed to perform reliable sensory evaluation work, the search for simple objective methods that can replace sensory evaluation will continue.

REFERENCES

Dugan, L., Jr. (1976). Lipids. *In* "Principles of Food Science," Part 1, Food Chemistry (O. R. Fennema, ed.). Marcel Dekker, New York.

Dutton, H. J., C. R. Lancaster, C. D. Evans, and J. C. Cowan (1951). The flavor problem of soy oil VIII linolenic acid. *J. Am. Oil Chem. Soc.* **28:**115.

Frankel, E. N. (1980). Soybean oil flavor stability. *In* "Handbook of Soy Oil Processing and Utilization" (D. R. Erickson, E. H. Pryde, O. L. Brekke, T. L. Mounts, and R. A. Falb, eds.). American Soybean Association, St. Louis, MO, and American Oil Chemists Society, Champaign, IL.

Hamerstrand, G. E., L. T. Black, and J. D. Glover (1981). Trypsin inhibitors in soy products: Modification of the standard analytical procedure. *Cereal Chem.* **58:**42.

Hill, B. S., H. E. Snyder, and K. L. Wiese (1982). Use of the pH stat to evaluate trypsin inhibitor and tryptic proteolysis of soy flours. *J. Food Sci.* **47:**2018.

Horan, F. E. (1974). Soy protein products and their production. *J. Am. Oil Chem. Soc.* **51:**67A.

Ingold, K. U. (1962). Metal catalysis. *In* "Lipids and Their Oxidation. H. W. Schultz, E. A. Day, and R. O. Sinnhuber, eds.). AVI Publ. Co., Westport, CT.

Jackson, H. W., and D. J. Giacherio (1977). Volatiles and oil quality. *J. Am. Oil Chem. Soc.* **54:**458.

Kakade, M. L., N. Simons, and I. E. Liener (1969). An evaluation of natural versus synthetic substrates for measuring the antitryptic activity of soybean samples. *Cereal Chem.* **46:**518.

Kakade, M. L., J. J. Rackis, J. E. McGhee, and G. Puski (1974). Determination of trypsin inhibitor activity of soy products: A collaborative analysis of an improved procedure. *Cereal Chem.* **51:**376.

Legendre, M. G., G. W. Fisher, W. H. Schuller, H. P. Dupuy, and E. T. Rayner (1979). Novel technique for the analysis of volatiles in aqueous and nonaqueous systems. *J. Am. Oil Chem. Soc.* **56:**552.

Lundberg, W. A. (1962). Mechanisms. *In* "Lipids and Their Oxidation" (H. W. Schultz, E. A. Day, and R. O. Sinnhuber, eds.). AVI Publ. Co., Westport, CT.

Min, D. B. (1981). Correlation of sensory evaluation and instrumental gas chromatographic analysis of edible oils. *J. Food Sci.* **46:**1453.

Mounts, T. L., and K. Warner (1980). Evaluation of finished oil quality. *In* "Handbook of Soy Oil Processing and Utilization" (D. R. Erickson, E. H. Pryde, O. L. Brekke, T. L. Mounts, and R. A. Falb, eds.). American Soybean Association, St. Louis, MO, and American Oil Chemists Society, Champaign, IL.

Olcott, H. S., and E. Einset (1958). A weighing method for measuring the induction period of marine and other oils. *J. Am. Oil Chem. Soc.* **35:**161.

Smouse, T. H. (1979). A review of soybean oil reversion flavor. *J. Am. Oil Chem. Soc.* **56:**747A.

Stinson, C. T., and H. E. Snyder (1980). Evaluation of heated soy flours by measuring tryptic hydrolysis using a pH stat. *J. Food Sci.* **45:**936.

5

Functional Properties of Soy Proteins

Proteins can interact with other food ingredients to form desirable food properties. This interaction is called functionality. For example, added protein can prevent fat or water from separating during heating of a meat product; can prevent staling by controlling moisture redistribution in baked goods; can form stable emulsions or foams. The study and control of these kinds of properties of proteins have become active research areas, and a lot of information is accumulating. It is the purpose of this chapter to organize and summarize that information as it applies to soy proteins.

Two ideas have provided much of the stimulus for research on functionality of food proteins. The first is that many of the better functional proteins in food systems are of animal origin, and because of population pressures in the world, we shall soon have to substitute plant proteins for scarce animal proteins. The second idea is that functional properties are a consequence of the inherent physical and chemical properties of the protein molecules, and with sufficient understanding of these basic molecular properties, we shall be able to predict, control, and possibly produce desirable functional properties of proteins. There are reasons to be skeptical of both of these ideas, but they have provided the stimulus to uncover much useful information about protein functionality.

Although considerable information exists, it is not well defined and organized. Problems exist with standardized terminology. For example, bound water may

163

mean only the water that is nonfreezable at $-104°F$ ($-40°C$) to some, while to others bound water may be all the water that is not separable by centrifugation at a given speed. Also, methods for measurement have not been standardized nor have methods for reporting measurement data.

The ultimate aim of terms and measurement techniques is to provide useful information to the user of the products. Since soy proteins can be and are used in many different kinds of foods, it is not surprising that differences in terminology and measurement exist. The student and practitioner need to be aware of this and to make sure that the details of the measurement systems are understood.

First we examine the interaction of soy proteins with water and how this can affect food properties. An important part of the discussion is the methods used to study the interactions. Second, we discuss soy protein–lipid interactions and their measurement. In the third section, the surface-active properties of proteins are discussed in relation to producing emulsions and foams.

Several good reviews of this general topic exist (Kinsella 1976, 1979; Cherry 1981). For extensive citations of the original literature and for in depth treatment of selected topics, these reviews and their references should be consulted.

INTERACTIONS OF SOY PROTEINS WITH WATER

The functionality of a protein added to a food system can best be tested in that food system. But to understand functionality better and to simplify the testing procedure, it is desirable to learn how proteins interact with single components of food systems. Perhaps the best studied of these kinds of interactions are those between proteins and water. Many types of measuring systems have been used to study protein–water interaction. Dry protein interacting with water vapor is expressed as a water sorption isotherm. Protein interacting with liquid water can be expressed as water adsorption, binding, swelling, solubility, viscosity, or gelation. We consider how some of these measurements are made and the different conditions that can influence them, but first a word about terminology.

Terms such as water binding, water adsorption and absorption, water holding capacity, constitutional water, interfacial water, multilayer water, and bulk water have been used by different investigators to try to express different degrees of interaction of proteins and water. Unfortunately there is no agreed upon standard terminology that can be recommended. The best guide is to be aware that the same term can mean different things depending upon who is using it, and to pay most attention to how the measurement is made rather than to the term used to express the protein–water interaction.

Water Sorption Isotherms

If dry protein is allowed to come to equilibrium with each of a series of different relative humidities (water activity or a_W in decimal terms), the result is

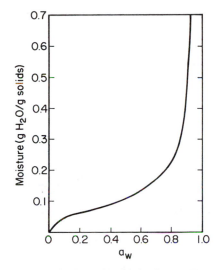

Fig. 5.1. General water sorption isotherm for protein. Source: Fennema (1976). Copyright by Marcel Dekker, Inc.

a water sorption isotherm as shown in Fig. 5.1. As implied in the term, the temperature has to be constant for all measurements in a series. Proteins are remarkably similar in displaying this same general shape of curve regardless of their solubility, amino acid composition, or state of denaturation. Note that the moisture content plotted on the ordinate is always on a dry rather than a wet weight basis.

There are several ways in which sorption isotherms can be developed. Known relative humidities can be produced by saturated solutions of different salts. If a small amount of the sample under investigation is incubated in a closed container over a saturated salt solution, it will come to equilibrium with that known relative humidity. The time needed for equilibrium may range from 2 days at intermediate humidities to 7 days at high humidities. After reaching equilibrium (determined by no further weight change), the sample can be analyzed for moisture content, and one has all the information needed for one point on the sorption isotherm.

A second approach to determining a sorption isotherm is to incubate a relatively large sample of known moisture content in a closed container with a small headspace. The headspace is fitted with a hygrometer to determine relative humidity. Again the sample has to come to equilibrium (determined by no further change in relative humidity), the relative humidity in the headspace is measured, and this measurement will establish one point of the isotherm. It is assumed because of the small headspace that moisture content of the sample does not change.

In both methods temperature has to be carefully controlled.

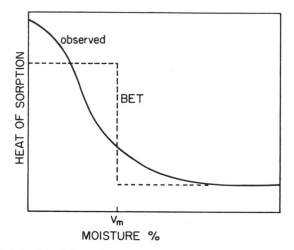

Fig. 5.2.   Relationship of heat of sorption and moisture content as actually observed and according to BET theory. Source: Labuza (1968).

The first section of the water sorption isotherm at low a_W shows rapid increase in water uptake by the protein with a relatively small change in a_W At intermediate relative humidities, the water uptake by protein is less rapid in relation to increases in a_W At high relative humidities ($a_W \geq 0.8$), the moisture content appears to be increasing very rapidly, but this is a consequence of the scale used for measuring moisture. On a dry weight basis the moisture content would be infinity at an $a_W = 1$.

Figure 5.2 shows how the energy of water sorption changes as the moisture content increases. The data of Figs. 5.1 and 5.2 have led to the idea that the first water molecules bind to protein most tightly (higher sorption energies) and subsequent molecules bind less tightly. Although there is an obvious continuum of sorption energies and amounts of moisture bound by proteins, regions of the curves have been characterized as representing different kinds of bound water. For example, Fennema (1977) has described the water bound in the first portion of the sorption isotherm (0.003 g water per gram of protein) as constitutional water, with any given water–protein structure having a half-life of about 10^{-6} sec. As water content increases (from 0.1 to 0.5 g water per gram of protein), the water is called interfacial, and any given water–protein structure persists for approximately 10^{-9} sec. Above 0.5 g water per gram of protein, the water is characterized as bulk water, which may be either free or entrapped, and structures persist for about 10^{-12} sec, the same time scale as for hydrogen bonding of water molecules in clusters.

Research Methods.   Several methods exist for studying the interaction between water and protein. We have just been describing one way in which protein modifies the vapor pressure of water. As a result of interaction of water molecules with protein, the vapor pressure is decreased, and the change in vapor pressure is used to gain information about the protein–water interaction.

Nuclear magnetic resonance spectroscopy (NMR) is an analytical method that gives information on the kind of environment in which nuclei are located. NMR is useful in differentiating between protons of water molecules having different mobilities. It is also useful in distinguishing between water molecules immobilized by interaction with protein and those immobilized by freezing. Hansen (1976) has studied moisture sorption isotherms of soy concentrates and made NMR measurements based on relaxation times and the difference between unfrozen and frozen samples. He found that the amount of water interacting with soy protein was 0.26 g per gram of protein.

Calorimetry is the study of the amounts of heat released or taken up as molecules go through a phase change. Since the water that interacts with protein does not freeze, calorimetry is useful in distinguishing total water from freezable water in protein samples. By freezing the sample to be studied and gradually raising its temperature in a calorimeter cell, one can measure the amount of heat taken up during melting. Dividing the total heat taken up by the heat of fusion for water (79.6 cal/g) gives the weight of freezable water in the sample. The sample can then be analyzed for total moisture, and total moisture minus the freezable moisture is a measure of the amount of water interacting with the protein in the sample. Muffett and Snyder (1980) used this technique to measure water interacting with soy protein and found not a single value but an increasing amount of unfreezable water as the moisture content increased. Others have also found that calorimetry studies give increasing bound water as total moisture increases (Ross 1978, Bushuk and Mehrota 1977, Biswas *et al.* 1975), but the reason for this has not been found.

If water is interacting with protein molecules, it would be reasonable to assume that the water would not be available to dissolve additional molecules. This idea has been used by measuring the solvation properties of protein–water mixtures. However, solvation ability seems to depend on the size of the added molecule, and so this technique has not been useful to measure water–protein interaction.

In addition to studies on overall interaction between proteins and water, studies have been made on the specific sites on a protein molecule where water would interact. The sites are equivalent to the amino acids that make up a protein. Some have tried to study water interaction with the individual amino acids, but such studies do not give useful information because the amino and carboxyl groups that are free in individual amino acids become peptide bonds in proteins; thus the sites are quite different in proteins as compared to free amino acids. Kuntz (1971) and Kuntz and Kauzmann (1974) solved this problem by studying polypeptides of individual amino acids using NMR and frozen samples. Their results are shown in Table 5.1 and indicate that a nonpolar amino acid side chain such as

Table 5.1

Proposed Amino Acid Hydrations Based on NMR Studies
of Polypeptides[a]

Amino acid residues	Bound water (moles water/mole amino acid)
Ionic	
Asp⁻	6
Glu⁻	7
Tyr⁻	7
Arg⁺	3
His⁺	4
Lys⁺	4
Polar	
Asn	2
Gln	2
Pro	3
Ser, Thr	2
Trp	2
Asp	2
Glu	2
Tyr	3
Arg	3
Lys	4
Nonpolar	
Ala	1
Gly	1
Phe	0
Val	1
Ile, Leu, Met	1

[a]Source: Fennema (1977).

that of alanine or valine would bind one water molecule. Polar side chains that develop charge asymmetries in water as a solvent will bind two or three water molecules per side chain, and ionic side chains such as exist in aspartic and glutamic acids and in lysine will bind four to seven molecules of water per molecule of amino acid.

Variables such as pH, salt concentration or ionic strength, solubility, and denaturation of the protein have relatively little effect on the moisture sorption isotherm (Fennema 1977). At high salt concentrations, ions will compete with protein and decrease the amount of water bound by a protein, but such concentrations are outside the range generally used for food products.

Information on moisture sorption isotherms for dry protein products is useful to determine what kind of packaging material would be best in maintaining a proper moisture content for the product. Also, the sorption isotherms can be useful in predicting how moisture will migrate if a dry soy protein product is mixed with other dry ingredients to form a new food product. The a_W of food products determines susceptibility to lipid oxidation, bacterial growth, and browning reactions, all of which are important quality aspects of dry protein products. More closely related to functional properties of proteins than the sorption isotherms is the amount of water that the protein can immobilize or hold as a tissue, gel, or dough. This kind of water is generally referred to as absorbed water.

Water Absorption or Swelling

When excess water is added to a food product or ingredient, some of the water will be immobilized, that is, it will not be apparent as free water. The amount of water absorbed in this way is a functional property of the food or ingredient as well as being related to other functional properties such as solubility, viscosity, and gelation. Water hydration capacity or water holding capacity (both abbreviated WHC) are other terms used for water absorption. There are several different experimental procedures for determining amounts of absorbed water, and a description of absorbed water can best be done in terms of the experimental procedures used.

Centrifugation. This is probably the most frequently used procedure for absorbed water. A severalfold excess of water is added to the food or ingredient, and by mixing or by high-speed dispersion, the food system is allowed to come to equilibrium with the moisture. The mixture is centrifuged at a low speed (1500 rpm), and the excess or free water is decanted.

Water absorbed is usually reported as weight increase in relation to the original dry weight of the sample. Hutton and Campbell (1977a) found values ranging from 200 to 500% for soy concentrate and from 200 to 1400% for soy isolate. Sometimes water absorption is reported on the basis of protein weight, and so the basis for the measurement is important to keep in mind.

In contrast to moisture sorption isotherms, absorbed water is strongly influenced by pH and solubility of the protein. As Quinn and Paton (1979) have emphasized, solubility of the protein is a complicating factor. With an excess of added water, soluble proteins will dissolve and will be decanted with the nonabsorbed water. For this reason Quinn and Paton (1979) have advocated a method for measuring absorbed water in which excess added water is avoided, and only enough water to saturate the protein is added. Data plotted in Fig. 5.3 show how the WHC varies with pH for a meat sample and for an extruded soy sample.

Ionic strength and the kind of salt will have an influence on the amount of water absorbed by proteins, but there is no easily discernible pattern.

Swelling. Another procedure for measuring the amount of water absorbed by soy proteins has been advocated by Hermansson (1979). It makes use of an

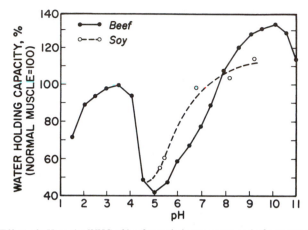

Fig. 5.3.   Effect of pH on the WHC of beef muscle homogenate and of soy extrudate. Sources: Rhee *et al.* (1981) and Fennema (1977).

apparatus in which a wetted filter paper is in contact with a reservoir of thermostatted water, which is connected with a glass capillary tube (Fig. 5.4). The sample is sprinkled on the filter paper, and as it absorbs water, water is withdrawn from the reservoir, which causes a change in the level in the capillary tube. A cover prevents evaporation. This empirical method provides useful information, but it depends on whether the sample is soluble in water (unlimited swelling, according to the terminology of Hermansson) or is insoluble (limited swelling). Also, the distribution of the sample on the filter paper achieved by sprinkling will affect the results.

Brabender Viscosity.   When a wheat flour dough is tested for viscosity changes as the dough is worked in a Brabender farinograph, the first step is to adjust the moisture content of the dough. Sufficient water is added so that the initial viscosity of the dough is 500 Brabender units. Thus the amount of water added is a measure of the water absorption of the dough. If the dough contains added soy protein, the amount of water absorbed at 500 Brabender units is much greater than without added soy protein, showing the water absorption capability of the added soy protein.

Solubility

The solubility of soy proteins is important in several different senses. As discussed in Chapter 2, the extraction and separation of individual soy proteins depend to a large extent on their solubilities and on how those solubilities change with pH, ionic strength, and reducing agents. In Chapter 4 we discussed another aspect of soy protein solubility: the quality of soy protein products as determined

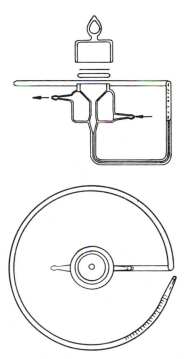

Fig. 5.4.   Equipment for measuring the spontaneous uptake of water. Source: Hermansson (1979).

by NSI and PDI. Now we are concerned with the interrelationships between solubility and functionality.

For some end uses a soluble protein preparation is an important aspect of product quality. In high-protein drinks, such as soy milk, or in soft drinks designed to have a high-protein content, solubility or dispersibility is all-important. Solubility keeps the viscosity low for a reasonable protein content of 3% in a drink, and solubility keeps the protein from settling out and causing a problem of redispersion.

For other uses as an ingredient, solubility may confer an advantage of ease of addition and ease of uniform distribution. If the functional need is for texture, water absorption, fat absorption, or gelation, then an insoluble protein preparation would be more useful. There are very few general rules that can be used to relate solubility to functionality.

Shen (1976) made an extensive study of the conditions that affect the solubility analyses of soy isolates. He studied blending conditions, centrifugation conditions, equilibration time, concentration, and temperature as they influence the amount of protein that goes into solution. Each variable did have an influence on the measured solubility of the sample soy isolate, showing that the analysis for solubility is not a simple, unambiguous procedure. The conditions chosen by

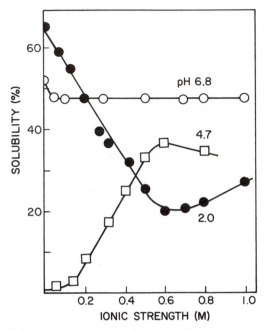

Fig. 5.5.   The effect of changing ionic strength on solubility of soy protein isolate at different pH. Source: Shen (1976).

Shen differ from those used for NSI, and consequently the measured solubilities differ depending on the procedure used. This is a specific example of the problem of measurement of functional properties. The measured value depends upon the method used, and for various, valid reasons investigators modify the methods to suit their particular aims. Consequently, reported data on functional properties must be considered in terms of the method used, making an inherently complicated situation even more complicated.

Aside from the analysis difficulties, soy isolates differ in their solubilities depending on the suspending medium. The most important variables are pH and ionic strength, as shown in Fig. 5.5. Increasing ionic strength causes the protein fraction that is insoluble at pH 4.7 to become increasingly soluble. Protein at pH 6.8 is influenced very little by changes in ionic strength, but if the pH is 2, protein solubility is lost with increasing ionic strength. As a general rule, the solubility of soy protein, and most other proteins, can be increased by adjusting the pH to about 10 or higher.

The effect of heat on soy protein solubility depends greatly on the purity of the protein preparation, the pH, the ionic strength, and the solvent system used (Wolf 1978). As noted in Chapter 2, Johnson and Snyder (1978) found that heating full-fat soy meal caused the protein bodies to become heat fixed and not lyse upon contact with water. If soy protein is first extracted from unheated soy meal, and the extract is heated at 212°F (100°C), the solubility pattern is quite different. The whey proteins, such as trypsin inhibitor and hemagglutinin, denature and precipitate, but the 11 and 7 S globulins dissociate into smaller subunits and remain soluble. With continued heating the small subunits may aggregate and precipitate.

Heating a soy protein solution of 8% or greater at 158°F (70°C) to 212°F (100°C) will cause the formation of gels, which will be discussed more fully in the section on gelation. The presence of reducing agents such as cysteine, bisulfite, or mercaptoethanol generally increases the solubility or inhibits gelation.

Time of storage of soy proteins can also change their solubility. Saio *et al.* (1982) and Nash and Wolf (1980) have studied the storage stability of isolated soy protein, soy meal, and whole soybeans, and they found that the proteins most likely to lose solubility during storage were the 7 and 11 S globulins. The intact soybeans were less affected than the meals or isolated protein. Figure 5.6 shows changes in solubility with storage time at relatively high temperature and high relative humidity for beans, defatted meal, and full-fat meal (Saio *et al.* 1982).

Viscosity

The flow properties of soy proteins are closely but not always directly related to solubility. Viscosity is an important functional property of foods that affects mouth feel, the textural quality of fluid foods such as beverages and batters, and the design of processing lines. For example, fluid flow through pipes, pumps, extruders, heat exchangers, and spray driers is a function of the viscosity of the material being processed.

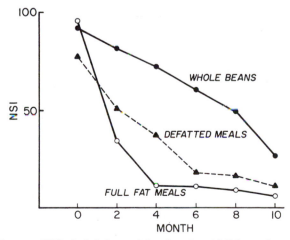

Fig. 5.6. Changes of NSI of whole beans, defatted meal, and full-fat meal stored at 96°F (35°C), 85% RH. Source: Saio *et al.* (1982).

Intrinsic viscosity is the viscosity of a dilute solution or dispersion extrapo-
lated to zero concentration. Intrinsic viscosity is a property of proteins that can
give information on shape and size of the molecules, and intrinsic viscosities
have been measured for dilute solutions of highly purified soy proteins. Unfortunately these highly reproducible measures of soy protein viscosities are useless
for determining functional aspects of protein viscosity. The main reason is that in
food systems concentrations of proteins and other ingredients are so high that
there is considerable molecular interaction, and that interaction cannot be pre-
dicted from information about the proteins only.

Measurement of viscosity that is of use in food systems is done by instruments
such as the Brookfield viscometer with a Helipath stand, so the spindle is constantly encountering new material. For batters the Brabender farinograph is
commonly used. The measured values are apparent viscosities, because the
mixtures being measured are non-Newtonian in their viscosity behavior, and the
apparent viscosities are valid for only that set of conditions under which they
were measured.

All of those variables affecting solubility will also affect viscosity. Usually,
the more material there is in solution the higher will be the viscosity. Figure 5.7
shows how pH affects solubility and viscosity of soy protein. If the pH is
sufficiently high (10 or above), protein molecules are denatured, and the loss of
tertiary structure has a larger effect on viscosity increase than is true for solubility
alone.

Shear rate at which the viscosity measurement is made will have an influence
on the apparent viscosity. This is shown for a soy isolate in Fig. 5.8, where the
effect of concentration is also shown. Apparent viscosity increases with concentration of the protein.

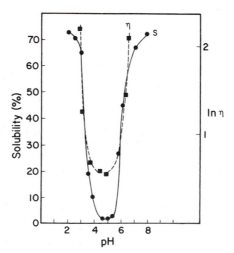

Fig. 5.7.   The relation of viscosity (η) and solubility (S). Source: Shen (1981).

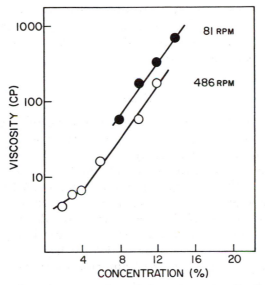

Fig. 5.8. Effect of protein concentration and shear rate on viscosity of soy protein isolate. Source: Hermansson (1972).

Table 5.2

The Effect of Aging on the Apparent Viscosity of Soy Protein
Isolate (12%)[a]

Method[b]	Spindle	Standing time (min)	Brookfield apparent viscosity (cP) at 3 rpm
A	1	15	2.6×10^3
	2	45	3.7×10^3
		75	4.5×10^3
		265	5.0×10^3
		455	8.0×10^3
	F	Weekend	8.0×10^6
B (90 sec)	1	15	7.1×10^2
		75	1.3×10^3
		135	2.1×10^3
	2	255	4.4×10^3
	F	Overnight	2.2×10^6
B (7.5 min)	UL	15	33.0
	UL	75	78.0
	1	135	1.9×10^2
		255	5.1×10^2
	F	Overnight	2.9×10^6

[a]Source: Shen (1981).

[b]A, Osterizer blender set at "blend" for 45 sec; B, ultrasonicator Braun-sonic 1510 at 150 W.

In accordance with the thixotropic behavior of soy protein, soy isolates show an increase in apparent viscosity with time. Table 5.2 gives data for a soy isolate dispersed in several different ways, and then stored for different times before making the viscosity measurement.

Gelation

If viscosity increases sufficiently, the soy protein can form a gel, and this is a useful functional property. Soy protein gels are important in comminuted sausage products and in soy curd products.

Gels produced by heating protein solutions involve a two-step process according to Ferry (1948). The first step is the loss of secondary and tertiary structure by heating (denaturation), which is irreversible. The actual formation of the gel, which occurs on cooling of the protein suspension, depends on a controlled aggregation of the protein molecules such that solution is trapped in the three-dimensional network. This step is reversible. If the aggregation occurs too rapidly, a weak, opaque gel may form, or the protein may simply precipitate with no gel formation.

Catsimpoolas and Meyer (1970) studied these reactions for gel formation using a soy isolate. They described the sequence of events as

$$\text{Sol} \xrightarrow{\text{heat}} \text{Progel} \underset{\text{heat}}{\overset{\text{cool}}{\rightleftarrows}} \text{Gel}$$
$$\underset{\begin{array}{c}\text{excess}\\\text{heat}\end{array}}{\big\downarrow}$$
$$\text{Metasol}$$

The sol to progel transition is brought about by heating sufficiently to denature the protein suspension. If other conditions are correct (see below), upon cooling the progel is converted to a gel. By heating the gel, the progel state is reformed. If the sol or progel is heated excessively, meaning a temperature of 257°F (125°C), a gel does not form upon cooling, and a metasol state is formed. Under the condition of excessive heat the protein structure has been modified, so that proper aggregation for gel formation is no longer possible.

There are several variables other than temperature that can affect the gelation process. Protein concentration has to be above a certain minimum, which is 8% for soy isolate, to form a gel. Also gel strength is dependent upon protein concentration, with higher concentrations forming stronger gels.

At extremes of pH, outside the range 2–10, gel strength is weak. This could be evidence of the importance of ionic cross-links in gel formation, since at extremes of pH the proteins would tend to be all negatively or positively charged. Ionic strength also has an influence on gel formation. Using sodium chloride the gel strength is inversely related to the concentration of salt.

The presence of reducing agents such as cysteine or mercoptoethanol will

inhibit gelation at low concentrations, indicating that disulfide cross links between protein molecules are important in the gelation process. However, the role
of disulfide bonds in gelation is not clear cut, because N-ethylmaleimide, a
sulfhydryl blocking reagent, has no effect on gelation, and gelation is promoted
by high concentrations of mercaptoethanol (a reducing agent that breaks disulfide
bonds).

The second type of soy protein gel starts with a heated soy milk preparation
that is gelled by the addition of calcium or calcium plus magnesium salts. The
protein concentration may be as low as 3%, but the aggregation is promoted by
the divalent cations. For proper gel strength, the amount and kind of calcium salt
are important. As the amount of added salt increases, the gel strength also
increases. For high-quality soy curd foods, 30–40 mM calcium sulfate is used,
and it is added slowly to avoid a rapid precipitation of the soy protein and a loss
of gel-forming ability.

An important variable in the gelation of soy proteins is the kind of food
mixture involved, and consequently the kinds of interactions taking place between soy proteins and the other food ingredients. In a relatively simple food
system such as soy curd, the interaction with lipid and with whey proteins will
affect the gel, as will the time and strength of pressing and the particle size of the
suspended solids in the original soy milk. In a more complicated system such as a
comminuted sausage product, there are even more interactions to consider. Peng
et al. (1982) have begun to study the interactions involved by considering only
soy 11 S proteins and myosin from the meat system, and they have found interactions. Whether the data obtained from such simplified systems will ever be of use
in interpreting or controlling the complex meat mixture is a serious question. It
may be that the degree of complexity is too great to justify a reconstruction based
on results from simplified systems.

Properties of the gels may be measured in a variety of ways. The gel strength
or hardness is the property of most interest, and it can be measured by compres-
sion, penetration, or viscosity in a variety of rheological instruments. The lack of
standardization of the measurements makes interpretation of published data difficult, but the use of different measuring techniques depending on instrument
availability and the degree of accuracy needed is perfectly understandable. In
addition to gel strength, elasticity properties, adhesiveness, and deformation
under a load may be of importance and can be measured, but with the same
difficulties of standardization noted above.

Texturization

Although we are including this subject under the general category "interactions of soy proteins with water," the functional property of texture does not
depend mainly on water–protein interactions. Texture, as it is built into soy
proteins, depends primarily on protein–protein interactions or interactions of soy

proteins with other components of the food system, including water. The functional property of texture does fit in the sequence of viscosity, gelation, and texture, and that is the reason for including it at this point.

Texture is a functional property that can be contributed to foods by soy proteins. Most obvious in this respect are the meat and cheese analogs made with soy as the only protein source. The processing technology for producing textured products from soy proteins has been covered in Chapter 3.

Variables that influence texture as generated by thermoplastic extrusion have been investigated by Rhee *et al.* (1981). They found that high solubility of the starting material was desirable for producing a textured matrix and that an NSI of 14 in the original defatted soy flour would not allow the generation of texture by extrusion. Also, they found that enzymatic digestion of the soy flour to decrease polypeptide molecular weights below 50,000 caused loss of texture upon hydration. On the other hand replacement of the soy flour with up to 15% sucrose or up to 60% soy isolate had very little effect on the texture generated by extrusion.

Methods for determining texture include those rheological methods already mentioned in connection with gelation and methods used traditionally in the meat industry such as force needed to shear a certain sized sample. Sensory evaluations of texture are needed, when analogs of traditional food products are made, to determine noticeable differences. Rhee *et al.* (1981) and Hermansson (1979) have emphasized the utility of microscopy in judging the influence of variables on texture. Both light and electron microscopy are effective tools to learn more about the basis for changes in texture.

INTERACTIONS OF SOY PROTEINS WITH LIPID

The interactions of soy proteins with water and with lipid are not mutually exclusive. In foods where both water and lipid are present, it would be difficult to determine the interaction of proteins with these other two components. Some of the properties of soy proteins just discussed, such as solubility, viscosity, and gelation, may differ in the presence of lipid. There are two ways in which protein and lipid may interact: lipid absorption and emulsions.

Lipid Absorption

This measurement is analogous to the measurement of water absorption, but it has been much less studied. The functional property that is involved is the prevention of lipid separation during heating of some lipid-containing product such as a comminuted sausage.

To determine lipid absorption, one stirs a small portion (0.5 g) of the protein to be tested with an excess of oil, and after allowing time for equilibration, centrifuges the mixture. The excess oil is decanted, and the fat absorption is the percentage weight increase of the protein sample. Obviously some of the oil will be trapped in the protein matrix, and there is no way to distinguish this fraction from oil that is more firmly bound to the protein.

Amounts of absorbed lipid range from 80 to 150%, with the larger amounts being absorbed by protein isolates. If results are expressed on the basis of protein weight rather than sample weight there is little difference between soy concentrates and soy isolates (Hutton and Campbell 1981).

The absorption of lipid by soy protein may be related to the hard to extract lipid fraction referred to in the section on solvent extraction of soy oil in Chapter 3. Oil that is absorbed in some way to protein surfaces or to hydrophobic sites in proteins may be difficult for the solvent to remove during extraction of flakes.

In contrast with water absorption, variables such as pH and temperature seem to have little effect on lipid absorption (Kinsella 1979).

Some consider color and flavor contributions of soy proteins as part of their functionality. In this respect, one of the problems with use of soy proteins in foods is the contribution of off-flavors. The off-flavors are considered to be lipid oxidation products that are bound to the protein.

A study has been made of the binding of low molecular weight alcohols, aldehydes, ketones, esters, and hydrocarbons to soy protein isolate (Aspelund and Wilson 1983). Using dry soy protein isolate as the column material in gas chromatography, homologous series of the classes of compounds noted above were passed over the column, and adsorption energies measured. Aspelund and Wilson found that binding energies increased with increase in chain length for each class of compound. Alcohols were most tightly bound, and hydrocarbons were least tightly bound. The other classes of compounds were approximately equal in binding energies. A chain length of nine carbons was needed to demonstrate binding of hydrocarbons, but four carbon alcohols were definitely bound. Chain lengths of at least six carbons were necessary to demonstrate binding of esters, aldehydes, or ketones.

Thus a start has been made in understanding the contribution of purified soy proteins to flavor. The conditions of preparation and storage of individual food products (rather than purified proteins) would add immeasurably to the complexity of the binding and release of flavor compounds.

Emulsions

Although water-in-oil food emulsions exist, the emulsions that make use of soy protein functionality are oil-in-water. Foods that utilize emulsifying properties of soy proteins are comminuted sausages, batters and doughs, coffee whiteners, frozen desserts, and salad dressings.

An oil-in-water emulsion consists of finely dispersed droplets of oil in a continuous phase of water or dilute aqueous solution. The size of the oil droplets may range from less than 1 to 20 μm or more. Protein is useful in an oil-in-water

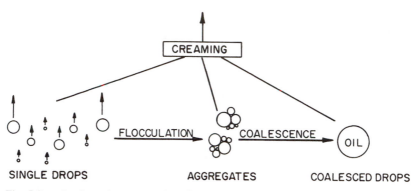

Fig. 5.9.   A schematic representation of emulsion instability. Source: Hermansson (1979).

emulsion because it can increase the amount of oil in such an emulsion (emulsion capacity), and it can stabilize the emulsion from creaming or from separating into two phases. If the oil droplets aggregate and rise to the top of the emulsion, this is known as creaming. If the oil droplets coalesce, and the two original phases of oil and water form, then the emulsion has broken. Figure 5.9 differentiates diagramatically between these two types of emulsion instability.

Proteins are effective in promoting emulsion formation because they have hydrophilic and hydrophobic regions based on their amino acid composition. Normally the hydrophilic region is exposed to the aqueous phase and the hydrophobic region is exposed to the interior of a globular protein in solution. When an oil–water interface becomes available as in an oil-in-water emulsion, protein molecules at the interface will unfold so that the hydrophobic region of the protein associates with the oil phase, and the hydrophilic region of the protein remains associated with the aqueous phase. Thus, protein that is soluble in the aqueous phase when the emulsion is being formed has a much better chance of orienting correctly at the interface to stabilize the emulsion. Solubility of soy protein preparations does correlate well with promoting emulsion capacity and stability.

Emulsion Capacity.   To measure emulsion capacity, a solution or suspension is made of the protein being tested, and it is stirred or sheared at a high rate in a vessel fitted to receive oil additions. Oil is added to the vessel at a constant rate so that an emulsion forms. Eventually, the emulsion capacity of the protein will be exceeded, and at that point some visible change will take place or there will be an abrupt change in viscosity or electrical conductivity of the continuous phase. The emulsion capacity of the test protein is then calculated as the milliliters of oil emulsified by 1 g of sample or by 1 g of protein. Soy proteins have emulsion capacities in the range 2–20 ml of oil per gram of protein (Hutton and Campbell 1977b).

As emphasized by Tornberg and Hermansson (1977), the measured emulsion capacity does depend strongly on the measuring system. That is the emulsifying ability of the test system (speed of the shearing element or power to a sonicator) will influence the measured emulsifying capacity of a protein. Consequently, absolute values for emulsion capacity should not be considered or compared without taking into account the measuring system.

Emulsion Stability.   Equally important with the emulsion capacity is the emulsion stability measurement. A large capacity with very little stability is of little use in most food systems. There are several approaches to measurement of emulsion stability. One can visually observe either creaming or breaking of the emulsion as a function of time. For this technique to be effective, the emulsion has to be weak, so that a visible change can be observed in a reasonable amount of time. More quantitative measures can be achieved by measuring an increase in lipid in the upper portion of the emulsion.

Tornberg and Hermansson (1977) devised a stability ratio (SR) based on a fat analysis in the lower portion of an emulsion. The analysis of the test sample (F_{test}) was divided by the analysis of the original emulsion (F_{orig}) and multiplied by 100:

$$SR = (F_{test}/F_{orig}) \times 100$$

The analyses are done after the samples have been centrifuged at a low speed (150 g).

Figure 5.10 compares SRs for a soy isolate (Promine-D), a whey protein concentrate, and a sodium caseinate. The protein isolate is superior to the other two proteins in emulsion stability, but more importantly Fig. 5.10 shows that the measured emulsion stability depends on the way in which the emulsion was made. In this instance the emulsion was formed by sonic energy, and emulsion stabilities differ widely, depending on the power setting. This dependence of stability on the way in which an emulsion is formed again makes it very difficult to compare data from different laboratories or to make generalizations about the variables affecting emulsion stability.

As noted earlier, solubility is important for good emulsion capacity and stability. Figure 5.10 shows that increased ionic strength enhances emulsion stability, and pH in the range of 7 to 9 generally increases emulsion capacity and stability perhaps as a result of solubility effects.

FOAMING

The functional property of foaming is closely related to emulsion formation but with air (or other gases) as the discontinuous phase rather than lipid. Both functional properties depend on the surface-active properties of the proteins involved.

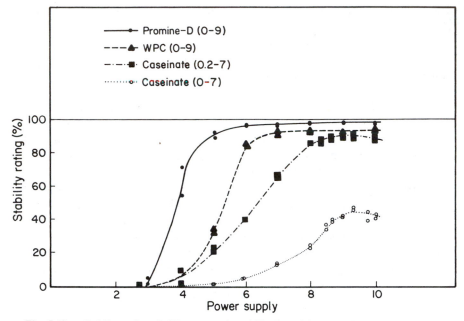

Fig. 5.10. Stability rating of different protein-stabilized emulsions, emulsified for one minute, as a function of the ultrasonic power supply. Numbers in parentheses are ionic strengh and pH, respectively. Source: Tornberg and Hermansson (1977). Copyright by the Institute of Food Technologists.

The foods in which foams are important and in which soy proteins can be useful are confections (nougats, marshmallow, divinity, and fudge), whipped toppings, icings, frozen desserts, meringues, and various types of cakes. The traditional protein for use in foam formation is egg albumen, but soy proteins have proven useful as whole or partial replacements for egg albumen.

As gas bubbles form in a foodstuff, functional protein concentrates at the surface of gas and liquid and unfolds, so that the surface tension is lowered, and viscosity is increased. These two effects tend to maintain the bubble and to minimize liquid drainage. For the protein to act in this way, it should be soluble initially to be able to diffuse quickly to surface sites, and it should be able to unfold, so that polar and nonpolar surfaces on the protein orient with the liquid and gas surfaces of the foam. The protein molecules must interact with each other sufficiently to form a surface film that resists breaking, but the interaction should not be so great that the protein tends to coagulate and thus break the foam (Cherry and McWatters 1981).

Many variables influence the ability of soy proteins to maintain this balance and thus perform well in generating and stabilizing foams. The pH should be above the isoelectric point to maintain solubility, but close to it so that molecular interaction is possible. A pH of 5 for soy proteins promotes foam functionality.

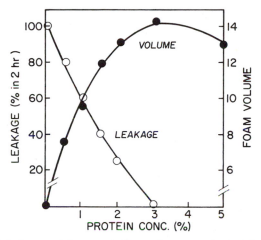

Fig. 5.11. Effect of soy protein concentration on foam expansion and foam stability. Source: Eldridge *et al.* (1963).

Figure 5.11 shows the influence of protein concentration on foam formation and stability. Since foam volume depends upon protein concentration, generally soy isolates are superior to concentrates, and soy concentrates are superior to flours.

Constituents other than protein can have a strong influence on the amount and stability of foams. Soy proteins frequently have some lipid remaining, and extraction of that lipid (probably phospholipid) with ethanol enhances foaming ability of soy proteins. Sucrose has the effect of decreasing slightly the foam volume but increases the foam stability, and so its presence is desirable. Sodium chloride (5%) will increase foam volume but decrease stability, and generally is not helpful.

As with other functional properties a wide range of methods is used to measure the foam volumes and stabilities. Foams may be generated by whipping, shaking, or sparging. The volumes generated are most frequently measured in a graduate cylinder. Stabilities are usually measured by the volume of liquid that drains from the foam with time or by the decrease in foam volume with time. The values found for these measurements are very much dependent upon the methods used for generating the foams initially and for this reason are difficult to use for generalizations.

Excessive agitation during the formation of a foam will cause the bubble size to decrease and may cause foam breakage. Since surface area increases exponentially as bubble size decreases, the need for more protein also increases exponentially. Foam breakage may be the result of lack of sufficient protein.

For best data on how the soy protein might perform in a particular food, it is always wise to make the food in question with the proteins being tested. But this is not foolproof for comparison of functional properties, since it is difficult to make an angel cake exactly the same way in various laboratories or to obtain exactly the same kind of protein isolate.

COMMENTARY ON FUNCTIONALITY

At the beginning of the chapter, we noted two ideas that are frequently cited by researchers interested in learning more about how vegetable proteins can substitute for animal proteins. The first of these ideas is that population pressures will mandate that vegetable proteins be substituted for animal proteins, simply because we shall not be able to produce sufficient animal protein products to supply the need. During 30 years of rapidly growing population, the world has expanded greatly its production and consumption of animal products, and there is good evidence that this trend will continue. At the same time the use of vegetable proteins in human foods has greatly expanded. We see no reason why the use of both animal and vegetable protein products cannot continue to grow as people become more affluent, and as we learn how to make economical use of vegetable protein substitutes for part of the animal proteins.

The second idea is that by accumulating sufficient basic information about the physical and chemical properties of proteins, we shall be able to predict the functional behavior of these proteins in a food system. There is an inherent complexity of protein functionality in foods that leads us to doubt the wisdom of pursuing basic physical–chemical information as a key to understanding functionality. Part of the complexity stems from the structure of the protein molecules, particularly a protein such as glycinin, the 11 S globulin from soy. The dodecahedron quaternary structure of glycinin with acidic and basic polypeptide chains that may vary in amino acid composition (based on the soybean variety) can give rise to a multiplicity of states that we try to describe with terms like native, denatured, or partially unfolded. Yet it is possible to gather basic information on solubility, water adsorption, and viscosity, for example, on glycinin under a variety of conditions of pH, ionic strength, temperature, and concentration.

The question is whether this kind of information is of any use in predicting functionality in a food system. Shen (1981) has argued that basic information, such as intrinsic viscosity of a purified protein solution, is meaningless to an understanding of functionality, and we agree with him. Adding greatly to the inherent complexity of protein functionality is the interaction of the added proteins with molecules of the food. Schoen (1977) has discussed the many causes of complexity in relating properties of proteins to functionality, but in the end he falls back on a plea for more and better basic understanding of individual proteins in simple model systems gradually working up to the complicated food systems. Schoen (1977) also makes a case for better use of sophisticated statistical techniques in dealing with the relationship of protein properties to functionality. As examples of how this should be done he cites the three papers by Akeson and Hermansson (Hermansson 1975, Hermansson and Akeson 1975a,b). Whether sophisticated statistical techniques can solve the problem of predicting functionality from basic protein chemistry still remains to be shown. We are not arguing against gathering basic information about proteins. This is necessary and worthwhile basic research. We are arguing against the idea that such information will be of use in predicting functionality of proteins in food systems.

Also, we are not arguing that it is futile to study functionality. Much has been learned about protein functionality in food systems that is useful, and more needs to be known. We are arguing that the farther one goes in the direction of simplified model systems, the less likely anything useful about food systems will be learned.

REFERENCES

Aspelund, T. G., and L. A. Wilson (1983). Adsorption of off-flavor compounds onto soy proteins: A thermodynamic study. *J. Agric. Food Chem.* **31:**539.

Biswas, A. B., C. A. Kumsah, G. Pass, and G. O. Phillips (1975). The effect of carbohydrates on the heat of fusion of water. *J. Solution Chem.* **4:**581.

Bushuk, W., and V. K. Mehrotra (1977). Studies of water binding by differential thermal analysis II. Dough studies using the melting mode. *Cereal Chem.* **54:**320.

Catsimpoolas, N., and E. W. Meyer (1970). Gelation phenomena of soybean globulins.I. Protein–protein interactions. *Cereal Chem.* **47:**559.

Cherry, J. P. (ed.) (1981). "Protein Functionality in Foods." *ACS Symp. Ser.* **147.** American Chemical Society, Washington, D.C.

Cherry, J. P., and K. H. McWatters (1981). Whippability and aeration. *In* "Protein Functionality in Foods" (J. P. Cherry, ed.). American Chemical Society, Washington, D.C.

Eldridge, A. D., P. K. Hall, and W. J. Wolf (1963). Stable foams from unhydrolyzed soybean protein. *Food Technol.* **17:**1592.

Fennema, O. (1976). Water and ice. *In* "Principles of Food Science," Part I, Food Chemistry (O. Fennema, ed.). Marcel Dekker, New York.

Fennema, O. (1977). Water and protein hydration. *In* "Food Proteins" (J. R. Whitaker and S. R. Tannenbaum, eds.). AVI Publ. Co., Westport, CT.

Ferry, J. D. (1948). Protein gels. *Adv. Protein Chem.* **4:**1.

Hamm, R. (1960). Biochemistry of meat hydration. *Adv. Food Res.* **10:**355.

Hansen, J. R. (1976). Hydration of soybean protein. J. Agric. Food Chem. 24:1136.

Hermansson, A. M. (1972). Functional properties of proteins for foods - Swelling. *Lebensm. Wiss. Technol.* **5**(1):24.

Hermansson, A. M. (1975). Functional properties of added proteins correlated with properties of meat systems. Effect and texture of a meat product. *J. Food Sci.* **40:**611.

Hermansson, A. M. (1979). Methods of studying functional properties of vegetable proteins. *J. Am. Oil Chem. Soc.* **56:**272.

Hermansson, A. M., and C. Akesson (1975a). Functional properties of added proteins correlated with properties of meat systems. Effect of concentration and temperature on water-binding properties of model meat systems. *J. Food Sci.* **40:**595.

Hermansson, A. M., and C. Akesson (1975b). Functional properties of added proteins correlated with properties of meat systems. Effect of salt on water-binding properties of model meat systems. *J. Food Sci.* **40:**603.

Hutton, C. W., and A. M. Campbell (1977a). Functional properties of a soy concentrate and a soy isolate in simple systems. *J. Food Sci.* **42:**454.

Hutton, C. W., and A. M. Campbell (1977b). Functional properties of a soy concentrate and a soy isolate in simple systems and in a food system. Emulsion properties, thickening function and fat absorption. *J. Food Sci.* **42:**457.

Hutton, C. W., and A. M. Campbell (1981). Water and fat absorption. *In* "Protein Functionality in Foods" (J. P. Cherry, ed.). American Chemical Society, Washington, D.C.

Johnson, K. W., and H. E. Snyder (1978). Soymilk: A comparison of processing methods on yields and composition. *J. Food Sci.* **43:**349.

Kinsella, J. E. (1976). Functional properties of proteins in foods: A survey. *Crit. Rev. Food Sci. Nutr.,* April, p. 219.

Kinsella, J. E. (1979). Functional properties of soy proteins. *J. Am. Oil Chem. Soc.* **56:**242.

Kuntz, I. D. Jr. (1971). Hydration of macromolecules. 3. Hydration of polypeptides. *J. Am. Chem. Soc.* **93:**514.

Kuntz, I. D., Jr., and W. Kauzmann (1974). Hydration of proteins and polypeptides. *Adv. Protein Chem.* **28:**239.

Labuza, T. P. (1968). Sorption phenomena in foods. *Food Technol.* **22:**263.

Muffett, D. J., and H. E. Snyder (1980). Measurement of unfrozen and free water in soy proteins by differential scanning calorimetry. *J. Agric. Food Chem.* **28:**1303.

Nash, A. M., and W. J. Wolf (1980). Aging of soybean globulins: Effect on their solubility in buffer at pH 7.6. *Cereal Chem.* **57:**233.

Peng, I. C., W. R. Dayton, D. W. Quass, and C. E. Allen (1982). Investigations of soybean 11 S protein and myosin interaction by solubility, turbidity and titration studies. *J. Food Sci.* **47:**1976.

Quinn, J. R., and D. Paton (1979). A practical measurement of water hydration capacity of protein materials. *Cereal Chem.* **56:**38.

Rhee, K. C., C. K. Kuo, and E. W. Lusas (1981). Texturization *In* "Protein Functionality in Foods" (J. P. Cherry, ed.). American Chemical Society, Washington, D.C.

Ross, K. D. (1978). Differential scanning calorimetry of nonfreezable water in solute–macromolecule–water systems. *J. Food Sci.* **43:**1812.

Saio, K., K. Kobayakawa, and M. Kito (1982). Protein denaturation during model storage studies of soybeans and meals. *Cereal Chem.* **59:**408.

Schoen, H. M. (1977). Functional properties of proteins and their measurement. *In* "Food Proteins" (J. R. Whitaker and S. R. Tannenbaum, eds.). AVI Publ. Co., Westport, CT.

Shen, J. L. (1976). Soy protein solubility: The effect of experimental conditions on the solubility of soy protein isolates. *Cereal Chem.* **53:**902.

Shen, J. L. (1981). Solubility and viscosity. *In* "Protein Functionality in Foods" (J. P. Cherry, ed.). American Chemical Society, Washington, D.C.

Tornberg, E., and A. M. Hermansson. (1977). Functional characterization of protein stabilized emulsions: Effect of processing *J. Food Sci.* **42:**468.

Wolf, W. J. (1978). Purification and properties of the proteins. *In* "Soybeans: Chemistry and Technology," Vol. 1, Proteins, 2nd ed. (A. K. Smith and S. J. Circle, eds.). AVI Publ. Co., Westport, CT.

6

Nutritional Attributes of Soybeans and Soybean Products

Recent advances in soybean processing technology have made possible deodorized and hydrogenated oils, lecithins, flours, protein concentrates, protein isolates, and textured products, in addition to the traditional Oriental foods, and thus have increased consumption of soybeans as foods. However, while the role of soybeans in human nutrition has increased over the last two decades throughout the world, only a small portion of soybean protein goes to human consumption.

From the standpoint of efficient use of soy protein in human nutrition, it would seem reasonable to produce palatable human foods from soybeans. Then soybeans and soybean protein would be consumed directly, and we would avoid the inefficiencies of feeding soybean protein to animals and eating the animal products. As will be explained in detail in Chapter 7, the direct consumption of soybean proteins by humans is a long-standing tradition in Eastern countries. In contrast, the tradition in the West is to feed soybean protein meal to animals being raised as food sources.

As countries develop economically, the trend seems to be for Eastern countries to adopt increasingly the practice of feeding soy protein to animals. Although the countries of the West have done a lot of experimenting with direct human consumption of soy protein products, there is no convincing evidence of a sizable change in this direction.

The nutritive quality of soybeans has to be judged not only from their nutrient content and physiological availability but also from the presence of inherent and derived constituents that may have undesirable impacts on acceptability and nutritive availability. Evaluation of these undesirable components must be included in the assessment of the overall nutritional quality of soybeans. This is especially important in the feeding of infants raised solely by a soybean-based formula.

Fortunately, some of the undesirable components, such as trypsin inhibitors, are readily inactivated by moist heat, which is often part of the processing of soybean foods. Therefore, the presence of potentially undesirable substances in raw soybeans has less practical significance when cooked soybeans or soybean products are consumed as a part of mixed diets.

INHERENT ATTRIBUTES OF SOYBEANS

Proteins and lipids along with some vitamins and minerals are the nutrients of soybeans. The carbohydrates, one of the major constituents of soybeans in terms of quantity, play a minor nutritional role. Digestibility of soybean protein is an essential factor in evaluation of nutritional availability.

Protein Digestibility

The texture of soybeans, as with other cereal grains and beans, is quite resistant to softening and thus digestibility of the whole bean is low. The digestibility of protein in different forms of traditional soybean products is listed in Table 6.1. The digestibility of protein in steamed beans and toasted beans is lower than that of soy curd and soymilk film. Soybean pulp has a digestibility between the two groups of soybean foods (Watanabe *et al.* 1971). In other words, the digestibility

Table 6.1
Digestibilities of Soybean Proteins
in Several Soybean Food Products[a]

Soybean product	Digestibility (%)
Steamed soybeans	65.3
Toasted soybeans	65.3
Soy curd	92.7
Soymilk film	92.6
Soy pulp	78.7

[a]Source: Watanabe *et al.* (1971).

Table 6.2

Comparison of Digestibilities of Soybean Protein Products
and Other Proteins by Adult Humans[a]

	Digestibility (%)[b]	
Protein source	Apparent	True
Soy flour, full fat	70	75–92
Soy flour, defatted, extruded	69–79	84–90
Soy protein isolate	81–82	93–97
Soy protein, spun fiber	83–88	101–107
Whole egg	73–86	93–100
Milk	69–77	90–98
Beef	73–82	91–99
Casein	71–78	94–97

[a]Source: Bressani 1981

[b]Apparent digestibility is given by $(I - F)/I$, where I is the ingested nitrogen and F the fecal nitrogen. True digestibility is given by $[I - (F - F_k)]/I$, where F_k is the fecal nitrogen on a nitrogen-free diet.

of soybean protein is low when the only treatment is heating, but further processing—such as soaking, grinding, and hot water extraction of protein—increases the digestibility considerably.

The digestibility of soybean protein also varies depending on the type of modern soybean product. In Table 6.2, data on protein digestibility by adults are compared for soybean products and animal products (Bressani 1981). Soy protein concentrates and isolates have a higher digestibility than soy flours.

One reason for this difference in digestibility is that in soy flours considerable cell structure is maintained, and it may be difficult for digestive enzymes to reach and react with the proteins in the form of protein bodies within cells.

Another reason is that constituents may be present in higher concentrations in flours than in concentrates or isolates, soybean trypsin inhibitors, for example.

Soybean Trypsin Inhibitors. The chemistry of trypsin inhibitors in terms of structure and reactivity has been discussed in Chapter 2. The fact that certain species of young animals do not grow well when fed raw soy meal as a sole source of protein has been known since 1917 at least. This biological activity of raw soy products has been referred to as growth inhibition, antinutritional activity, or rather vaguely as toxic factors. The cause remained a mystery until the 1940s when Kunitz discovered a trypsin-inhibiting protein in raw soybeans (Kunitz 1946).

Then it was thought that the mystery had been solved. If a trypsin inhibitor prevents trypsin from acting, then protein would not be hydrolyzed, and ob-

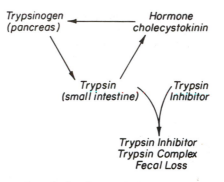

Fig. 6.1.   Feedback control mechanism for trypsin in the small intestine and disturbance by trypsin inhibitor.

viously an animal could not grow well when cut off from its supply of essential amino acids. However, a simple experiment showed that this was not the correct interpretation of events. When trypsin inhibitor was fed to young animals along with completely hydrolyzed soy protein, growth inhibition still occurred.

As scientific investigation of the situation continued, other complicating factors became known. It was discovered that supplementing a raw soy diet with sulfur-containing amino acids would overcome the growth inhibition. Also, investigators learned that young animals fed raw soy diets had greatly enlarged pancreases. The pancreatic hypertrophy only occurred in those species in which the pancreas made up a substantial proportion of the weight of the young animal (mainly small animals).

The generally accepted explanation for this series of observations is outlined in Fig. 6.1. Trypsin levels in the small intestine are monitored by the hormone cholecystokinin. When levels are sufficient a signal is sent to the pancreas to stop production, and when levels are insufficient the pancreas is called upon to produce more trypsinogen. Trypsin inhibitor disturbs this control system by complexing with trypsin. The complex is no longer active, and the feedback system calls on the pancreas to produce more trypsinogen. This constant need for more trypsinogen causes the hypertrophy of the pancreas. Furthermore, the overproduction of trypsin (an enzyme rich in sulfur-containing amino acids) and its loss in the feces cause a serious depletion of sulfur-containing amino acids in the young animal. This is particularly true in animals whose pancreas weight is large in proportion to total body weight (small animals such as chicks, rats, and mice). This explains why effects of trypsin inhibitor can be overcome by supplementing the diet with sulfur-containing amino acids.

The practical way of dealing with the trypsin inhibitors in raw soy meal has been to subject the meal to moist heat. For soy meals used in animal feeds, the heating occurs in the desolventizer–toaster. For soymilk and soy curd the heating step is an integral part of the normal processing (see Chapter 7). No heating step

Table 6.3

Trypsin Inhibitor Activity of Raw and Toasted Soy Flour[a]

Sample	Trypsin inhibitor activity		
	TIU/mg sample[b]	mg TI/g sample[c]	Raw flour (%)
Raw soy flour			
A	92.5	48.7	—
B	105.5	55.5	—
Average	99.0	52.1	
Toasted soy flour[d]			
A	5.5	2.9	6
B	9.4	5.0	10
Average	7.5	4.0	8
Commercial soy flour[e]			
Toasted	9.7	5.2	10

[a]Source: Anderson et al. (1979). Copyright by Academic Press Inc.
[b]TIU, trypsin inhibitor units as defined by Kakade et al. (1969).
[c]Calculated on the basis of 1.9 TIU = 1 g TI.
[d]Live steam at 100°C (212°F) for 30 min.
[e]Several lots; heat treatment conditions unknown.

may be involved in the processing of some types of soy concentrates and isolates in order to retain protein functionality and solubility. In some of these products, the trypsin inhibitors may be partially removed (with the whey fraction in producing protein isolates), or may be inactivated by heating food products (to which the concentrates or isolates have been added) just before consumption.

The amounts of trypsin inhibitor activity in raw soy flour and the decrease due to heating are shown in Table 6.3. Approximately 90% of trypsin inhibitor activity is inactivated by conventional desolventizer–toaster heating. In the laboratory, steaming for 30 min at 100°C (212°F) will accomplish the same effect. The residual trypsin inhibitor activities for several different soy products are shown in Table 6.4.

In applying heat to soy products, it is essential to use an optimum amount to minimize trypsin inhibitor activity but to avoid loss of the essential amino acid lysine. Too much heat can decrease nutritional value. Figure 6.2 shows how nutritional value of soy meal changes with time of steaming at 100°C (212°F), and the concurrent change in trypsin inhibitor. The measure of nutritional value is protein efficiency ratio (PER, weight gain/weight of protein ingested), and it begins to decline after 10 min heating. As heating time continues, the PER declines further due primarily to the loss of lysine in nonenzymatic browning reactions, but destruction of sulfur amino acids is also very likely. The presence of some moisture during heating greatly improves the inactivation of trypsin inhibitor or of any other protein.

Table 6.4
Trypsin Inhibitor Activity of Various Commercially Manufactured
Soy Protein Products[a]

	Trypsin inhibitor activity	
Product	TIU/mg	Raw soy flour (%)
Raw soy flour	99.0	100
Toasted soy flour[b]	6–15	6–15
Soy concentrate A	12.0	12
Soy concentrate B	26.5	27
Soy isolate A	8.5	9
Soy isolate B	19.8	20
Soy isolate C	20.9	21
Soy food fiber	12.3	12
Chicken analog	6.9	7
Ham analog	10.2	10
Beef analog	6.5	7
Textured soy flour	9.8	10

[a]Source: Anderson *et al*. (1979). Copyright by Academic Press, Inc.
[b]Several lots were analyzed.

Not all of the increase in PER shown in Fig. 6.2 is due to inactivation of trypsin inhibitor. Rackis *et al*. (1975) have shown that there is no pancreatic hypertrophy in feeding experiments with rats when only 54% of the trypsin inhibitor activity has been lost due to heating. Presumably, if there is no pancreatic hypertrophy, there would be no growth inhibition.

Kakade *et al*. (1973) removed Kunitz trypsin inhibitor from soy extracts using

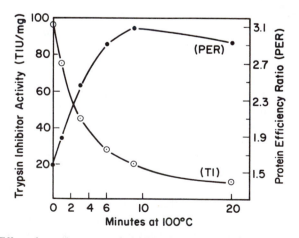

Fig. 6.2.   Effect of steaming on trypsin inhibitor activity and protein efficiency ratio (PER) of soy meal. Source: Anderson *et al*. (1979). Copyright by Academic Press, Inc.

Table 6.5
Effect of Trypsin Inhibitor Removal on the Protein Efficiency Ratio (PER)
and Rat Pancreas Weights[a]

Soy extract	TIU/mg protein	PER[b]	Pancreas wt (g) per 100 g body wt
Original	125.1	1.4	0.74
Heated original	13.2	2.7	0.52
Inhibitor free[c]	12.9	1.9	0.65

[a]Source: Anderson *et al.* (1979). Copyright by Academic Press, Inc.
[b]Data significantly different at $p < 0.03$.
[c]Trypsin inhibitor removed by affinity chromatography.

affinity chromatography, and compared those extracts without trypsin inhibitor to others with trypsin inhibitor in rat feeding experiments. The results shown in Table 6.5 show that PER is only partially restored to the heated value by removal of trypsin inhibitor. The other factor involved in heating raw soy flour is probably heat modification of soy proteins making them more digestible.

People have realized for some time that more than Kunitz trypsin inhibitor is involved in the growth inhibition caused by feeding raw soy flour. This realization led to the search for other factors and as a result the trypsin inhibitor of Bowman and Birk was found. As already discussed in Chapter 2, the Bowman–Birk inhibitor will inhibit both trypsin and chymotrypsin. Also, it is strongly cross-linked by disulfide bonds and consequently is resistant to heat denaturation. The low molecular weight of the Bowman–Birk inhibitor gives it strange solubility characteristics for a protein such as having solubility in ethanol. Usually scientists studying soy protein nutrition assume that the trypsin inhibitor activity remaining after heating is due to the Bowman–Birk inhibitor. Its physiological significance is not known.

Just as the search for another factor influencing the availability of soy protein led to the discovery of the Bowman–Birk inhibitor, so did that search lead to the discovery of soybean hemagglutinin.

Soybean Hemagglutinin.   Soybeans contain several hemagglutinins (Liener 1974) comprising 1–3% of defatted soy flour (Turner and Liener 1975). A seven-fold variation in hemagglutinating activity among 108 soybean varieties has been reported (Kakade *et al.* 1972). Soy hemagglutinins are glycoproteins of about 110,000 molecular weight, containing 5% carbohydrate (mainly mannose and N-acetyl-D-glucosamine). When first discovered, soybean hemagglutinins were thought to be responsible for approximately one-half of the rat growth inhibition attributable to raw soybean meal (Liener and Rose 1953). Soy hemagglutinins are readily inactivated by heat and by pepsin (Liener 1958). Inactivation is complete when 12% of the hemagglutinin peptide bonds are cleaved. Hemagglutinin activity of a crude raw soybean extract was selectively removed by affinity chromatography on Sepharose bound concanavalin A (Kakade *et al.* 1973). The growth rate of young rats was not changed by hemagglutinin removal, leading to the current concept that hemagglutinins are not involved in the growth inhibition from raw soy flour.

Protein and Amino Acids.   The predominant proteins in soybean products are glycinin and β-conglycinin, the 11 and 7 S storage proteins. Hence, the nutritional value of soybean products with respect to protein is determined primarily by the quantity and quality of these proteins.

Soybean protein has been used in the past and continues to be used today principally as a feedstuff for animals. Both the quantity and quality of soybean protein are factors that give it a competitive edge over other protein feedstuffs.

Dehulled soy flakes are sold on the basis of a minimum of 47.5–49.0% protein content, and soybean flakes with hulls added have a minimum of 44% protein. The high protein content of both products allows blending of soy flakes with other ingredients such as corn or a starchy tuber like cassava to dilute the protein to the 16–24% needed in the final feedstuff.

In addition to the advantage of a large quantity of protein, soybeans have high-quality protein, which means that the amino acid content is good for meeting the essential amino acid requirements of animals. The limiting amino acid in soy protein is methionine, or more generally the sulfur-containing amino acids. Hence, soy protein is complementary to cereal protein in which lysine generally is the limiting amino acid.

There is relatively little controversy surrounding the use of soy protein in animal feeding, but considerable controversy exists about soy protein in the human diet. Much of the controversy stems from the program of the U.S. FDA to inform consumers of nutrient content of foods through its nutritional labeling program. The U.S. government has the responsibility of devising a system by which consumers are kept informed of the nutrient quality of certain foodstuffs. If a company decides to change an ingredient in a product and that change results in a less nutritious product, there should be some way of alerting the consumer to that fact.

One way of doing this is by the name given to products with soy protein as an ingredient. This subject will be covered in Chapter 11 on grades and standards for soy products.

The other system in use in 1986 is a complicated set of rules known as Nutritional Labeling, and protein is evaluated by PER measurements.

Protein, as with other nutrients in nutritionally labeled foods, is given as a percentage of the USRDA (recommended daily allowance) for a certain serving size. Since proteins vary in quality based on their amino acid content, the USRDA is set at 45 g if PER is equivalent to casein or higher and at 65 g if PER is lower than for casein. Part of the controversy about use of soy protein in human foods is over the PER method used to evaluate protein quality.

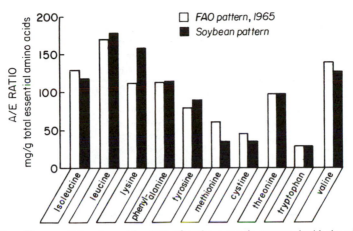

Fig. 6.3. The essential amino acid pattern of soybean protein compared with the whole egg protein. Source: Liener (1978).

Methods for evaluating the quality of proteins fall into several categories. Simplest to use are those based on amino acid analysis of the protein. A comparison of the amino acid analysis with the amino acid needs of humans or with proteins that are considered to be nutritionally adequate such as eggs leads to the concept of chemical score.

Figure 6.3 shows the essential amino acid content of soybeans compared to the whole egg standard used by FAO in 1965. The data are given as an A/E ratio, which is the amount of any particular amino acid (mg) divided by the amount of total essential amino acids (g). Comparing the value for the limiting amino acid for soybean protein (cysteine + methionine) of 74 to 107 for whole egg protein gives a chemical score of $^{74}/_{107} = 69\%$. The chemical score thus indicates that soy protein would be roughly 70% as good as whole egg protein in meeting human requirements.

There are several shortcomings of this chemical scoring procedure: First, comparing with egg protein is not as useful as comparing with actual human requirements, since whole egg protein is particularly rich in essential amino acids. Second, chemical score gives no measure of the extent to which the amino acids are available. If there is a digestibility problem, the total quantity of amino acids may not be biologically available. This is more likely to be a problem with plant proteins than with animal proteins. Third, the emphasis on limiting amino acids obscures the very positive aspect of soy protein, its high lysine content, which is readily evident from Fig. 6.3. In spite of these shortcomings, chemical score can be a useful tool in evaluating nutritive quality of proteins.

To avoid some of the shortcomings of the chemical score, biological systems are used to evaluate nutritional quality of proteins. Table 6.6 compares several types of foods using different biological and chemical evaluation systems.

PER is a biological test system in which the test protein is fed to young rats

Table 6.6

Nutritional Quality of Proteins Based on Biological Evaluation and on Amino Acid Composition

Protein	PER	BV[a]	NPU[b]	Limiting amino acid	Chemical score	E/T ratio[c] (g/g total N)
Whole egg	3.8	87–97	91–94	None	100	3.22
Cows milk	2.5	85–90	86	S amino acids	60	3.20
Beef muscle	3.2	76	71–76	S amino acids	80	2.79
Salmon	—	72	71	Tryptophan	75	—
Soybean	0.7–1.8	58–69	48–61	S amino acids	69	2.58
Peanut	1.7	56	43–54	S amino acids	70	2.08
Rice	1.9	75	70	Lysine	57	—
Corn	1.2	60	49–55	Lysine	55	—
Wheat	1.0	52	52	Lysine	57	2.02

[a]Biological value.
[b]Net protein utilization.
[c]Total essential amino acids/total amino acids.

under carefully controlled conditions of total diet, housing, etc., and growth of the rats is measured as weight gain over 28 days. PER is then determined by dividing weight gain by weight of protein ingested. The protein content of the diet is calculated as N $\times$ 6.25, but for soybean protein N $\times$ 5.7 would be better.

Biological value (BV) and net protein utilization (NPU) are evaluation systems based on measurement of protein ingested (I), protein excreted in feces (F), and protein (nitrogen converted to protein) excreted in urine (U). The excretion values are corrected for excretion of nitrogen when no protein is being fed to correct for normal protein turnover. Digestibility is defined as $(I - F)/I$ (protein absorbed) as discussed in Table 6.2. BV is defined as $[I - (F + U)]/(I - F)$ or nitrogen retained/nitrogen absorbed. NPU is $[I - (F + U)]/I$ or nitrogen retained/nitrogen ingested. From Table 6.6 it is obvious that soybeans occupy an intermediate position between animal proteins and cereal proteins based on both biological and chemical evaluation systems.

Nutritionists find fault with the BV, NPU, and PER evaluation systems because they are in effect single-point assays based on a single level of protein fed (Hegsted 1977). Response of animals to protein fed does change with the amounts fed, and differentiation between proteins is greatest at low levels when the animal is in negative nitrogen balance.

With respect to evaluation of soy protein, a more serious complication than the single-level question is the use of rats in PER measurements. Table 6.7 shows the amino acid requirements for rats and for both infant and adult humans. Clearly the requirement for rats is appreciably higher for sulfur amino acids and for lysine. Since sulfur amino acids are limiting in soy proteins, the rat assay does not give valid information on how humans would respond to soy protein foods. Table 6.8 shows how different protein sources are evaluated by chemical

Table 6.7

Comparison of Amino Acid Requirement Patterns[a]

Essential amino acids	Amino acids as percentage of protein		
		Human	
	Rat	FNB[b]	FAO/WHO
Histidine	2.5	1.7	1.4[c]
Isoleucine	4.6	4.2	4.0
Leucine	6.2	7.0	7.0
Lysine	7.5	5.1	5.5
Total sulfur amino acids	5.0	2.6	3.5
Total aromatics	6.7	7.3	6.0
Threonine	4.2	3.5	4.0
Tryptophan	1.25	1.1	1.0
Valine	5.0	4.8	5.0
Arginine	5.0		

[a]Source: Steinke (1979). Copyright by Academic Press, Inc.
[b]For infants.

score when rat requirements are used instead of human requirements. Those proteins that are deficient in sulfur amino acids are not well evaluated by PER with rats (Torun *et al.* 1981).

An even greater difficulty in evaluating nutritive quality of proteins than the shortcomings of the specific methods is the fact that proteins are evaluated singly. Yet we eat mixed diets with many proteins being ingested in each meal,

Table 6.8

Amino Acid Score Based on Rat
and Human Requirements[a]

Protein source	Rat	Human[b]
Casein	71	100
Egg whole	95	100
Lactalbumin	65	100
Beef	90	100
Pork	87	100
Isolated soy protein	58	100
Sesame flour	36	54
Oat flour	52	77
Wheat flour	33	49

[a]Source: Steinke (1979). Copyright by Academic
Press, Inc.
[b]Based on FNB pattern.

and the nutritional well-being of humans with respect to protein depends on the mixture ingested. There is no reasonable way to use the evaluations of proteins made singly to estimate how they will behave in a complicated mixture. It is reasonably certain that the mixture will be better than any sort of additive combination of individual evaluations, because the amino acid compositions of proteins will complement each other rather than be less than expected.

In Tables 6.6 and 6.8, wheat protein is shown to be fairly poor in nutritional value by any of the methods used. Nevertheless, wheat is a valuable source of protein at reasonable cost to people throughout the world. Thus there is a fallacy in relying too heavily on nutritive evaluation by any of the suggested methods for protein. Use of soy protein in human foods is diminished by both problems of underestimating nutritive value by PER and by relying on data from single protein analyses.

Soybean protein can provide positive nitrogen balance for animals and humans (Mitchell 1950, Liener 1978). Nitrogen balance is another concept in protein nutrition that can be explained in previously described terms as $I - (F + U)$. The fact that sulfur-containing amino acids are limiting in soy protein raises the question of whether soy protein for human nutrition should be supplemented with methionine. This question has been explored for adult nutrition (Scrimshaw and Young 1979, Jansen 1979) making use of nitrogen balance over a range of protein levels fed.

Scrimshaw and Young (1979) found that in feeding several levels of a soy protein isolate to young men, methionine supplementation improved nitrogen balance when low levels of soy protein were used and when the subjects were in negative nitrogen balance. However, when soy protein was fed at a level to achieve positive nitrogen balance, methionine supplementation showed no significant improvement over nonsupplementation.

Working with preschool-age children, Torun (1979) found that feeding soy protein isolate at 1 g of protein per kilogram of body weight per day achieved normal growth and nitrogen retention. The soy protein isolate without methionine supplementation compared quite adequately with cow's milk. Nitrogen retention values for the two isolates fed were 96 and 86% when compared to cow's milk, and a composite score for nitrogen retention, absorption, and balance gave 98 and 107% for the two isolates in comparison with cow's milk.

The most stringent test of soy protein nutritive quality is achieved by feeding human infants, and comparisons of soy protein and cow's milk in infant feeding have been done. Fomon and Ziegler (1979) reported that methionine-supplemented soy isolate formulas promote retention of nitrogen and growth in infants to the same extent as milk-based formulas. Furthermore, at a level of 9.3% of energy from soy protein isolate, supplementation by L-methionine did not change retention of nitrogen or rate of growth by infants. The infants' serum concentration of urea nitrogen was significantly higher when the formula was not supplemented with methionine. This indicates less efficient utilization of soy protein by infants in the absence of methionine supplementation.

Table 6.9

Summary of Nitrogen Balance Data and Protein Quality Indices
in Soy-Beef Replacement Study[a,b]

Soy/beef ratio:	100/0	75/25	50/50	27/75	0/100
Nitrogen balance (mg N/kg/day)	−2.3	−3.2	−0.9	−1.1	−1.7
Biological value	53	52	55	53	53
Digestibility (%)	97	99	98	98	98

[a]Source: Scrimshaw and Young (1979). Copyright by Academic Press, Inc.

[b]None of the diets showed significant ($p > 0.05$) difference.

Torun *et al.* (1981) summed up the situation of methionine supplementation for soy protein in human diets, by saying that methionine supplementation may be useful when total dietary protein intake is marginal, but that there is little justification for supplementation when soy protein is used as part of a mixed diet, and protein intake is adequate.

Another important aspect of the nutritive quality of soy protein is the extent to which it can be used as a substitute for meat. Scrimshaw and Young (1979) studied this question, and part of their results are shown in Table 6.9. When mixtures of soy and beef were fed at levels close to those necessary for nitrogen balance, no significant differences were found in the entire range of beef–soy mixtures.

In contrast Kies and Fox (1971) had found that beef was superior to textured vegetable protein (soy protein isolate) in nitrogen balance studies with adults. The difference was due to levels fed. When the level of protein fed gives negative nitrogen balance, beef is superior to soy protein, but at levels that give exact nitrogen balance or positive nitrogen balance, differences between beef and soy protein are not apparent.

Hopkins and Steinke (1981) reviewed some of the nutritional data on special food blends containing soy protein. In Chapter 8 we explore in detail the various products and nutritional implications of blends between soy proteins and other plant protein sources.

Soy Protein and Atherogenicity. It has been observed that the type of protein in the diet can have an influence on blood cholesterol levels and on atherosclerosis. Most work has been done with the rabbit as an experimental animal. When fed low-fat, cholesterol-free diets with casein as the protein source, rabbits develop high plasma cholesterol (Fig. 6.4). If the diet is prolonged for 10 months, the rabbits also develop atherosclerosis. In contrast, substituting soy protein isolate for the casein keeps the plasma cholesterol relatively low and avoids development of atherosclerosis.

Much experimentation has been done to determine the cause of the effects of

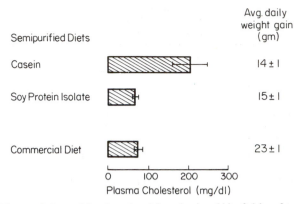

Fig. 6.4.   Plasma cholesterol levels and weight gains in rabbits fed low-fat, cholesterol-free, semipurified diets or a commercial diet. Results are given as mean + SEM for groups of 6 rabbits fed the diets for 28 days. Source: Carroll *et al.* (1979). Copyright by Academic Press, Inc.

different proteins on blood lipids. Amino acid differences have been investigated (Huff *et al.* 1977), and feeding mixtures of amino acids corresponding to casein will also generate elevated plasma cholesterol. Feeding amino acids corresponding to soy protein does lower plasma cholesterol but not to the extent that intact soy protein does.

Another complicating factor is that the type of carbohydrate in the diet has an influence on the results. Kritchevsky (1979) reported that when raw potato starch is substituted for dextrose, plasma cholesterol levels with either casein or soy protein are low (about 50 mg/dl).

Of more interest than the blood lipid changes in rabbits due to diet are the differences that may occur in humans. Carroll *et al.* (1979) and Carroll (1982) reported that statistically significant lowering of plasma cholesterol could be achieved by replacing animal protein with soy protein in the diets of young women. Also, Sirtori *et al.* (1977) found lowered plasma cholesterol when hypercholesteremic patients were shifted from a diet predominant in animal protein to a diet essentially devoid of animal protein and containing predominantly soy protein.

Other researchers have not been able to duplicate these results with humans, and the reasons are unknown (Carroll 1982), although an exact duplication of diets and conditions with human patients is always difficult. Bodwell and Hopkins (1985) concluded that only cholesterol levels in type II hyperlipoproteinemic individuals (cholesterol level greater than 300 mg/dl) are lowered as a result of using soy protein in the diet.

Lipids

The nutritional value that is inherent in soybean oil comes from calories, essential fatty acids, and fat-soluble vitamins.

Calories.   In the United States the main preoccupation seems to be to avoid foods with calories. But the body's demands for energy take precedence over other nutrient demands, and a palatable, concentrated source of calories, such as soybean oil, is a valuable nutritional resource.

About 41 lb (18.6 kg) of soybean oil was consumed per capita in the United States in 1982 (Anon. 1983). This was approximately 70% of all visible fat and oil consumption. Visible fat and oil consumption is consumption of those products that are primarily lipid, such as salad oils, shortenings, and margarines as opposed to invisible fat and oil consumption, which is the fat and oil in food products such as meats, eggs, and dairy products. Total fat and oil consumption in the United States in 1982 was 135 lb/capita (61.4 kg/capita) of which 58 lb/capita (26.4 kg/capita) or about 40% is visible fat and oil. In rat studies, soybean oil was found to be 98.5% digestible (Emken 1980). Thus, soybean oil is a valuable calorie source in the United States and to a lesser degree in Western Europe and Japan.

Nutritional questions about fat and oil consumption have been raised because of the disease condition of atherosclerosis, plaques of fatty deposits on the lining of arteries. These plaques are dangerous because they restrict blood flow, and they can loosen to form emboli that may lodge in a coronary artery and cause a heart attack. The plaques have a high content of saturated fatty acids and cholesterol; consequently trying to minimize the saturated fatty acids and cholesterol in blood lipids would seem to be a good protective measure. Also, there is evidence from epidemiological studies and direct experimental evidence that high concentrations of saturated lipids and cholesterol in blood tend to promote atherosclerosis.

It is possible to control to some extent the saturated lipid and cholesterol content of the blood by diet. Eating a diet low in saturated fat and low in cholesterol is helpful. Furthermore, eating a diet high in polyunsaturated fatty acids tends to lower blood cholesterol. With respect to the atherosclerosis problem, soy oil in the diet seems to be beneficial.

Essential Fatty Acids.   Linoleic acid is an essential fatty acid for humans and there is also evidence that linolenic acid is essential (Anon. 1980). Soy oil can serve as a source of both fatty acids, and even partially hydrogenated soy oil still contains about 25% linoleic and 3% linolenic acids. The supply of essential fatty acids in the normal diet in the United States is not a problem, since many foods will supply the small amounts of essential fatty acids needed.

Fat-Soluble Vitamins.   Any fat or oil can serve as a carrier of fat-soluble vitamins. Soybean oil is a good source of vitamin E. At the current rate of consumption in the United States, soybean oil provides about 5 mg/day of α-tocopherol, which satisfies half of the highest RDA. The other tocopherols would also have vitamin E activity, and so soy oil provides more than half the RDA for vitamin E. Soy oil does contain some β-carotene as a vitamin A

Table 6.10
Fiber Content of Soybean Hulls[a]

Type of fiber	Content (% dry basis)
Total crude fiber	36–47
Total dietary fiber	87–88
Acid detergent fiber	35–50
Buffered acid detergent fiber	54
Neutral detergent fiber	49–67
In vitro fiber	70
Crude cellulose	41–53
Crude hemicellulose	14–33
Crude lignin	0.7–1.3

[a]Source: Erdman and Weingartner (1981).

precursor, but the amounts are small, particularly after bleaching, and soy oil is not considered a useful source of vitamin A.

Fiber and Oligosaccharides

The carbohydrates of soybeans, containing little starch and hexose, are largely polysaccharides with some oligosaccharides. The soybean polysaccharides are considered unavailable to humans mainly due to difficulties in digestion. Aside from its own indigestibility, crude fiber depresses the digestibility of other nutrients to a considerable extent.

In recent years, however, the possible beneficial effects of dietary fiber have received much attention. Particle size, density, hydration capacity, and ion exchange capacity are four major physical properties of dietary fiber. These properties not only are instrumental in the role a particular fiber source may play in the rheology of a food product, but they may also determine the physiological role of the specific dietary fiber source. An important benefit of dietary fiber for humans is increasing the water-holding capacity of the stools. Increased volume and softness of stools parallels the water-holding capacity. Increased stool bulk, softness, and decreased transit time may reduce diverticular disease, hemorrhoids, and possibly other diseases of the lower gastrointestinal tract.

Soybean hulls are reported to contain about 87% dietary fiber, consisting of 40–53% crude cellulose, 14–33% crude hemicellulose, and 1–3% crude lignin on a dry basis (Erdman and Weingartner 1981) as shown in Table 6.10. Dehulled soybean flour contains 6.2% neutral detergent fiber, 5.7% acid detergent fiber, 4.6% crude cellulose, 0.5% crude hemicellulose, and 1.3% lignin. Soy protein concentrates contain slightly higher levels of dietary fiber. In a practical sense, soy protein isolates have no fiber.

Soybean hulls tend to disintegrate into a uniform powder, while the other cereal bran particles have a tendency to form elongated flakes. Therefore, even though all products pass through the same screen, their sizes are different, and the dietary fiber prepared from soybean hulls has relatively low density and relatively high hydration capacity compared to fibers extracted from cereal sources. The hydration capacity of fiber appears to be correlated with the uronic acid concentrations in the fibers. The relatively high uronic acid content in soybean hull fiber may also offer an explanation for its distinctively high cation exchange capacity. This high cation exchange capacity suggests that this fiber source may have the capacity to reduce mineral bioavailability if the fiber is consumed in large enough quantities in the diet.

In order to investigate this assumption, diets based on dehulled or whole soybean flours were fed to rats, and growth and zinc deposition in bone were measured (Weingartner *et al.* 1979). Results showed that inclusion of soybean hulls, at least at low levels in the diet, had no effect on the bioavailability of zinc native to the bean. The presence of soybean hulls in diets also did not affect bioavailability of added calcium. Dintzis *et al.* (1979) investigated changes in composition and morphology of soybean hulls along with other brans after passage through the human alimentary tract. They found that soybean hulls could be greatly disrupted by the human gastrointestinal tract depending on individual transit times. In a person with short transit time, cellulose and lignin were almost fully recovered. In a constipated individual, recovery of lignin was about 33%, and cellulose and hemicellulose were only about 10% recovered in stool. Mean fecal weight increased from 68 to 128 g ($p < 0.05$) as did plasma triglycerides. Feeding the men finely ground soybean hulls improved glucose tolerance.

No adverse effects of soybean hulls were noted on mineral balance of zinc, copper, or iron. The overall results suggest beneficial effects on glucose and lipid metabolism with moderate intakes of soybean hulls as a source of dietary fiber. The metabolic effect of soybean fiber for humans is still a scientific curiosity and not a practical consideration.

Due to the absence of α-galactosidase in the human small intestine, galactoside-containing oligosaccharides such as raffinose and stachyose in soybeans are not digested. However, fermentation of such sugars by intestinal microorganisms in the large intestine produces CO_2 and H_2 to cause flatus.

VanStratum and Rudrum (1979) fed healthy male students soy protein products to determine the effect on flatus production. As could be expected, the subjects showed great individual variability, but the composite results showed a clear trend (Fig. 6.5). The control meal caused a mean flatus egestion of 400 ml, while defatted soy in the test meal caused a high awareness of intestinal movements and a doubling of gas production as compared to the control. The refined soy products caused intermediate reactions.

Flatus activity of soy protein products in humans is further illustrated in Table 6.11. Both full-fat and defatted soy flours cause increased flatulence, whereas

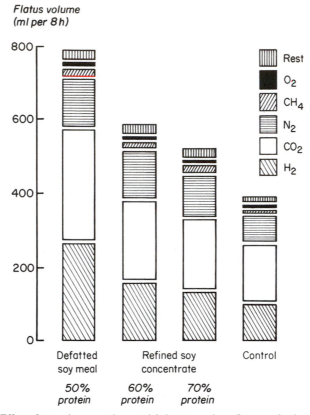

*Flatus volume
(ml per 8 h)*

Fig. 6.5. Effect of several soy protein materials in test meals on flatus production in men. Meals contained 75 g (dry basis) of the test materials. Source: VanStratum and Rudrum (1979).

protein isolate and high molecular weight polysaccharides show no more flatus activity than the basal diet. When soy flour is extracted with 80% ethanol to produce a protein concentrate, the flatulence effects are reduced. Alcohol extractives and soy whey solids, which contain most of the oligosaccharides, are responsible for large amounts of flatus.

The amount of oligosaccharides that can be removed from mature whole soybeans by leaching is summarized in Table 6.12. Germination also reduces soybean oligosaccharide content (Lee *et al.* 1959, Anderson *et al.* 1979), and the changes during germination of soybeans are given in Table 6.13. However, germinated soybean sprouts still have flatus activity. The polysaccharides, which normally do not cause flatulence, may have been partially hydrolyzed during germination. These modified polysaccharides may now become substances for the formation of flatus by the intestinal microflora and thereby compensate for the loss in raffinose and stachyose (Anderson *et al.* 1979).

Table 6.11

Effects of Soy Products on Flatus in Man[a]

Product[b]	Daily intake (g)	Flatus volume (cm³/hr)	
		Average	Range
Full-fat soy flour	146	30	0–75
Defatted soy flour	146	71	0–290
Soy protein concentrate	146	36	0–98
Soy proteinate	146	2	0–20
Water-insoluble residue[c]	146	13	0–30
Whey solids[d]	48	300[e]	—
Ethanol extractive 80%[d]	27	240	220–260
Basal diet	146	13	0–28

[a]Source: Anderson *et al.* (1979). Copyright by Academic Press, Inc.

[b]All products were toasted with live steam at 100°C (212°F) for 40 min.

[c]Fed at a level three times higher than that present in the defatted soy flour diet.

[d]Amount equal to that present in 146 g of defatted soy flour.

[e]One subject; otherwise four subjects per test.

Trace Minerals

The first observations that soybean protein products affect trace element availability arose from zinc requirement studies. Chicks fed soybean protein isolate had a higher zinc requirement than those fed casein or gelatin. Furthermore, autoclaving the soy protein isolate or addition of ethylenediaminetetraacetic acid

Table 6.12

Protein Loss and Oligosaccharide Removal from Whole Soybeans by Various Treatments[a]

Treatment	Protein loss[b]	Oligosaccharide removal[c]
Soak, room temp. 15 hr[d]	—	8
Boil		
20 min, water	1.0	33
60 min, water	2.6	59
20 min, 0.5% NaHCO₃	1.3	21
60 min, 0.5% NaHCO₃	6.8	60
60 min, pH 4.3	2.0	46

[a]Source: Anderson *et al.* (1979). Copyright by Academic Press, Inc.

[b]Grams of protein/100 g of protein in original dry bean.

[c]Grams of oligosaccharide/100 g oligosaccharide in original dry bean.

[d]Ratio of beans to water was 1:10 in all treatments.

Table 6.13
Effect of Germination on Autolysis of Sucrose,
Raffinose, and Stachyose[a]

	Loss of oligosaccharides (%)		
Germination period (hr)[b]	Sucrose	Raffinose	Stachyose
48	0	30	50
96	80	100	96

[a]Source: Anderson *et al*. (1979). Copyright by Academic Press, Inc.
[b]After imbibition.

(EDTA) reduced the zinc requirement of turkey poults. In general, the zinc in animal products is more available than that in plant proteins. Soybean zinc is 50–60% available (O'Dell 1979).

There is overwhelming evidence that soluble phytate added to purified diets decreases zinc availability in experimental animals and human subjects (Reinhold *et al*. 1973). Not only does phytate bind zinc and make it less readily absorbed, but as in the case of soybean protein, excess calcium aggravates the situation. In the absence of phytate, excess calcium has no effect on zinc availability. It has been postulated, therefore, that calcium, zinc, and phytate interact to form a highly insoluble complex, which reduces the absorption of zinc to a greater extent than phytate alone. Bioavailability of zinc depends, among other things, on the presence of chelating agents and on the calcium concentration in the diet, and thus absolute values of bioavailability cannot be obtained.

Erdman *et al*. (1980) observed that zinc bioavailability changes, depending on the type of soy isolate being fed. Zinc is more readily available from an acid isolate than from an isolate that has been neutralized. They postulated that zinc–phytate–protein complexes may form in the neutralized products due to the negative charge on the protein. These complexes may confer resistance to proteolysis as a result of drying and therefore reduce the availability of zinc.

Trace amounts of iron are extremely important for the nutritional well-being of people, but there are many complicating factors in the bioavailability of iron. For example, heme iron is much more readily available than nonheme iron. Absorption of nonheme iron by adults is promoted by consumption of animal protein but is not promoted by consumption of plant protein. The amount of iron absorbed varies, depending on the stores of the individual. An anemic person or a menstruating woman would absorb more iron from the same foods than would an iron-replete man. Until 1981 it was assumed that soy protein had no different effect on iron bioavailability than did other plant proteins (Erdman and Forbes 1981). This concept was based mainly on animal experiments. In 1981 data were published on the effect of soy protein ingestion on bioavailability of nonheme

Table 6.14

Nonheme Iron Absorption in Adult Men[a]

Product	Iron absorbed (%)
Study 1[b]	
Egg albumen	2.49 ± 0.4–0.5
Casein	2.74 ± 0.6–0.7
Soy isolate	0.46 ± 0.1–0.2
(Supro 710)	
Study 2[c]	
Egg albumen	5.50 ± 1.1–1.3
Full-fat soy flour	0.97 ± 0.2–0.5
Textured soy flour	1.91 ± 0.3–0.4
(Supro 50A)	
Soy isolate	0.41 ± 0.1
(Supro 710)	

[a]Source: Bodwell (1983).
[b]Mean ($n = 15$) + S.E.
[c]Mean ($n = 10$) + S.E.

iron based on human experiments (Cook *et al.* 1981, Morck *et al.* 1981). They found that iron absorption by adult males was markedly decreased in the presence of soy products as compared to animal proteins (casein and egg albumen). Table 6.14 shows some of the data for iron absorption for different kinds of protein in the diets.

Because of the widespread use of soy protein in the school lunch program and in military feeding, these effects on iron bioavailability have been reexamined. Preliminary results (Bodwell 1983) show that with women, men, and children the amounts of soy protein used in feeding programs do not adversely affect iron bioavailability.

The results of Cook *et al.* (1981) and of Morck *et al.* (1981) have also raised questions about the utility of using rats as a model for iron absorption in humans. Schricker *et al.* (1983) have found that rats differ sufficiently from humans in their relative response to different soy foods based on iron status that they should not be used as a model for human bioavailability of iron.

Fiber also has been implicated in decreasing mineral absorption, but the low concentration of crude fiber in soy protein isolate in particular suggests that fiber is not responsible for the low zinc availability associated with these products.

Mineral availability may not be of practical importance when there is only partial substitution of soy protein for animal protein in the diet (O'Dell 1979). Although many researchers believe that soy protein plays a causal role in reduced bioavailability of minerals from soybeans, there is no evidence that soy protein per se directly affects the bioavailability of minerals (Erdman and Forbes 1981).

Table 6.15

Goitrogenicity of Soy Products[a]

Product	Iodine content (g/100 g diet)	Thyroid weight (mg/100 g body weight)
Raw soy flour	1.0–2.3	37
Toasted soy flour	0.7	19
Soy isolate	0.9	16
Soy infant food, iodized	40	7
Raw soy flour + 10 μg I_2/g protein	196	8
Casein	30	7

[a]Source: Block *et al.* (1961). Copyright by Academic Press, Inc.

Goitrogenic Substances

It has been apparent for some time that goiters occur in rats consuming large quantities of soybean meal. Small quantities of iodine added to the diet effectively prevent thyroid enlargement, but the low iodine content of soybean is not the only reason for soy goitrogenicity (Block *et al.* 1961). Raw full-fat soy flour produces greater thyroid hypertrophy when fed to rats than does toasted and defatted soy flour or isolated soy protein, even though iodine content was higher in the raw flour than in the toasted and defatted flour or isolate (see Table 6.15).

The thyroid enlargement due to soybeans is completely reversible, and the goitrogenic agent appears to be a peptide comprised of two or three amino acid residues or a glycopeptide containing one or two amino acid residues linked to a sugar residue (Konijn *et al.* 1972, 1973).

The soybean goitrogen mode of action may involve the prevention of intestinal thyroxine readsorption (Beck 1958). When raw soy flour is fed to rats with radioactive iodine, an increased thyroid iodine uptake coupled with increased fecal iodine loss is observed.

Estrogenic Compounds

The isoflavone glucosides genistin, daidzin, and glycitein-7-O-β-glucoside are the major soybean estrogenic compounds. The concentration of each glucoside and its relative estrogenicity are shown in Table 6.16 along with diethylstilbestrol estrogenic activity for comparison. Other isoflavones, namely, 6,7,4′-trihydroxyflavone and coumestrol, have been found in soybeans (Knuckles *et al.* 1976, Friedlander and Sklarz 1971).

Diethylstilbestrol is 10^5 times more active than genistin, and daidzin is three-fourths as active as genistin (Bickoff *et al.* 1962), and so estrogenic activity due to the soy isoflavones is minimal. Even though coumestrol is 35 times more

Table 6.16
Soy Meal Estrogenicity[a]

Estrogen	Concentration in soy meal (ppm)	Relative estrogenicity	Concentration × estrogenicity
Diethylstilbestrol[b]	—	1×10^5	—
Genistin	1644	1.0	1644
Daidzin	581	0.75	436
Glycitein 7-O-β-glucoside	338	c	—
Coumestrol	0.4	35	14

[a]Source: Anderson *et al.* (1979). Copyright by Academic Press, Inc.
[b]Included for comparison purposes.
[c]Estrogenicity unknown.

estrogenic than genistin, coumestrol is much less a factor in soybean estrogenicity because of its presence in only minute amounts (120 μg/100 g). The estrogenic activities of the glycitein glucoside and 6,7,4'-trihydroxyflavone are not known.

Most of the estrogenic activity of soybean meal is attributable to genistin, which is stable to autoclaving. Soybean meal concentrates prepared by alcohol extraction display no estrogenic activity due to genistin extraction in aqueous ethanol, and soy protein isolates contain only small amounts of isoflavones (Nash *et al.* 1967).

Both genistin and genistein cause significant decreases in weight gain when fed to rats at the 0.5% level, whereas no effect is noted when genistin or genistein is included in the diets at the 0.1% level (Magee 1963). Soybean meal, when used at a dietary protein level of 19%, would provide only a 0.1% level of genistin plus genistein and a 0.16% level of total isoflavonoids, and so estrogenic effects should be negligible.

CHANGES DUE TO PROCESSING

The processing of soybeans into the many food and feed products that exist does cause changes in their nutritive value. The changes can be both positive and negative. The purpose of this section is to examine those changes that occur for both proteins and lipids. The chemical and physical changes will be noted, but a precise judgment as to their influence on nutritional values is more difficult to achieve. It should be noted at the outset that the changes known to occur during processing do not in any measurable way adversely affect the nutritive value of soy products now being used.

Proteins

The processing of soy proteins may subject them to heat, alkali, or mild oxidation, all of which can have an influence on nutritive value.

Heat.   Although soy proteins are exposed to mild heat during solvent extraction, this has little influence on their nutritive value. The main heating step in soy protein processing is desolventizing–toasting, and this has a predominantly beneficial effect on nutritive value. The heat inactivation of trypsin inhibitors and the heat denaturation of soybean globulins, making them more susceptible to proteolysis, improve soy proteins for either human foodstuffs or animal feeding. If the heating is excessive, for example, autoclaving at 130°C (246°F) for 24 hr, the nutritive value is impaired, and the mechanism is thought to be cross-linking of peptide chains by acylation of free amino groups. One such cross-link would be ϵ-N(γ-glutamyl)-lysyl amide (Cheftel 1977):

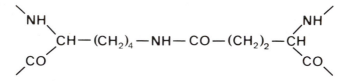

Such a cross-link could impair nutrition by making lysine unavailable due to the acylation, or by making the peptide bonds difficult to hydrolyze. The latter seems to be the mechanism whereby nutritive value is lost.

Acylation of lysyl amino groups has been experimented with as a means of possibly improving functional properties of soy proteins (Franzen and Kinsella 1976). Table 6.17 shows that acylation of lysine does not necessarily interfere with its nutritional availability.

Table 6.17
Nutritional Availability of Acylated Lysine Derivatives[a]

Derivative	Utilization as a source of lysine (%)	Animal
ϵ-N-formyl-L-lysine	~50	Rat
ϵ-N-acetyl-L-lysine	~50	Rat,chick
ϵ-N-(τ-glutamyl)-L-lysine	~100	Rat,chick
ϵ-N-(α-glutamyl)-L-lysine	~100	Rat
ϵ-N-glycyl-L-lysine	~80	Rat
α-N-glycyl-L-lysine	~100	Rat
ϵ-N-(N-acetylglycyl)-L-lysine	0	Rat
ϵ-N-propionyl-L-lysine	0	Rat
ϵ-N-propionyl-L-lysine	~70	Chick

[a]Source: Cheftel (1977).

Severe heating of soy proteins can also cause loss of hydrogen sulfide from cystine, thereby destroying one of the essential amino acids. This reaction is also catalyzed by alkali and will be discussed further in connection with generation of lysinoalanine.

We need to emphasize that the kind of heating involved in protein cross-linking or loss of hydrogen sulfide is not normally encountered in soy protein processing. Also, acylation of proteins is not an approved process and has been studied only to determine potential benefits.

Alkali.   Soy protein is exposed to alkali during the processing of soy isolates. As mentioned above, hydrogen sulfide can be lost from cystine in the presence of heat and alkali:

$$R-CH_2-S-S-CH_2R + H_2O \rightarrow R-CH_2SH + R-CH_2-SOH$$

$$R-CH_2SOH \rightarrow RCHO + H_2S$$

Also, cystine can react in alkaline conditions to yield dehydroalanine, which in turn can react with lysine to form lysinoalanine (Fig. 6.6).

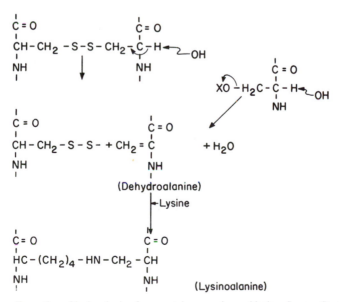

Fig. 6.6.   Formation of lysinoalanine from cysteine or serine and lysine. Source: Struthers *et al.* (1979). Copyright by Academic Press, Inc.

Soon after lysinoalanine was first discovered as an unusual amino acid in 1964, it also was discovered to cause renal lesions in rats. Further studies quickly showed that lysinoalanine is widely distributed in proteins although present at very low levels, and this was cause for concern. As evidence continued to accumulate, it became clear that the renal lesions were species specific for rats and not even all rat strains were susceptible (Struthers *et al.* 1979). Also, the amounts of lysinoalanine needed to produce the lesions were equivalent to the amounts found in proteins treated with alkali for industrial uses. The amounts found in proteins for food use could not produce the lesions. At present there is no reason to believe that the trace amounts of lysinoalanine found in food proteins constitute a health hazard.

All naturally occurring amino acids are in the L form, and it is the L form that is nutritionally valuable. Alkali treatment of proteins can cause isomerization of the L to D isomer, and as a result make the amino acid unavailable. While this isomerization reaction is known to occur, it is not considered a problem because of the very small amounts of amino acids that are isomerized.

Alkali has a further effect on soy proteins that can affect their nutritional value. In the presence of carbohydrates with free reducing groups, protein will react with the carbonyl groups as the first step of the Maillard or nonenzymatic browning reaction. The ε-amino group of lysine is involved and lysine can become unavailable as a nutrient. Alkali catalyzes the browning reaction as well as heat and low moisture. The loss of lysine through the Maillard reaction is the main reason one has to be careful about overheating soy meal, and nutritional value is lost if soy proteins are excessively heated.

Oxidation. Heating proteins in air, exposure to peroxides formed in lipid oxidation, or exposure to H_2O_2 can oxidize some of the amino acids. Particularly susceptible to oxidation are the sulfur-containing amino acids and tryptophan. Since sulfur-containing amino acids are limiting in soy protein, any loss is of potential nutritional significance.

Cystine and cysteine are oxidized in stages as shown in Fig. 6.7. Cysteic acid is not capable of replacing cysteine nutritionally, but less severely oxidized forms such as cysteine sulfenic acid and cystine mono- and disulfoxides can substitute for the amino acids. Probably these less oxidized amino acids can be reduced to cystine and cysteine for use in protein biosynthesis.

Methionine can also be oxidized sequentially to methionine sulfoxide and then to methionine sulfone:

$$HOOC-\underset{\underset{NH_2}{|}}{CH}-(CH_2)_2-S-CH_3 \rightarrow R-\overset{\overset{O}{\|}}{S}-CH_3 \rightarrow R-\overset{\overset{O}{\|}}{\underset{\underset{O}{\|}}{S}}-CH_3$$

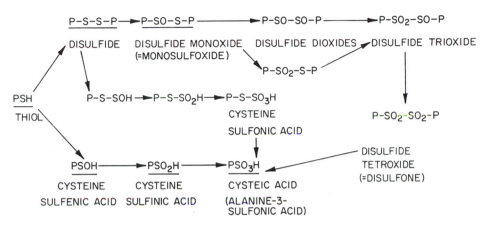

Fig. 6.7. Oxidation derivatives from cysteine and cystine residues (stable derivatives are underlined). Source: Cheftel (1977).

Methionine sulfoxide can be used nutritionally, but methionine sulfone cannot. Marshall *et al.* (1982) in a study of the stability of sulfur-containing amino acids during legume processing found that in the processes used for soy isolate processing there was no loss of sulfur amino acids. However, in one of the four commercial isolates they examined they found that over 50% of the methionine had been converted to methionine sulfoxide. They guessed that this might have been due to the use of H_2O_2 to bleach pigments.

Lipids

The refining of soy oil does have some minor effects on its nutritive value. Carotenoids are lost on bleaching, which means some loss of vitamin A potential. Vitamin E is also lost during refining, but the amount remaining still makes soy oil an excellent source of vitamin E. The most significant changes in soy oil that might have some influence on nutritive value are changes from cis to trans configuration of fatty acid double bonds during hydrogenation and changes in frying oils after exposure to high temperatures for long times.

Hydrogenation. During partial hydrogenation of soy oil, 12% of the oleic acid and 4–6% of the linoleic acid may be in the trans form. Since such a change could conceivably present some problems in metabolism of the trans fatty acids, considerable research effort has been expended to learn about the health effects of the modified fatty acids.

The rates of breakdown to CO_2 and H_2O are measurably different for the trans fatty acids compared to cis fatty acids, but the differences are minor and not thought to be a problem. Trans fatty acids are not converted to biologically active prostaglandins. Feeding studies with rats and swine have shown no effect on growth, longevity, or reproduction due to trans fatty acids. Also, histological examinations of tissues from these feeding studies have shown no abnormalities. The evidence gathered thus far cannot definitively answer the question about health effects due to trans fatty acids, but the evidence has shown that there is no cause for alarm. Trans fatty acids are found naturally in food products from ruminants, and so humans have been exposed to them for thousands of years. In the past 30 years during which the consumption of partially hydrogenated soy oil has steadily increased, there has been no evidence of short- or long-term toxic effects due to partially or fully hydrogenated soy oil.

Heat.   Cooking of food is known to cause complex chemical changes that are not fully understood. Since fats and oils are often exposed to high temperatures for prolonged periods during frying of foods, the effects of heat on fats have been studied extensively. Studies on extremely abused fats have shown that they can cause many symptoms when fed to animals. For example, fatty livers, decreased fat absorption, diarrhea, and interference with reproduction in female rats have been noted. However, fats causing these kinds of symptoms are so badly oxidized that they would never be consumed willingly by humans as food. Studies on less drastically abused fats, which conceivably would be consumed by humans, have not shown the same kinds of toxic symptoms.

In summary, after many years of extensive research there are no known health hazards from consumption of hydrogenated or heated oils.

REFERENCES

Anderson, R. L., J. J. Rackis, and W. H. Tallent (1979). Biologically active substances in soy products. *In* "Soy Protein and Human Nutrition" (H. L. Wilcke, D. T. Hopkins, and D. H. Waggle, eds.). Academic Press, New York.

Anon. (1980). "Recommended Dietary Allowances," 9th ed. National Academy of Sciences, Washington, D.C.

Anon. (1983). Oil Crops Outlook and Situation Report. U.S. Department of Agriculture—Economic Research Service, August, No. 2, Washington, D.C.

Beck, R. N. (1958). Soy flour and fecal thyroxine loss in rats. *Endocrinology* **62**:587.

Bickoff, E. M., A. L. Livingston, A. P. Hendrickson, and A. N. Booth (1962). Relative potencies of several estrogenlike compounds found in forages. *J. Agric. Food Chem.* **10**:410.

Block, R. J., R. H. Mandl, H. W. Howard, C. D. Bauer, and D. W. Anderson (1961). The curative action of iodine on soybean goiter and the changes in the distribution of iodoamino acids in the serum and in thyroid gland digests. *Arch. Biochem. Biophys.* **93**:15.

Bodwell, C. E. (1983). Effects of soy protein on iron and zinc utilization in humans. *Cereal Foods World* **28**:342.

Bodwell, C. E., and D. T. Hopkins (1985). Nutritional characteristics of oilseed proteins. *In* "New Protein Foods," Vol. 5, Seed Storage Proteins (A. M. Altschul and H. W. Wilcke, eds.). Academic Press, New York.

Bressani, R. (1981). The role of soybeans in food systems. *J. Am. Oil Chem. Soc.* **58**:392.

Carroll, K. K. (1982). Hypercholesterolemia and atherosclerosis: Effects of dietary protein. *Fed. Proc.* **41**:2792.

Carroll, K. K., M. W. Huff, and D. C. K. Roberts (1979). Vegetable protein and lipid metabolism. *In* "Soy Protein and Human Nutrition" (H. L. Wilcke, D. T. Hopkins, and D. H. Waggle, eds.). Academic Press, New York.

Cheftel, J. C. (1977). Chemical and nutritional modifications of food proteins due to processing and storage. *In* "Food Proteins" (J. R. Whitaker and S. R. Tannenbaum, eds.). AVI Publ. Co., Westport, CT.

Cook, J. D., T. A. Morck, and S. R. Lynch (1981). The inhibitory effect of soy products on nonheme iron absorption in man. *Am. J. Clin. Nutr.* **34**:2622.

Dintzis, F. R., L. M. Legg, W. L. Deatherage, F. L. Baker, G. E. Inglett, R. A. Jacob, S. J. Reck, J. M. Munoz, L. M. Klevay, H. H. Sandstead, and W. C. Shuey (1979). Human gastrointestinal action on wheat, corn, and soy hull bran—preliminary findings. *Cereal Chem.* **56**:123.

Emken, E. A. (1980). Nutritive value of soybean oil. *In* "Handbook of Soy Oil Processing and Utilization" (D. R. Erickson, E. H. Pryde, O. L. Brecke, T. L. Mounts, and R. A. Falb, eds.). American Soybean Association, St. Louis, MO, and American Oil Chemists Society Champaign, IL.

Erdman, J. W., Jr., and R. M. Forbes (1981). Effects of soya protein on mineral availability. *J. Am. Oil Chem. Soc.* **58**:489.

Erdman, J. W., Jr., and K. E. Weingartner (1981). Nutrition aspects of fiber in soya products. *J. Am. Oil Chem. Soc.* **58**:511.

Erdman, J. W., Jr., K. E. Weingartner, G. C. Mustakas, R. D. Schmutz, H. M. Parker, and R. M. Forbes (1980). Zinc and magnesium bioavailability from acid-precipitated and neutralized soybean protein products. *J. Food Sci.* **45**:1193.

Fomon, S. J., and E. E. Ziegler (1979). Soy protein isolates in infant feeding. *In* "Soy Protein and Human Nutrition" (H. L. Wilcke, D. T. Hopkins, and D. H. Waggle, eds.). Academic Press, New York.

Franzen, K. L., and J. E. Kinsella (1976). Functional properties of succinylated and acetylated soy protein. *J. Agric. Food Chem.* **24**:788.

Friedlander, A., and B. Sklarz (1971). Catecholic flavanoids from soybean flakes. *Experientia* **27**:762.

Hegsted, D. M. (1977). Protein quality and its determination. *In* "Food Proteins" (J. R. Whitaker and S. R. Tannenbaum, eds.). AVI Publ. Co., Westport, CT.

Hopkins, D. T., and F. H. Steinke (1981). Uses of soy protein in mixed protein systems to meet nutritional needs. *J. Am. Oil Chem. Soc.* **58**:452.

Huff, M. W., R. M. G. Hamilton, and K. K. Carroll (1977). Plasma cholesterol levels in rabbits fed low fat, cholesterol-free semi-purified diets: Effects of dietary proteins, protein hydrolysates and amino-acid mixtures. *Atherosclerosis* **28**:187.

Jansen, G. R. (1979). The importance of protein quality in human nutrition. *In* "Soy Protein and Human Nutrition" (H. L. Wilcke, D. T. Hopkins, and D. H. Waggle, eds.). Academic Press, New York.

Kakade, M. L., N. R. Simons, and I. E. Liener (1972). An evaluation of natural vs. synthetic substrates for measuring the antitryptic activity of soybean samples. *Cereal Chem.* **46**:518.

Kakade, M. L., N. R. Simons, I. E. Liener, and J. W. Lambert (1972). Biochemical and nutritional assessment of different varieties of soybeans. *J. Agric. Food Chem.* **20**:87.

Kakade, M. L., D. E. Hoffa, and I. E. Liener (1973). Contribution of trypsin inhibitors to the deleterious effects of unheated soybeans fed to rats. *J. Nutr.* **103**:1772.

Kies, C., and H. M. Fox (1971). Comparison of the protein nutritional value of TVP, methionine enriched TVP and beef at two levels of intake for human adults. *J. Food Sci.* **36**:841.

Knuckles, B. E., D. Defremery, and G. O. Kohler (1976). Coumestrol content of fractions obtained during wet processing of alfalfa. *J. Agric. Food Chem.* **24**:1177.

Konijn, A. M., S. Edelstein, and K. Guggenheim (1972). Separation of a thyroid active fraction from unheated soya bean flour. *J. Sci. Food Agric.* **23**:549.

Konijn, A. M., B. Gershon, and K. Guggenheim (1973). Further purification and mode of action of a goitrogenic material from soybean flour. *J. Nutr.* **103**:378.

Kritchevsky, D. (1979). Vegetable protein and atherosclerosis. *J. Am. Oil Chem. Soc.* **56**:135.

Kunitz, M. (1946). Crystalline soybean trypsin inhibitor. *J. Gen. Physiol.* **29**:149.

Lee, K. Y., C. Y. Lee, T. Y. Lee, and T. W. Kwon (1959). Chemical changes during germination of soybean. (1) Carbohydrate metabolism. *Seoul Univ. J. (Biol. Agric.)* **8**:35.

Liener, I. E. (1958). Inactivation studies on the soybean hemagglutinin. *J. Biol. Chem.* **233**:401.

Liener, I. E. (1974). Phytohemagglutinins: Their nutritional significance. *J. Agric. Food Chem.* **22**:17.

Liener, I. E. (1978). Nutritional value of food protein products. *In* "Soybeans: Chemistry and Technology," Vol. 1, Proteins, 2nd ed. (A. K. Smith and S. J. Circle, eds.). AVI Publ. Co., Westport, CT.

Liener, I. E., and J. E. Rose (1953). Soyin, a toxic protein from soybean III. Immunochemical properties. *Proc. Soc. Exp. Biol. Med.* **83**:539.

Magee, A. C. (1963). Biological responses of young rats fed diets containing genistin and genistein. *J. Nutr.* **80**:151.

Marshall, H. F., K. C. Chang, K. S. Miller, and L. D. Satterlee (1982). Sulfur amino acid stability. Effects of processing on legume proteins. *J. Food Sci.* **47**:1170.

Mitchell, H. H. (1950). Nutritive factors in soybean products. *In* "Soybeans and Soybean Products," Vol. 1 (K. S. Markley, ed.). Wiley (Interscience), New York.

Morck, T. A., S. R. Lynch, B. S. Skikne, and J. D. Cook (1981). Iron availability from infant food supplements. *Am. J. Clin. Nutr.* **34**:2630.

Nash, A. M., A. C. Eldridge, and W. J. Wolf (1967). Fractionation and characterization of alcohol extractables associated with soybean proteins. Nonprotein components. *J. Agric. Food Chem.* **15**:102.

O'Dell, B. L. (1979). Effect of soy protein on trace mineral availability. *In* "Soy Protein and Human Nutrition" (H. L. Wilcke, D. R. Hopkins, and D. H. Waggle, eds.). Academic Press, New York.

Rackis, J. J., J. E. McGhee, and A. N. Booth (1975). Biological threshold levels of soybean trypsin inhibitors by rat bioassay. *Cereal Chem.* **52**:85.

Reinhold, J. G., K. Nasr, A. Lahingarzadeh, and H. Heydayati (1973). Effects of purified phytate and phytate rich bread upon metabolism of zinc, calcium, phosphorous and nitrogen in man. *Lancet* **I**:283.

Schricker, B. R., D. D. Miller, and D. Van Campe (1983). Effects of iron status and soy protein on iron absorption by rats. *J. Nutr.* **113**:996.

Scrimshaw, N. S., and V. R. Young (1979). Soy protein in adult human nutrition: A review with new data. *In* "Soy Protein and Human Nutrition" (H. L. Wilcke, D. T. Hopkins, and D. H. Waggle, eds.). Academic Press, New York.

Sirtori, C. R., E. Agradi, F. Conti, O. Mantero, and E. Gatti (1977). Soybean protein diet in treatment of type II hyperlipoproteinemia. *Lancet* **I**:275.

Steinke, F. H. (1979). Measuring protein quality of foods. *In* "Soy Protein and Human Nutrition" (H. L. Wilcke, D. T. Hopkins, and D. H. Waggle, eds.). Academic Press, New York.

Struthers, B. J., R. R. Dahlgren, D. T. Hopkins, and M. L. Raymond (1979). Lysinoalanine: Biological effects and significance. *In* "Soy Protein and Human Nutrition" (H. L. Wilcke, D. T. Hopkins, and D. H. Waggle, eds.). Academic Press, New York.

Torun, B. (1979). Nutritional quality of soybean protein isolates:Studies in children of preschool age. *In* "Soy Protein and Human Nutrition" (H. L. Wilcke, D. Y. Hopkins, and D. H. Waggle, eds.). Academic Press, New York.

Torun, B., F. E. Viteri, and V. R. Young (1981). Nutritional role of soya protein for humans. *J. Am. Oil Chem. Soc.* **58**:400.

Turner, R. H., and I. E. Liener (1975). The effect of the selective removal of hemagglutinins on the nutritive value of soybeans. *J. Agric. Food Chem.* **23**:484.

Van Stratum, P. G., and M. Rudrum (1979). Effects of consumption of processed soy proteins on minerals and digestion in man. *J. Am. Oil Chem. Soc.* **56**:130.

Watanabe, D. J., H. O. Ebine, and D. O. Ohda (1971). "Soybean Foods." Kohrin Shoin, Tokyo (in Japanese).

Weingartner, K. E., J. W. Erdman, Jr., H. M. Parker, and R. M. Forbes (1979). Effect of soybean hull upon the bioavailability of zinc and calcium from soy flour based diets. *Nutr. Rep. Int.* **19**:223.

7

Oriental Soy Food Products

Traditional foods based on the soybean have been limited to the Orient where soybean foods have been eaten by everyone—young and old, rich and poor. The kinds of soybean foods, the ways to prepare them, and the manner of consuming them are deeply embedded in the different individual cultures. Although such traditional foods are increasingly popular in most of the Asiatic countries today, they have been most extensively developed and used in China, Korea, and Japan, and to a lesser extent in Indonesia.

Soybean cooking is a fine art in the Orient. Many simple and yet effective processes have been developed to make a great variety of wholesome and nutritious soybean foods. There are fresh, dried and powdered, liquid and moist, and paste products available, and they are used in meals throughout the year as cooking materials and seasonings. Considering that soybean foods are nutritious and highly digestible, it is clear that the soybean has played an important nutritional role for the Oriental population in the past and will continue to do so in the future.

For convenience, soybean foods may be categorized as nonfermented and fermented. Most of them are heat-treated sufficiently during processing, or by cooking before consumption, that the inherent antinutritional substances in the soybean are inactivated. Fresh soybeans, toasted soy powder, soy sprouts, soymilk, soymilk film, and soy curd are the principal nonfermented foods. The fermented products are fermented whole soybeans, soy sauce, soy paste, fermented soy curd, and fermented soy pulp.

Traditionally, the majority of products have been made at home or on a village scale. Those food products and the details of preparing them have been kept as a family tradition and are not widely known outside the Orient. In recent years, industrial mass production of such foods has become popular to meet changes in life styles (mainly urbanization). Such commercial products increasingly are being introduced to other parts of the world.

Nonfermented soybean products of high moisture content are perishable and thus difficult to transport for long distances. The scale of production for such foods is relatively small in spite of large demand. In contrast, the scale of industries to produce fermented soybean products is large since the fermented products have a long shelf-life.

Most traditional soybean foods (fermented and nonfermented) are known to have originated in China, and then gradually have been introduced or have spread into Korea, Japan, and other Asiatic countries. During the centuries-long process of such technology transfer, considerable modification and adaptation of processes have been made. To suit specific environmental conditions and cultural practices, significant variations in the commodities, in the processes, and in the ways of consumption were inevitable. The most extensive research and developmental efforts on such traditional foods have been accomplished in Japan. The findings, based on modern science and technology, promptly were applied to mass production. Thus it has been possible to meet the increases in demand following the recent urbanization in Japan and to promote commercialized soybean foods in other Asiatic countries.

In Table 7.1 annual per capita consumption of soybeans in several Oriental countries is shown. Direct consumption of soybeans as food is small, ranging from 3 to 5 kg per person per year. The large increase in consumption in recent years in Japan and Korea is a result of use of soybean meal for animal feeding.

Considering the heavy reliance on a starchy cereal diet in the Orient, a diet that is low in protein and lipid, incorporation of soybeans has been significant in terms of the nutritional well-being of the Oriental population. Protein derived from soybeans may range from 3 to 5 g per person per day and thus contributes about 10% of total protein intake. Furthermore, since rice is deficient in lysine, and the soybean contains relatively large amounts of this amino acid, soybean foods not only augment the quantity, but also the quality of the protein intake. In addition, fermented soybean foods often supply vitamins that may be lacking in a cereal diet.

For all of the processes used and the end products obtained, except for the toasted soybean and its powder, the first and most important step in making soybean foods is to soak the dry bean in water. This soaking process has long

Table 7.1

Per Capita Annual Consumption of Soybeans
in Selected Oriental Countries[a]

Country	Quantity (kg)	
	1964–1966	1975–1977
China	6.7	4.7
Indonesia	2.8	4.2
Japan	5.1	17.1
Korea	5.0	8.5
Malaysia	2.6	1.6
Philippines	0.2	0.2
Singapore	4.3	2.7
Thailand	0.6	0.6

[a]Source: Wang *et al.* (1979) and FAO (1980).

been believed to reduce the cooking time as well as to increase the wholesomeness of the products.

Because of its high fat, low carbohydrate, negligible starch, and compact texture, the soybean does not soften as readily as many other seeds upon cooking without prior soaking in water. The bean is rinsed several times with fresh water, and the rinse water is discarded. The clean bean is soaked in an excess of water.

Depending on the temperature, the soaking time may vary from several hours to overnight. Overnight soaking, however, is widely used in the household to prepare for further processing the next day. Under controlled laboratory conditions, the soybean reaches maximum water absorption after soaking at 20°C (68°F) for 18 hr or at 28°C (82°F) for 8 hr. The thoroughly soaked bean usually weighs about 2.2–2.4 times the original weight, depending on the variety and has slightly less than 3 times the original volume.

The soaked bean, after draining and rinsing, is ready for the next step of processing. If soaked and dehulled soybeans are needed, the dehulling can easily be accomplished after soaking by rubbing the beans against each other. The seed coats, freed from the bean by rubbing, can be floated away with water.

Some soybeans do not take up water when soaked, and these have been labeled hard beans. Hard beans are a problem because subsequent heating is not effective in inactivating their trypsin inhibitor or lipoxygenase and they affect the texture of whole-bean products such as natto and tempeh.

Based on studies of the effect of alcohols on converting hard beans to normal beans, Arechavaleta-Medina and Snyder (1981) concluded that the barrier to moisture is the cuticle, the outermost layer of the seed coat.

The major classes of soybean foods that can be produced from soaked soybeans are described in detail in the following sections.

Table 7.2

Traditional Nonfermented Soybean Food Products

Food items	Names used by major consumers	Description	Uses
Fresh soybeans	Mao-tou (China), Put kong (Korea), Edamame (Japan)	Picked green, large, soft beans	Cooked in pod, served as fresh vegetable or snack, pod removed before eating
Toasted soy powder	Tou-fen (China), Kong ka au (Korea), Kinako (Japan), Bubuk kadele (Indonesia)	Toasted dry bean ground into powder, nutty flavor, dark yellow color	Coating for rice cakes or sprinkled over cooked rice
Soy sprouts	Huang-tou-ya (China), Kong na mool (Korea), Daizu no moyashi (Japan)	Germinated with H_2O in dark, yellowish cotyledon with white hypocotyl (5 cm)	Cooked as vegetable or in soup
Soymilk	Tou-chiang (China), Kong kook, Doo Goo (Korea), Tonyu (Japan)	Milklike water extract of soybeans; heated and filtered	Served hot for breakfast (China); served cold with noodles (Korea)
Soymilk film	Tou-fu-pi (China), Kong Kook (Korea), Yuba (Japan), Fu chok (Malaysia)	Creamy yellow film from surface of boiling soymilk; sheets, sticks, or flakes	Cooked with meat, soups, vegetables, edible packaging material; not eaten daily
Soy curd	Tou-fu (China), Doo bu (Korea), Tofu (Japan), Tahu (Indonesia), Tau foo (Malaysia), Tokua (Philippines)	White protein curd from soymilk; bland taste; can be dried, frozen, or fried	Served with seasoning or after cooking as part of main meal or in soups

TRADITIONAL NONFERMENTED SOYBEAN FOOD PRODUCTS

The processes involved in making traditional, nonfermented soybean food products are cooking, sprouting, toasting, grinding, extraction, coagulation, or combinations of these. Through such processes digestibility is increased, antinutritional substances are inactivated, and the objectionable beany flavor is either decreased or is converted into some acceptable flavor. Soybeans are good sources of protein, lipid, and other nutrients, and through such processes are made into palatable foods. Simultaneously, those foods become perishable, particularly high-moisture and lipid products, such as soy sprouts, soymilk, soy curd, and soymilk film.

Table 7.2 describes the major food products in this category, giving the names used by major consumers, description of the products, and the uses. Table 7.3 gives proximate chemical composition of the same food products.

Table 7.3
Proximate Chemical Composition (%) of Nonfermented Soybean Food Products[a]

Product	Moisture	Protein	Lipid	Carbohydrate (fiber)	Ash
Fresh soybean	68	13	6	11 (2)	2
Toasted soy powder	5	38	19	32 (3)	5
Soy sprouts	82	8	2	8 (1)	1
Soymilk	94	3	1	1 (0)	0.3
Soymilk film	9	52	24	12 (0)	3
Soy curd	88	6	3	2 (0)	0.6

[a]Source: Watanabe *et al.* (1971) and Wang *et al.* (1979).

Since preferences vary depending on the culture and locality, it is difficult to judge the quantity consumed for each food item. However, we estimate the quantities consumed in decreasing order as follows: soy curd, soy sprouts, fresh beans, soymilk, toasted soy powder, soymilk film. Soymilk, soymilk film, and soy curd may be further processed into a wide range of products including fermented products.

Many different names for the same commodity are used by different major consumers. Although such names are important and essential, their use in international publications and meetings often brings misunderstanding and confusion. To avoid such possible confusion, the Eighth Association for Science Cooperation in Asia (ASCA) Conference (held in Medan, Indonesia, in February 1981) resolved to compile a publication in which the raw materials, responsible microorganisms, form, and use of products will be coded to achieve a universal nomenclature. Local names in different languages will be listed for reference in parentheses. Therefore, in this book the principle of the newly proposed system has been adopted where possible.

Fresh Soybean (Mao-Tou, China; Put Kong, Korea; Edamame, Japan)

When soybeans are picked at about 80% maturity in the green-yellow pod, the beans are still green, soft, and large. Fresh beans, cooked in the pod with salt for 10–15 min until tender, can be served as a delicious hors d'oeuvre with beer or rice wine and as a nourishing snack between meals. The pod is removed before eating.

The beans often are shelled, cooked like other beans, and served as a seasonal vegetable from summer to fall. The shelled beans also can be cooked along with rice to add a delicate taste and complementary protein to the cooked rice. The steamed fresh bean has the highest NPU value among all soybean food products (Standal 1963).

Presoaked dry beans are often roasted and may be further fried in deep oil and consumed as a snack. Presoaked dry beans are also cooked with meat and vegetables in a number of ways, and sometimes the presoaked beans are cooked with rice just as with fresh beans.

Toasted Soy Powder (Tou-Fen, China; Kong Ka Ru, Korea; Kinako, Japan; Bubuk Kedele, Indonesia)

Dry beans are toasted for about 30 min until they become brown and acquire the characteristic toasted flavor. The beans are then cooled by spreading in the open air, ground into a powder, and sieved to collect only the fine powder. This process was originally carried out in individual households, but the process is improved by mechanization today. However, the production scale is still small compared with production of other soybean foods.

The powder is used by sprinkling it on cooked rice or on rice cakes. The powder also may be mixed with lard and sugar and used as a filling or coating material for pastry (China). After mixing with spices, such as garlic and chili powders, it is served with *longtong,* which is boiled rice wrapped in banana leaves (Indonesia).

Soy Sprouts (Huang-Tou-Ya, China; Kong Na Moal, Korea; Daizu No Moyashi, Japan)

Soybeans require moisture to sprout, but with too much water the beans may rot, and with not enough water the beans will not germinate. Usually, the presoaked beans are poured into a container with a drain and a cover in a dark room. They are covered to avoid light and to prevent evaporation. The beans must be watered as frequently as needed to reduce the heat generated during sprouting. It usually takes 5–10 days for the sprout to reach a desired length of about 5 cm (2 in.) depending on the temperature. At that time, the sprouts can be removed from the container and are ready for cooking. Although any soybean may be used for sprouts, smaller varieties are preferred. Soybean sprouts resemble mungbean sprouts, but the cotyledons of soy sprouts are more distinct because of their large size, yellowish color, and beany odor (Fig. 7.1).

A special feature of soybean sprouts is the considerable content of ascorbic acid synthesized during sprouting (Weakley and McKinney 1957). Usually, no ascorbic acid is found in the dry bean. The galactose-containing oligosaccharides (stachyose and raffinose) are metabolized during sprouting (Lee *et al.* 1959). At the same time, trypsin inhibitor activity is decreased and protein digestibility is increased (Suberbie *et al.* 1981).

The sprouts are eaten as a cooked vegetable throughout the year. They are used as are other green vegetables in soups, salads, and side dishes. During cooking, it is desirable to minimize heating to maintain the inherent crisp texture and distinct taste and to minimize the destruction of ascorbic acid.

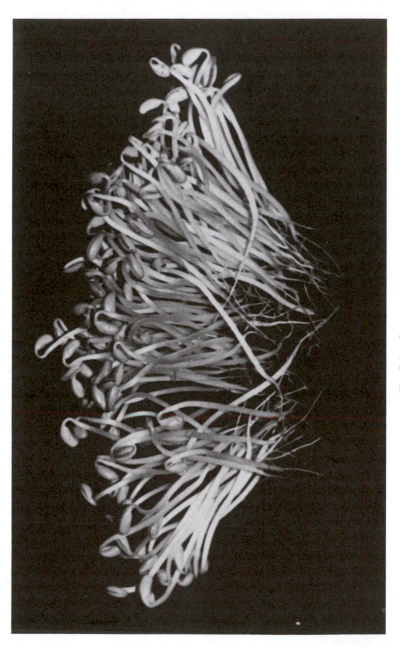

Fig. 7.1. Soy sprouts.

Soymilk (Tou-Chiang, China; Kong Kook or Doo Yoo, Korea; Tonyu, Japan)

In the traditional way of making soymilk, presoaked soybeans are ground with a stone mill while a small stream of water is added. The ground mass is heated to boiling for 15–30 min with constant stirring and additional water. The heating improves nutritive quality and flavor and pasteurizes the soymilk. The cooked, ground mass is strained through cloth, and the soymilk, a highly stable oil emulsion, passes through the cloth. Today, in the laboratory or at home, an electric blender is used for grinding the beans. To insure the maximum protein recovery, a water to dry bean ratio of 10:1 is usually recommended. From 200 g of soybeans, approximately 1 L of soymilk is obtained. Although any variety of soybean may be used, soybeans that have large uniform size, light colored hilum, thin seed coat, and high protein are preferred for soymilk.

A typical flow sheet for a modern soymilk production plant is shown in Fig. 7.2. Soybeans are thoroughly dried to facilitate dehulling, dehulled by physical means such as impact mills, ground thoroughly with added hot water, vacuum deodorized, and mixed with additives, such as sugar, vegetable oil, or vitamins. The process continues with homogenizing, pasteurizing, filling in bottles and capping, and finally retorting and packaging for shipping (Fig. 7.3). When paper containers are used, the sterilized soymilk is filled and sealed aseptically, and then packaged for shipping. Such products have a long shelf-life at ambient temperatures, but storage stability is extended by refrigeration. Flavor of such products can be varied, and vanilla, apple, and chocolate flavors have been used. Currently manufacturing plants exist for producing one million bottles of soymilk per day.

The only by-product from making soymilk is the insoluble residue known as soy pulp (tou-cha, China; be jee, Korea; okara, Japan). The pulp still has considerable nutritive components (Hackler *et al.* 1963, 1967) and is used as a food after further cooking with and without fermentation or as an animal feed.

In comparison with cow's milk, soymilk is low in lipid, carbohydrate, calcium, phosphorus, and riboflavin, but high in iron, thiamin, and niacin. Soymilk contains approximately the same amount of protein as cow's milk. Soymilk is deficient in sulfur-containing amino acids compared to cow or human milk. Animal experiments have shown that the nutritive quality of soymilk ranges anywhere from 60 to 90% of that of cow's milk. Methionine supplementation raises soymilk nutritive quality to essentially the same level as that of cow's milk (Liener 1978). However, the deficiency of sulfur amino acids appears to be of little consequence in practical infant feeding. The difference may be due to a less intense requirement for sulfur amino acids by growing children compared to growing rats (Liener 1978).

Soymilk has been of considerable interest as a substitute for cow or human milk, particularly in the feeding of infants who are allergic to cow's milk. Also, soymilk would be useful when cow's milk is either too expensive or unavailable.

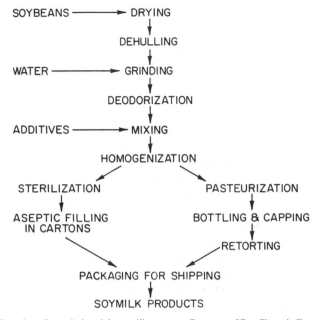

Fig. 7.2.   Flow sheet for an industrial soymilk process. Courtesy of Dr. Chung's Foods Co., Ltd,
Seoul, Korea.

Soymilk can be converted into a dried powdered form by either spray or roller
drying and used similarly to dry milk solids from cow's milk.

In China, soymilk is traditionally made fresh daily and sold on the street or in
small cafes. Soymilk is served hot after boiling, often as part of breakfast. Some
prefer it flavored with sugar, while others prefer a salty taste achieved by the
addition of soy sauce, a little bit of green onion, a dash of sesame oil, and even
small salt pickled vegetables.

In Korea, traditionally soymilk is not consumed directly but is eaten with
cooked noodles and salt. This dish is served often in summer as a cold lunch.

With increasing awareness of the nutritive quality of the soybean, soymilk is
becoming more popular. Soymilk is now bottled or packed in sterile cartons and
sold as a cold drink throughout Asia.

Numerous attempts have been made to produce and market soymilk in the
United States with relatively little success. Nevertheless, more research work has
been done on soymilk in the United States than has been done on the other
traditional Oriental soy foods.

In the 1960s research at Cornell University established that lipoxygenase activity was a problem in generating painty off-flavors in soymilk. Wilkens *et al.*
(1967) developed a "hot grind" process to heat the beans sufficiently to inactivate
lipoxygenase but not to reduce drastically the amount of protein extracted. This

Fig. 7.3. A view of the production line in a modern soymilk plant with a capacity of 550 bottles per minute.

process was used to produce PhilSoy, a soymilk, at the University of the Philippines in Los Banõs.

Extensive research on soymilk has been done at the University of Illinois on
soymilk production processes and quality. To avoid the loss of valuable nutrients
during the filtration step, Nelson *et al.* (1976) recommended heating soaked
intact beans to inactivate lipoxygenase, extraction with dilute sodium bicarbo-
nate, and homogenization of the ground whole bean. The solids stay in suspension with this procedure, but the soymilk is viscous and has a chalky mouth feel.

The quality of the soymilk produced by the Illinois process has been improved
by studies on the viscosity, calcium addition, and off-flavors.

The loss of protein during filtration can be explained by the observation of
Johnson and Snyder (1978) that heating intact soybeans causes heat fixation of
protein bodies at the same time that lipoxygenase is inactivated. If soybeans are
disrupted before heating, lipoxygenase acts to generate off-flavors, but protein
bodies are disrupted. Subsequent heating does not cause soy proteins to
precipitate.

Aside from the off-flavors in soymilk generated by lipoxygenase, an astringency is frequently, but not always, detected. Chien and Snyder (1983) found that
mixing soymilk with cow's milk was an effective way to minimize the astringency sensation.

It is sometimes difficult to distinguish between soymilk and infant formulas
based on soy protein. Infant formulas will be discussed further in Chapter 9, but
usually soy infant formulas make use of isolated soy protein with carbohydrates,
lipids, vitamins, and minerals added.

Soymilk Film (Tou-Fu-Pi, China; Yuba, Japan; Fu Chok, Malaysia)

Soymilk film is prepared traditionally by a surface concentration process.
Soymilk is heated to boiling in a flat, shallow open pan and maintained at a
temperature just below boiling. A film forms at the surface just as with heated
cow's milk. The films are removed successively from the milk surface by passing
a rod underneath and lifting them free. Films are removed continuously until no
more form. The films are air dried or heated to speed drying. The resulting
cream-yellow film is converted to sheets, sticks, or flakes for further use.

The chemical composition of the soymilk film varies not only with the composition of the soymilk, but also with the stage at which the film forms. In
general, the protein and lipid contents of successively formed films decrease,
while the carbohydrate and ash contents gradually increase. Even though the
moisture content is low, the shelf-life of the product is short because of its
relatively high lipid content.

Soymilk film is quite brittle and is always softened by soaking in cold water
before cooking. The film can be used in soups, or as an edible wrapping for

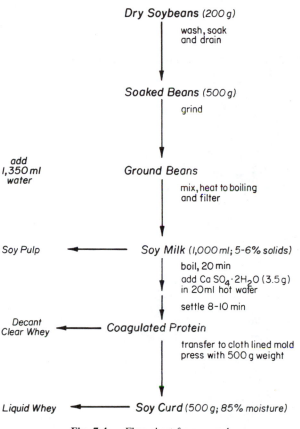

Dry Soybeans (200 g)

wash, soak
and drain

Soaked Beans (500 g)

grind

add
1,350 ml
water

Ground Beans

mix, heat to boiling
and filter

Soy Pulp   ◄————   *Soy Milk (1,000 ml; 5-6% solids)*

boil, 20 min

add Ca SO$_4$·2H$_2$O (3.5 g)
in 20 ml hot water

settle 8-10 min

Decant
Clear Whey   ◄———— *Coagulated Protein*

transfer to cloth lined mold
press with 500 g weight

Liquid Whey   ◄———— *Soy Curd (500 g; 85% moisture)*

Fig. 7.4.   Flow sheet for soy curd.

vegetables and meats for further cooking. Since it is an expensive specialty item, it is not eaten everyday.

Soy Curd (Tou-Fu, China; Doo Bu, Korea; Tofu, Japan; Tahu or Tau Foo, Indonesia and Malaysia; Tokua, Philippines)

Soy curd is closely associated with soymilk, as the initial step in making the curd is to make the milk. A laboratory procedure for soy curd making is illustrated in Fig. 7.4. When a divalent cation is added to the milk as a coagulating agent, the protein along with lipid is precipitated, and the mass is placed into a molding box to form the soy curd. The resulting curd has a bland taste, a soft texture, and a white color (Fig. 7.5).

Traditionally, the concentrated liquid obtained during the process of salt making from sea water, containing mainly magnesium sulfate and magnesium chloride (yen-lu, China; kan soo, Korea; nigari, Japan), is used as the coagulat-

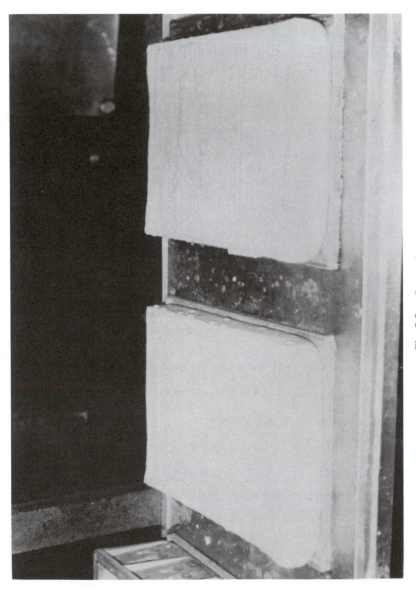

Fig. 7.5. Soy curd.

ing agent. Today, calcium salts, such as sulfate, chloride, lactate, gluconate, or citrate, and acids such as citric and acetic are used. However, calcium sulfate appears to be superior to the others, since it coagulates the protein slowly to give the maximum precipitation and the smoothest texture.

The calcium salt (3.5 g) should first be dissolved in hot water (20 mL) and then gently mixed with 1 L of hot (70–80°C, 158–176°F) soymilk obtained from 200 g of dry soybeans. After the curd settles, the supernatant fluid is removed, and the precipitate is transferred to a molding box with small holes in the sides and bottom and lined with coarse filter cloth, The ends of the filter cloth are folded over and a board placed on top. The curd is then subjected to pressure from the top with weights of 500 g to squeeze out the remaining supernatant whey. When draining has stopped, the curd is ready to be removed from the box. From 200 g of soybeans, 450–500 g of the curd with a moisture content of 85% is obtained (Wang *et al.* 1979).

The curd is usually made, sold, and consumed in the same day, although it can be kept in cold water for a couple of days without spoiling. As the curd is perishable and difficult to transport for long distances, production is on a relatively small scale.

Development of production machinery and packaging techniques in recent years have made possible an increased production scale. Large plants have begun to produce soy curd that is water packed in polyethylene containers, sealed with a sheet of transparent film, and pasteurized by immersion in hot water for one hour to give a shelf-life of up to one week. The yield of curd varies with the variety of soybean used and the processing method employed, but yields commercially are 4 to 4.5 times the soybean weight (Watanabe 1969).

The chemical composition of curd also may vary depending on the variety of soybean and the method used for curd production. However, soy curd is rich in protein, lipid, and minerals, and is easily digestible because all insoluble matter is removed in processing. Perhaps it is the most important nonfermented soybean food in the Orient in terms of the quantity consumed and in terms of its nutritional significance.

There are numerous ways of eating soy curd as soups or separate dishes cooked with meat and vegetables. It can also be served seasoned without further cooking.

Depending on the amount and type of coagulant used and the pressure applied, curds of different consistency can be made. Soft-textured product results from adding less coagulant, or by not applying pressure. For example, tou-fu-nao (China) or kinugoshi tofu (Japan) has a silky texture, is soft, and has a white color. It can be made by adding only half the amount of coagulant to a twofold concentrated milk and eliminating pressing. Since supernatant whey is not removed, this soy curd contains more of the nutrients originally present in the soybean than traditional soy curd.

Frying, drying, or freezing fresh curd also produces numerous processed products. These processes drastically change the fragile smooth texture of the fresh curd into a hard and chewy product. For example, frozen curd (tung-tou-fu, China; kori tofu, Japan) is like a resilient, absorbent sponge having a chewy texture. Because of the lack of freezing facilities, the frozen curd is traditionally made by exposing the fresh curd to extremely cold weather. The fresh curd is exposed to severe weather until frozen solid and kept for weeks to develop the spongy texture. Then it is thawed in warm water, pressed to expel all the water, and finally dried for longer shelf-life. The dried product is rehydrated by placing it in cold water before cooking. In Japan, there are large-capacity production plants based on these general procedures (Watanabe 1969).

TRADITIONAL FERMENTED SOYBEAN FOOD PRODUCTS

Fermented soybean food products use whole soybeans or other nonfermented soybean products as substrates, sometimes along with other cereals. Although some destruction of protein and other nitrogen compounds may occur during fermentation, protein and carbohydrate fractions are hydrolyzed into smaller constituents, which increase digestibility. Also, vitamins accumulate in the fermented products. The major food items in this category are listed in Table 7.4, and their proximate chemical compositions are listed in Table 7.5. Fermented foods are produced by specific microorganisms or by several such organisms acting in a sequential manner. The characteristic flavors derive mainly from the specific microorganisms used but partly from the substrate and partly from the process.

Fermented foods usually contain salt and have relatively long shelf-life. Although it takes longer to produce fermented soy foods than nonfermented, fermented products are produced on a larger scale in mechanized, modern factories and marketed in larger areas than nonfermented soy products.

Soy foods produced by fermentation are easily digested. The quantity consumed and the frequency of use are, in general, greater for fermented soy food products than for nonfermented products. The quantity consumed may be listed in decreasing order as follows: soy sauce, soy paste, fermented whole soybeans, fermented soy pulp, and fermented soy curd.

Fermented Whole Soybeans

Whole soybeans can be fermented into wholesome foods without any prior processing. After presoaking and heating, the beans are allowed to ferment with a starter culture of specific microorganisms, and after completion of fermentation, such products may be consumed as they are or after cooking. Depending upon the type of microorganism and fermentation process used, the resulting

Table 7.4

Traditional Fermented Soybean Food Products

Food items	Names used by major consumers	Microorganism involved	Description	Uses
Fermented whole soybeans	Tempeh (Indonesia and Malaysia)	*Rhizopus*	Soft bean bound by white mycelia, cake-like, nutty aroma	Fried as part of main meal, snack, and in soups
	Natto (Japan)	*Bacillus*	Soft bean covered by viscous, sticky polymer	Seasoned and eaten with cooked rice
	Hamanatto (Japan)	*Aspergillus*	Black and salty bean fermented with wheat flour	Cooked with meat and vegetables, eaten with rice or gruel
Soy sauce	Chiang-yu (China), Kang jang (Korea), Shoyu (Japan), Kecap (Indonesia and Malaysia), Tayo (Philippines)	*Aspergillus, Pediococcus, Torulopsis, Saccharomyces*	Dark brown liquid, salty and meaty taste	All-purpose seasoning agent in cooking
Soy paste	Chiang (China), Doen jang (Korea), Miso (Japan), Tauco (Indonesia and Malaysia), Tao si (Philippines)	*Aspergillus, Pediococcus, Saccharomyces*	Light to dark brown paste, salty and soy sauce flavor	Soup base and seasoning agent
	Ko chu jang (Korea)	*Aspergillus*	Reddish color with hot, salty flavor	Seasoning agent as normal soy paste
Fermented soy curd	Su-fu (China)	*Actinomucor, Mucor*	Creamy cheese, mild flavor, salty	Relish, also cooked with meat or vegetables
Fermented soy pulp	Tempeh gembus (Indonesia)	*Rhizopus*	Like Tempeh, nutty texture and aroma	Like Tempeh, popular in Central and Eastern Java
	Oncom ampas tahu (Indonesia)	*Neurospora*	Like Tempeh, orange red in color	Like Tempeh popular in Western Java

Table 7.5

Proximate Chemical Composition (%) of Fermented Soybean Food Products[a]

Product	Moisture	Protein	Lipid	Carbohydrate (fiber)	Ash
Fermented whole soybeans					
Tempeh	64	18	4	13	1
Natto	59	17	10	12 (2)	3
Hamanatto	36	26	12	14 (3)	12
Soy sauce	72	7	0.5	2 (0)	18
Soy paste	50	14	5	16 (2)	15
Ko chu jang	48	9	4	19 (4)	20
Fermented soy curd	60	17	14	0.1	9
Fermented soy pulp					
Tempeh gembus	81	5	2	11	1
Oncom ampas tahu	84	4	2	8	2

[a]Sources: Watanabe (1969); Watanabe *et al.* (1971), Mheen *et al.* (1981).

products vary in color, texture, and flavor. The way of serving also differs. Since no component is removed from the beans and since some vitamins accumulate in such products during fermentation, they are quite nutritious. Therefore, converting whole soybeans into fermented products appears to be a most efficient and economical way of utilizing soybeans. Some of the typical food products in this category are tempeh (or tempeh kedele) in Indonesia and natto in Japan.

Tempeh.   Making tempeh in Indonesia is a household art of long tradition. The procedure may vary in detail from one household to another, but the principal steps are as follows. Presoaked and dehulled beans are boiled with excess water for about 30 min, drained, and spread for surface drying. A small piece of tempeh from a previous fermentation is mixed with the beans as a starter. The inoculated beans are wrapped with banana leaves and allowed to ferment at ambient temperature for 1 or 2 days depending on the air temperature. By this time, the beans are covered with white mycelium and bound together as a cake. A general procedure for tempeh making is illustrated in Fig. 7.6.

Any variety of soybean is suitable for making tempeh. Recently, a pure starter mold has been marketed under the name ragi tempeh. The optimal temperature for the fermentation is between 30 and 32°C (86 and 90°F), and about 20 hr is sufficient time to produce a palatable product. A simple incubator can be constructed from a styrofoam picnic basket with a 7½ W light bulb maintaining the temperature at about 30°C (86°F). Instead of banana leaves, a shallow (2 cm or 0.8 in.) wooden, plastic, or metal tray with pin-sized perforations in the bottom and cover can be used for a fermentation container (Wang *et al.* 1979). In Malaysia, tempeh is made without removal of the soybean seed coat, and thus the product contains more fiber (Karim 1981).

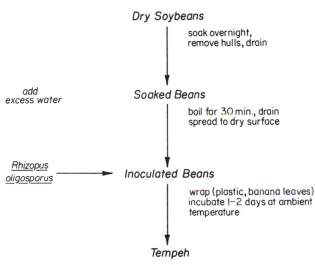

Fig. 7.6.   Flow sheet for making tempeh.

Tempeh is not only easily made, but it also has a universally acceptable pleasant texture and flavor. However, tempeh is perishable and is usually consumed the day it is made. Its shelf-life can be prolonged by either sun drying after cutting into slices or blanching to inactivate mold enzymes and then freezing. During tempeh fermentation, *Rhizopus oligosporus* produces an antibacterial agent that is especially active against some gram-positive bacteria (both microaerophilic and anaerobic). This helps to minimize possible bacterial contamination during tempeh fermentation.

There are many attractive ways to serve tempeh. The simplest and most popular is to cut into thin slices, dip into a salt solution, and fry in coconut oil. Sliced tempeh can also be baked or added to soup. It is served daily as a side dish with cooked rice and is also consumed as a snack in Indonesia. Tempeh is consumed less at main meals in Malaysia and more as a snack item.

Natto.   Making natto is a simple procedure that traditionally was done by each household for its own consumption. Before the responsible microorganism was isolated, natto was made by incubating cooked soybeans covered with rice straw in a warm room for a couple of days. Rice straw was credited not only with supplying the fermentation organism, but also with providing the aroma of straw and simultaneously in absorbing unpleasant odors generated during fermentation.

Natto is one of the few products in which bacteria predominate during fermentation. The responsible bacterium has been identified as *Bacillus natto*, an aerobic gram-positive rod closely related to *B. subtilis*. Now a pure culture fermentation has been adopted for making natto in Japan both in the household and commercially. The presoaked beans are cooked until tender, drained, cooled to 40°C (104°F), inoculated with a water suspension of *B. natto*, packed in a wooden box or polyethylene bag, and incubated at 40–43°C (104–109°F) for 12–20 hr. The natto produced has a characteristic odor and musty flavor and has a slimy appearance since it is covered with a viscous and sticky polymer of glutamic acid produced by the microorganism. The product has a short storage life, partly because of a moisture content of more than 50%. To improve its shelf-life and to broaden its application in food uses, natto can be converted into dry powder. In Japan, natto is eaten with soy sauce and mustard and often used for breakfast and dinner along with cooked rice.

Hamanatto is a special kind of natto made by fermenting whole soybeans with a strain of *Aspergillus oryzae*. Although its production is limited in Japan, similar products are widely available and consumed in other Oriental countries. The methods of preparation may vary from country to country, but the essential features are similar. Presoaked beans are cooked, drained, cooled, mixed with parched wheat flour, and inoculated with the starter culture. After incubation, the product is packed with salt and other seasonings and aged for several weeks or months. The final product is blackish, salty, and has a flavor resembling soy paste.

Soy Sauce (Chiang-Yu, China; Kan Jang, Korea; Shoyu, Japan; Kecap, Indonesia and Malaysia; Tayo, Philippines)

Making soy sauce was a somewhat complicated and delicate household art. It involved three major steps: the starter culture mass preparation (a solid-state fermentation); the fermentation and ripening in brine; and the finishing process. The starter mass is usually made in autumn by allowing the slow fermentation of a cooked soybean mash. Molds grow on the surface and bacteria grow inside the mash during drying in air. The typical molds found are *A. oryzae* and *A. sojae*, and the bacterium is *B. subtilis*. In the spring the dried culture mash is soaked in brine, and further fermentation and ripening continue for several months. The supernatant dark brown and salty liquid separated from the brine mixture is soy sauce, and the brown and salty paste remaining is soy paste. In this way homemade soy sauce and paste are obtained simultaneously.

The two products available in the market today are manufactured separately in modern factories. Although there has been a wide range of diversification in the ingredients, processes, and the final fermented products through a long and slow process of development in different localities, the principal procedures for making soy sauce and paste have remained the same even in the modern mass production of today. The industrial manufacture of such fermented products has been achieved primarily in Japan through the introduction of modern microbiology and process technology. Therefore, we shall describe in some detail the Japanese process of soy sauce production.

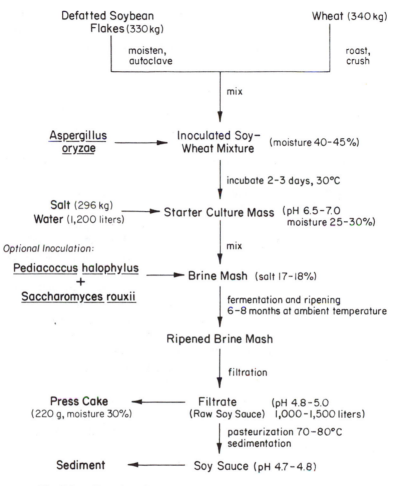

Fig. 7.7. Flow sheet for making soy sauce. Source: Yokotsuka (1981).

The flow sheet of a modern soy sauce manufacturing process is shown in Fig. 7.7. The first step is the solid-state fermentation to make the starter mass used in the brine fermentation. The soybeans in the form of defatted flakes are moistened and cooked in a continuous pressure cooker, while the wheat is roasted and cracked. About equal amounts of the two are mixed and inoculated with the seed mold, a pure culture of *A. oryzae* or *A. sojae*. The mixture is placed into a shallow perforated vat, and forced air is circulated through the mass. After about three days incubation at 30°C (86°F) the green-yellowish starter mass (about 26% moisture) is harvested. This starter culture mass (koji, Japan) serves as the source of enzymes for hydrolyzing starch into sugars and protein into amino acids and peptides. These compounds are fermented in the subsequent brine fermentation.

The harvested starter mass is transferred into deep fermentation tanks containing salt solution of about 22%. The two are mixed thoroughly, and the brine mash, now having about 18% salt, is held for 6–8 months at ambient temperature with occasional aeration.

The salt concentration is critical because putrefaction may occur at levels below 16% salt, whereas too high salt content reduces enzyme activity. The 18% salt also limits growth of the microorganisms to a few desirable osmophilic types. Initially, *Pediococcus halophilus* grows and produces lactic acid to drop the pH from 6.5–7.0 down to 4.7–4.8. Next, *Saccharomyces rouxii* generates a vigorous alcoholic fermentation, and lastly, *Toroulopsis* strains produce phenolic aroma compounds in the brine fermentation (Yokotsuka 1981).

The final step is finishing, which includes filtration and pasteurization. The fermented brine mash is filtered or pressed, and the resulting raw soy sauce is heated to 70–80°C (158–176°F). This heating process is necessary to develop the color and aroma, and to inactivate most of the residual enzymes. After clarifying the treated sauce by sedimentation or filtration the clear supernatant is bottled and marketed.

Owing to the remarkable advances made in recent years in soy sauce technology, in particular with soybean treatments and process mechanization, the yield of soy sauce has been increased from 65 to 90% on the basis of total nitrogen recovered. Furthermore, the quality of the product has greatly improved (Fukushima 1979). The use of defatted soybean flakes in industrial soy sauce production is fully justified, since the oil of the soybean is usually lost as a part of wastes during finishing.

Soy sauce is a dark brown liquid with a salty taste and a meaty flavor. It is an all-purpose seasoning agent used in the preparation of a wide range of cooked foods as well as a table condiment. Its use is still greatest in the Orient, but it is rapidly being accepted in many other countries throughout the world.

Soy Paste (Chiang, China; Doen Jang, Korea; Miso, Japan; Tauco, Indonesia and Malaysia; Tao Si, Philippines)

As described already, soy paste formerly was obtained after removal of the soy sauce from the fermentation of cooked, whole soybean mash in brine. Today, soy paste is produced commercially as a sole product without separation of soy sauce. Also, other cereals are added to soybeans as fermentation substrates. Although innumerable varieties are possible based on the kinds of substrates, amount of salt, and length of fermentation and aging, the basic processes include the production of starter culture mass, fermentation in brine solution, and finishing.

The starter culture mass for soy paste is made similarly to that of soy sauce, but using either rice or barley alone. At the final stage, salt is mixed with the starter mass to stop further growth of the mold. The salted starter mass is mixed with the cooked whole soybean mash, and a calculated amount of water is added to bring the moisture content to about 48%. At this time, an inoculum of salt-tolerant yeast and lactic acid bacteria may be added to accelerate the fermentation. The mixture is allowed to ferment and ripen in a tank under pressure at about 30°C (86°F) for 1–3 months. At least twice during the fermentation, the fermenting mash is transferred from one tank to another to accelerate the fermentation and to make the mash more uniform. The ripened paste is finally blended, pressed, and pasteurized. Pasteurization is either through a tube heater before packing, or in hot water after packing (Yokotsuka 1981).

The soy paste is light to dark brown in color and resembles peanut butter in consistency and texture. The paste is used by dissolving it in water as a base for various types of soup. It also can serve as a seasoning for cooked meat and vegetable dishes.

Hot soy paste (ko chu jang, Korea) is a unique fermented product popular only in Korea, reflecting the hot spice preference of Koreans. The first step is to make the starter culture mass using only cooked soybean mash as the substrate. The starter mass is ground, mixed with cooked rice and red pepper in brine solution and allowed to ferment and ripen for several months. The finished product is similar to tomato ketchup in appearance and is used in a variety of dishes in the same way as soy paste.

Fermented Soy Curd (Su-Fu, China)

Su-fu is made from soy curd by a mold fermentation and a brining process. However, soy curd for this purpose requires a hard consistency with less than 70% moisture. The hard curd is cut into small cubes, is heated for pasteurization (which further reduces surface moisture), and is inoculated with the selected mold. *Actinomycor elegans* is the seed mold of choice but other molds such as *Mucor* and *Rhizopus* strains are used. The time of fermentation differs depending upon the mold used, ranging from 3 to 7 days at 20°C (68°F). The fermented cubes are placed in brine containing 12% salt and rice wine for several months. Finally, the product is bottled with the brine and sterilized for marketing.

Su-fu is a creamy cheese-type product with a mild flavor and a salty taste. It is consumed directly as a relish or is cooked with vegetables and meat (Hesseltine and Wang 1972).

Fermented Soy Pulp

There are two fermented soy pulp products available in Indonesia, and they differ from each other by the microorganism involved. The insoluble pulp obtained from making soymilk or soy curd is washed several times with cold water, pressed to remove excess water, cooked, and inoculated with small pieces of tempeh or with ragi tempeh, the starter culture of *R. oligosporus*. The inoculated pulp is placed in a wooden tray, covered with banana leaves, and allowed to ferment for 24 hr at ambient temperature. The resulting tempeh gembus is soft like a sponge and easily sliced. When fried, it has a texture, aroma, and taste similar to french fried potatoes. It is consumed just as tempeh kedele and is very popular in Central and Eastern Java. Oncom ampas tahu is another fermented product made from soy pulp, but with *Neurospora sitophila* as the starter culture. This orange-pink product is similar to tempeh gembus and is served in the same way in Western Java.

REFERENCES

Arechavaleta-Medina, F., and H. E. Snyder (1981). Water imbibition by normal and hard soybeans. *J. Am. Oil Chem. Soc.* **58:**976.

Chien, J. T., and H. E. Snyder (1983). Detection and control of soymilk astringency. *J. Food Sci.* **48:**438.

FAO (1980). "Food Balance Sheets and per Capita Food Supplies." FAO, Rome.

Fukushima, D. (1979). Fermented vegetable (soybean) protein and related foods of Japan and China. *J. Am. Oil Chem. Soc.* **56:**357.

Hackler, L. R., D. B. Hand, K. H. Steinkraus, and J. P. VanBuren (1963). A comparison of the nutritional value of protein from several soybean fractions. *J. Nutr.* **80:**205.

Hackler, L. R., B. R. Stillings, and R. J. Polimeni (1967). Correction of amino acid indices with nutritional quality of several soybean fractions. *Cereal Chem.* **44:**638.

Hesseltine, C. W., and H. L. Wang (1972). Fermented soybean food products. *In* "Soybean: Chemistry and Technology," Vol. 1, Proteins (A. K. Smith and S. J. Circle, eds.). AVI Publ. Co., Westport, CT.

Johnson, K. W., and H. E. Snyder (1978). Soymilk: A comparison of processing methods on yields and composition. *J. Food Sci.* **43:**349.

Karim, M. I. (1981). Preliminary studies on nutritional qualities of Malaysian Tempeh. *Pertanika* **4:**129.

Lee, K. Y., C. Y. Lee, T. Y. Lee, and T. W. Kwon (1959). Chemical changes during germination of soybean. II. Carbohydrate metabolism. *Seoul Univ. J.* **8:**35.

Lie, G. H. (1981). Nutritional aspects of fermented foods in Indonesia. Paper presented at 8th ASCA Conf., Medan, Indonesia.

Liener, I. E. (1978). Nutritional value of food protein products. *In* "Soybeans: Chemistry and Technology," Vol. 1, Proteins, 2nd ed. (A. K. Smith and S. J. Circle, eds.). AVI Publ. Co., Westport, CT.

Mheen, T. I., T. W. Kwon, and C. H. Lee (1981). Traditional fermented food products in Korea. Paper presented at 8th ASCA Conf., Medan, Indonesia.

Nelson, A. I., M. P. Steinberg, and L. S. Wei (1976). Illinois process for preparation of soymilk. *J. Food Sci.* **41:**57.

Standal, B. R. (1963). Nutritional value of proteins of Oriental soybean foods. *J. Nutr.* **81:**279.

Suberbie, F., D. Mendizabal, and C. Mendizabal (1981). Germination of soybeans and its modifying effects on the quality of full-fat soy flour. *J. Am. Oil Chem. Soc.* **58:**192.

Wang, H. L., G. C. Mustakas, W. J. Wolf, L. C. Wang, C. W. Hesseltine, and E. B. Bagley (1979). "Soybeans as Human Food—Unprocessed and Simply Processed." *Util. Res. Rep.* **5,** USDA, NRRC, Peoria IL.

Watanabe, T. (1969). Industrial production of soybean foods in Japan. UNIDO Expert Group Meeting on Soya Bean Processing and Use. USDA, Peoria, IL.

Watanabe, D. J., H. O. Ebine, and D. O. Ohda (1971). "Soybean Foods." Kohrin Shoin, Tokyo (in Japanese).

Weakley, F. B., and L. L. McKinney (1957). Modified indophenol-xylene method for the determination of ascorbic acid in soybeans. *J. Am. Oil Chem. Soc.* **34:**281.

Wilkens, W. F., L. R. Mattick, and D. B. Hand (1967). Effect of processing method on oxidative off-flavors of soybean milk. *Food Technol.* **21:**1630.

Yokotsuka, T. (1981). Industrial application of proteinous food fermentation. Paper presented at 8th ASCA Conf., Medan, Indonesia.

8

Soybean-Supplemented Cereal Grain Mixtures

No single cereal grain contains protein with an ideally balanced amino acid composition for human nutrition, but mixing cereals, or cereals with oilseeds, usually improves the protein quality for nutrition. The principle is simple: combine cereals that are deficient in one or more of the amino acids with foods that are rich in them.

The advantages of mixing cereals or cereals and oilseeds have been recognized instinctively for centuries in many parts of the world. For example, in south Asia, pulses (*dal*) commonly are eaten with the wheat-based *chapatti*; in the Middle East, ground chickpea (*hummos*) or sesame paste (*tahini*) is eaten with round wheat breads (*pita*); and in Latin America, beans (*frijoles*) are eaten with flat corn breads (*tortillas*) (Berg 1973). These kinds of combinations increase the quality of protein in the mixture and often increase quantity as well.

Although the quality of soybean protein is not equal to that of animal protein, it is considerably better than cereal protein. Also, soybeans contain more protein than cereals. Proteins in soybeans are often deficient in sulfur-containing amino acids that are present in excess in cereals. Conversely, soybeans contain essential amino acids lacking in cereals (lysine and tryptophan); hence, the mixture becomes a nutritional bargain with respect to protein.

As a particular example, soybeans are mixed with rice and cooked in the Orient. Both fresh and dry soybeans can be used. This practice may have started to improve the flavor of cooked rice or to save rice by partially replacing it. Generally, rice is low in protein (about 8%) but has good quality. Still rice is deficient in lysine. Soybean has a high protein content (about 40%) and is rich in lysine. Thus, mixing soybeans with rice efficiently and economically increases the protein content and improves the quality of rice.

Protein malnutrition was recognized internationally at the first meeting of the Food and Agriculture Organization (FAO) Nutrition Committee in Baguio, Philippines in 1948. Surveys conducted by the World Health Organization (WHO) and FAO in 1950 and 1951 made clear that protein malnutrition prevailed in most of the developing countries. From that beginning, there have been innumerable national and international meetings on specific aspects of the protein nutrition problem, and these gatherings stimulated interest and concern in national governments, international relief organizations, research institutes, industries, and UN agencies in combatting dietary protein deficiency in developing countries. As a result, widespread interest developed in the international nutrition community about the possibility of developing low-cost nutritious foods from nonconventional sources. There was interest from nutritionists due to the complicated nature of the particular problem, and there was interest from food scientists to develop a palatable and nutritious food from low-cost ingredients.

In the 1950s, the Institute of Nutrition for Central America and Panama (INCAP) in Guatemala developed Incaparina, after *atole*, a corn-based beverage commonly used in Central America. The ingredients were locally produced commodities at a cost substantially less than that of cow's milk. In INCAP scientists had successfully demonstrated that such foods were good nutritionally, and also that children ill with protein malnutrition could recover through use of these food mixtures. The success of Incaparina stimulated the production of many new formulated foods by blending cereals, and these foods were eagerly sought by the international community.

In 1955, WHO appointed the Protein Advisory Group (PAG) to advise on the safe use of novel protein sources. In 1960, PAG became a tripartite organ of FAO, WHO, and the United Nations International Childrens' Emergency Fund (UNICEF), advising all three organizations on their joint program of protein-rich foods (Kapsiotis 1969). PAG was replaced in 1977 by the Advisory Group on Nutrition, and the PAG Bulletin has been superseded by the United Nations Food and Nutrition Bulletin.

The research stimulated by PAG produced evidence that proteins from unconventional sources, such as groundnuts, cottonseed, soybean, sesame, and coconut, when properly processed, could be used safely for feeding humans.

However, when such proteins were considered for direct human consumption, many questions emerged concerning the nutritive value, sanitary and micro-biological status, and toxicological safety. To help improve this situation, PAG published several "Tentative Quality and Processing Guides" for research institu-tions and industries engaged in processing such new formulated food mixtures. Items such as the identification of raw materials, type of processing, bac-teriological and chemical analysis, shelf-life, toxicity in animals, limited feeding trials on selected human beings, and large-scale trials in the field were included in the guidelines (DeMayer 1969 and Kapsiotis 1969).

Although protein–calorie malnutrition may affect all age groups, attention has been focused on the most vulnerable group—infants and children. Various gov-ernments and institutes with and without the assistance of UN organizations have developed and introduced protein products for this vulnerable group in many developing countries throughout the world.

Concurrently, in the period following World War II, the use of wheat for human consumption increased considerably, even in countries where soil and climate are not suitable for wheat production and consumption depends on im-ports. Due to the popularity of breads and pastas, this increase in wheat con-sumption at the expense of indigenous products is expected to continue.

To moderate this increased demand for wheat, which for many countries constitutes an economic drain, FAO launched its "Composite Flour Program" in 1964. The program's objective was to seek raw materials that could replace part of the wheat for production of bread, pasta, and similar wheat flour foods. This replacement required the formulation of flour mixtures from indigenous materials that would combine optimal nutritive value with good processing characteristics. To minimize dependency on wheat imports, emphasis was put on use of the ingredients that were already being produced or could be produced in the country of need. The quality of baked goods and pasta products from such mixtures should be comparable to products made from wheat. The addition of extra nutrients such as vitamins, minerals, and amino acids had the potential for enhancing the nutrition of the foods made from these flour mixtures (DeRuiter 1978).

During the 1950s, milk from the United Stated had been the mainstay of institutional feeding programs in a number of developing countries. Authoriza-tion for the milk donation program came from the U.S. Agricultural Act of 1949. Other U.S. laws that continued the authority were Public Law (PL) 480 of 1954 and the Food for Peace Act in 1966. The donation program authorized under Title II of PL 480 is administered by the Agency for International Development (USAID) with the U.S. Department of Agriculture (USDA) handling commodity availability and purchasing. Foreign distribution is done by U.S. voluntary agen-cies, international relief organizations, or recipient governments and is monitored by USAID.

In many countries, institutional feeding was based on the expectation that milk donations from the United States would continue. However, the milk was available primarily because of U.S. price support policies that created surpluses rather than in response to needs of developing countries (Berg 1973).

In 1964 the realization came that the milk surplus would decrease drastically and that the prospects for an immediate recovery in future years were poor. Recognizing that it would be difficult for recipient countries to secure adequate milk, program administrators in the United States sought alternative ideas for supplementing international feeding programs. The initial objective was to develop an economical milk substitute made with low-cost nonconventional proteins. The favored raw materials were cereals and oilseeds in a nutritious mixture.

In response to the need for a low-cost milk substitute, USDA in cooperation with USAID and the National Institutes of Health (NIH) developed guidelines for the nutrient composition and properties of these new formulated foods. The new foods were meant to serve as supplements in the diets of children and pregnant and lactating mothers or in the emergency feeding of adults, but were were not intended to serve as a sole source of food. In 1966 cereal blends were introduced into the U.S. donation program to substitute for the surplus milk that had been used by developing countries.

As described briefly at the beginning of this chapter, mixing of cereal grains as a part of the diet has been practiced instinctively for centuries in different parts of the world. Such practices with the help of modern nutritional sciences and food technology can lead to food mixtures to serve the needs of the developing world. In formulating such mixtures, the soybean is an important ingredient.

The history of deliberately compounded food mixtures goes back at least to the 1940s. They were initiated by different organizations with different backgrounds, motivations, and concepts. However, a certain similarity existed between them. Such products had and still have many different names: enriched food mixture, fortified food mixture, plant protein mixture, enriched cereal mixture, supplementary food mixture, weaning food mixture, high-nutrition low-cost food mixture, and others. We shall describe only those products containing soybeans as an ingredient. They fall into three categories: protein-rich food mixtures, composite flours, and cereal blends.

PROTEIN-RICH FOOD MIXTURES CONTAINING SOY FLOURS

Most of the diets in developing countries, where protein-rich food mixtures are needed, have the disadvantage of being low in total protein and deficient in lysine. Thus, the logical supplement for such diets is an economical protein abundant in lysine. To serve this purpose vegetable protein concentrates are most commonly used. They can be from soybean, cottonseed, groundnut, or sesame

Table 8.1
Some Nutritional Properties of Defatted Oilseed Flours

Defatted flour	Protein content (%)	Limiting amino acid	Excess amino acid
Soybean	47	Methionine	Lysine
Cottonseed	40	Lysine	
Groundnut	46	Methionine	
Sesame	48	Lysine	Methionine

Source: Bressani (1969).

seed. The protein content of the flours made from these seeds and the limiting amino acids are shown in Table 8.1.

The best material is soybean, based on its high lysine content, but cottonseed, sesame, and groundnut are also useful protein sources. Protein-rich food mixtures are formulated to contain approximately 25% protein, which meets the human infant requirement for quantity. The 25% protein can be achieved by dilution of defatted oilseed flours with low-protein cereal grains or tuber flours.

Achieving the correct level of protein in the mixture is much easier than achieving a mixture that will be optimal in available amino acids. Bressani (1969) outlined two approaches to choosing correct proportions of protein ingredients: (1) The choice could be based strictly on amino acid compositions of the protein components of the mixture compared to a reference protein for optimal nutrition. This method is lacking in the same way that chemical score is lacking as an evaluation procedure. That is, the availability of amino acids in the proteins

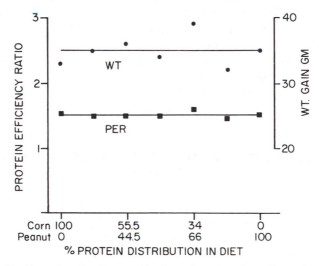

Fig. 8.1. Nutritive value of various combinations of peanut and corn flours. Source: Bressani (1969).

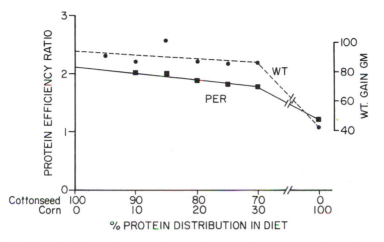

Fig. 8.2.   Nutritive value of various combinations of cottonseed and corn flours. Source: Bressani (1969).

is not known. (2) The two protein sources can be mixed in several proportions and fed. A measure of protein evaluation such as biological value or PER is used to determine the response to the various mixtures. Four different patterns of response can be obtained:

1. Type I (Fig. 8.1) results from the mixture of two protein ingredients with similar amino acid deficiency. The protein quality of the mixture is equal to that of either component. An example of type I would be mixtures of groundnut and corn proteins. Since both are deficient in lysine, the mixture does not improve protein quality. Therefore, mixtures of these two ingredients can have different total protein concentrations but the protein quality of all proportions of the two proteins will be similar.

(2) The type II pattern (Fig. 8.2) comes from a mixture of two protein sources with a deficiency of the same amino acid; however, one of the proteins contains more of the amino acid than the other. An example of this type is given by mixtures of cottonseed flour and corn. Cottonseed protein can be diluted 70–80% with corn meal without decreasing its protein quality. Cottonseed protein is deficient in lysine but contains more of this amino acid than corn protein. However, cottonseed flour cannot supply the lysine needed to cover the deficiency in corn at high levels of corn.

(3) An example of the type III pattern (Figs. 8.3 and 8.4) is given by black beans and cottonseed or by corn and soy flours. Black bean protein and cottonseed in a ratio of 40:60 give a good PER. The essential amino acids of one protein complement the essential amino acid pattern of the other protein, resulting in a mixture with a higher protein quality than either ingredient.

(4) An example of the type IV pattern (Fig. 8.5) is given by a mixture of cottonseed and sesame flours. The mixture shows no complementary effect, and

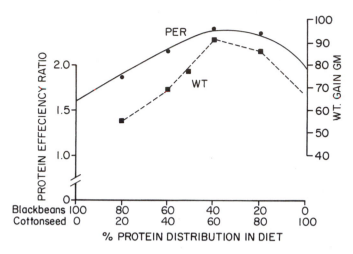

Fig. 8.3.   Nutritive value of various combinations of cottonseed flour and cooked black beans. Source: Bressani (1969).

the protein quality depends on the amount of each ingredient in the mixture. Sesame is more deficient in lysine than cottonseed but sesame is a better source of methionine than cottonseed.

Formulation of a novel food mixture to supply optimum protein nutrition is an essential but minor part of the whole process of making a new food available.

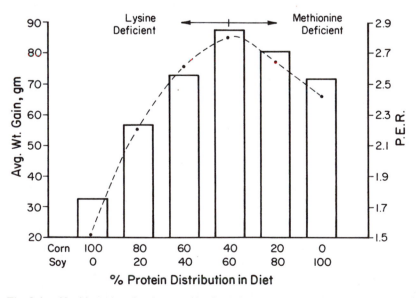

Fig. 8.4.   Nutritive value of various combinations of corn and soybean flour. Bars are for weight gain and points for PER. Source: Bressani (1969).

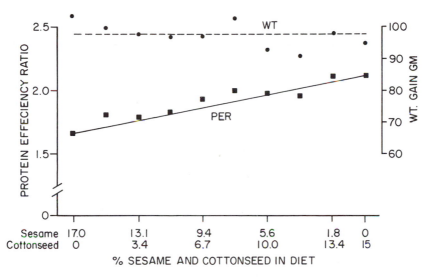

Fig. 8.5.   Nutritive value of various combinations of cottonseed and sesame flours. Source: Bressani (1969).

Guidelines have been written (Anon. 1983) for testing procedures to be done before any human feeding trials are started. The categories in which information is needed include specifications on raw materials and manufacturing procedures, nutritional value, sanitation both from a hygienic and pathogenic standpoint, toxicological safety, and technological and physical properties.

Soybeans have been valuable ingredients in new food mixtures because of their protein quantity and quality but for other reasons as well. Soybeans have been produced in large quantity for many years, and so they represent a known and stable supply with known processing parameters and composition.

Soybeans used in protein-rich food mixtures are generally in the form of defatted or full-fat flours with adequate heat treatment. These flours may be obtained from a large soybean mill in defatted form or may be produced on a village scale by simple equipment operated by hand (Wang et al. 1979). Today extrusion cooking increasingly is applied to obtain high-quality soy flours for local production of mixtures (Smith 1969, Harper et al. 1978).

Recognizing the many advantages of properly processed soy flour as a protein supplement, considerable effort has gone into the formulation of blends of soy flour with cereals. These mixtures, when suitably fortified with vitamins and minerals, have great potential for the feeding of infants, children, and people of all ages in developing areas of the world.

A vegetable protein mixture known as multiple-purpose food (MPF) with toasted soy flour providing the major source of protein was developed by the Meals for Millions Foundation, Santa Monica, California, in 1944 (Sterner 1976). It is evident from the data presented in Table 8.2 that MPF and other

Table 8.2

Effect of Soybean-Containing Food Mixtures on Low-Protein Cereal Diets[a]

Basal diet	Food mixture	Protein content of diet (%)	Gain in weight[b] (g/week)
Rice	No supplement	8.1	5.0
	MPF (soy flour)	12.4	16.1
	Groundnut/soy (50/50)	10.1	10.7
	Groundnut/soy (50/50) plus 0.8% methionine	10.2	15.6
Rice plus tapioca	No supplement	5.5	3.0
	Groundnut/sesame/soy (40/20/40)	10.5	12.5
	Groundnut/coconut/soy (30/30/40)	15.5	18.1
	Skim milk	15.5	18.1
Corn plus tapioca	No supplement	6.5	4.2
	Groundnut/chick pea/skim milk/soy (25/20/25/30)	17.2	14.2
	Groundnut/sesame/chick pea/soy (30/10/30/30)	17.2	14.9
	Skim milk	16.2	15.6

[a]Source: Liener (1978).

[b]Weanling mice fed over an 8-week period.

soybean-containing food mixtures markedly increase the growth-promoting properties of diets.

The most practical approach to the formulation of vegetable protein mixtures has been to blend soybean flour with the cereals that are a traditional part of the diet in a particular country or region. Scientists at the Central Food Technological Research Institute (CFTRI), Mysore, India, have been particularly active in development of mixtures using soybean protein to supplement the protein of groundnut and other indigenous legumes. They developed mixtures composed of equal amounts of full-fat soybean flour and groundnut flour and fortified the mixtures with 1% L-lysine and DL-methionine as well as vitamins and minerals. Other mixtures of soybean flour, groundnut flour, and coconut meal (4:3:3) and of soybean flour, groundnut flour, and sesame flour (2:2:1) have demonstrated fully their nutritional merits for growing children (Liener 1978).

Incaparina is the name given to a series of vegetable protein mixtures developed by INCAP. These mixtures were formulated to provide 25% or more of the daily protein requirement and to be comparable in quality to animal protein. Originally, Incaparinas were mixtures of corn and cottonseed flour, but in later formulations, cottonseed flour was replaced wholly or partially by soybean flour. Incaparina 14, for example, consists of 59% corn, 38% toasted soybean flour, 3% torula yeast, 1% $CaCO_3$ and 4500 IU vitamin A per 100 g. Its biological value of 78.6 and digestibility of 91.8 are comparable to skim milk values

(Bressani and Elias 1966). Such mixtures may be used to prepare a beverage or gruel by simply adding water, or they may be incorporated into other foods such as soups, puddings, cookies, or precooked baby foods.

In Korea, a precooked weaning food composed of rice, soybean, and dried milk fortified by vitamins and minerals was developed by scientists at the Korea Institute of Science and Technology (KIST) in Seoul (Kwon *et al*. 1976). The product has been produced successfully on a commercial scale since 1975. Also, the same institute developed a supplementary food mixture (Wooryang A) utilizing locally available corn grits/defatted soy flour/sesame flour (68/20/2), corn oil, sugar, vitamins, and minerals. The mixture is precooked by a locally fabricated low-cost extrusion cooker, and its production (90 kg/hr) has been a model program for local nutrition improvement since 1978 (Cheigh 1978).

Similar food mixtures containing soybean have been produced in Colombia (Incaparina and Bienestarina), Bolivia (Maisoy), Costa Rica (Torti-Rice), Chile (Supercil and Fortesan), Ecuador (Leche Avena), Guyana (Corex), Ethiopia (Faffa), South Africa (Pro Nutro), Tanzania (Licha), Thailand (Kaset infant food), and Sri Lanka (Thriposha). In addition to those mixtures already being produced, many formulations containing soybean are still in the process of evaluation.

COMPOSITE FLOURS CONTAINING SOY FLOUR

Substituting nonwheat flour for wheat flour in making bread is feasible at 5–10% without involving major technological difficulties or impairment of bread quality. A further increase in nonwheat flour causes undesirable changes such as loaf volume decrease and a dense crumb texture. A limited substitution for wheat flour does not involve sufficient economic savings to serve as a solution to the problem of wheat imports in the non–wheat-producing countries. Therefore, under its composite flour program, FAO originally stressed the desirability of exploring preparation of bread chiefly from starchy tubers and defatted oilseeds, because these products are readily available in most developing countries. By the request of FAO, the Institute for Cereal Flour and Bread TNO (ICFB-TNO) in The Netherlands launched a program in 1964 to investigate the problem of making bread entirely from nonwheat flours. Flour derived from tubers has a high starch content and is poor in protein. To obtain a nutritious flour, the protein content has to be improved considerably by adding defatted oilseed flours. Since nonwheat composite flour is devoid of gluten, special measures have to be devised to produce acceptable baked products.

The search for suitable composite flours did not commence in 1964. Since time immemorial, bread produced in Germany and other European countries included mixtures of wheat and rye, and many special dietary products contain flour mixtures. During World Wars I and II, bean, potato, and barley flours were used as supplements to wheat flour in bread making (DeRuiter 1978). In the 1950s, CFTRI did research and development on composite flours for products such as unleavened pancakes (*chapatties*), fermented and steamed cupcakes (*idlis*), protein biscuits, baby food preparations, rice substitutes, and pasta products.

The initiative taken by FAO in 1964 emphasized research and development. Thus, many wheat-producing countries and wheat-importing developing countries initiated research programs on composite flours. The experience gained so far has clearly demonstrated that for both ease of production and consumer acceptance, wheat still is the essential component in composite flours (DeRuiter 1978). The exact percentage of wheat flour required to bake an acceptable product with a composite flour depends on the quality and quantity of wheat protein in the flour and on the nature of the product.

Composite flours are intended for use as raw materials in baked goods and pasta products and for the preparation of family meals. The bulk of the research and development work centers around the production of bread, biscuits, and pastas.

There are two distinct lines of approach in the production of composite flours. One is the development of completely new flour mixtures without wheat. The other is based on wheat flour and tries to establish to what extent wheat flour can be replaced by other flours without causing major changes in the products and production procedures. As a general rule, addition of protein concentrates of neutral taste in small amounts will not present problems to the production process or product quality. Generally the nutritive value of the product will improve.

If the principal aim is to promote the use of indigenous flour, the final flours must contain a minimum of wheat flour. As wheat content decreases, the problems regarding both process and product increase and become more complicated. With no wheat flour at all, a completely different mode of production may be required, and the new products may be quite different from wheat products in appearance, flavor, and eating quality. Thus, the introduction of nonwheat composite flour poses far greater problems than for a composite flour with wheat, and only a small proportion of the research and development has been focused on nonwheat composite flours.

Development and Application of Nonwheat Composite Flours

It is generally accepted that wheat flour is outstanding for bread making because of its special protein, gluten. Gluten is essential for the retention of carbon dioxide formed during fermentation and for temporarily binding the water required to gelatinize the starch. With nonwheat composite flours devoid of gluten, the air entrapped in the dough during mixing and the carbon dioxide produced by the yeast are not retained. Thus a low-volume crumb with irregular cells is produced. Fortunately, it was found that the addition of emulsifiers to such gluten-free dough improves gas retention and bread structure (Rotsch 1954,

Jongh 1961). Addition of emulsifiers such as glycerylmonostearate (GMS), calcium stearoyl lactylate (CSL), sodium stearoyl lactylate (SSL), pentosans, and xanthan gum induce attraction between starch granules in a concentrated suspension. When this attraction is strong enough, the dough acquires properties sufficiently cohesive for retention of carbon dioxide.

In this type of research, the choice of raw materials is limited, and one depends on what is available in a given country. The composite flour must have good nutritive quality, and the preparation of bakery products should allow reasonable latitude in procedures while ensuring quality. Most important from the consumer's viewpoint are that taste and eating characteristics be acceptable. ICFB-TNO has tested numerous combinations of starch and protein flours with

Fig. 8.7.   Section through a loaf of cassava/soy bread. Source: DeRuiter (1978).

Fig. 8.6. Breads made from cassava/soy (80/20) composite flour. Source: Kim and DeRuiter (1969).

different gluten substitutes to establish their bread-making suitability (Kim and DeRuiter 1968). The best results were obtained with a composite flour containing 80 parts of cassava starch and 20 parts of defatted soy flour, with GMS and CSL as dough improvers. Table 8.3 shows the formula and process. A collection of different kinds of loaves and rolls made from cassava/soy (80/20) composite flour is shown in Fig. 8.6, and the cross section of the bread in Fig. 8.7.

For a good product with cassava/soy composite flour, it is important to (a) add enough water to make a semiliquid batter, (b) add GMS as an emulsion, (c) mix the components intensively and remix them after a short fermentation, and (d) insure the protein in the soy flour has an NSI of approximately 60. These factors all facilitate a large amount of air being incorporated into the dough during mixing. The cassava/soy bread obtained is slightly soggy and rubbery immediately after baking, but after half a day, the crumb becomes drier and less rubbery, and closely resembles cake. It is palatable, and consumer acceptance is best at this time.

DelValle *et al.* (1976) and Franz (1975) have developed formulas of corn and soybean flours for tortillas. The Instituto de Investigaciones Technologias (IIT) of Bogota, Colombia, also has developed a similar composite flour from pregelatinized corn flour mixed with soy flour and cottonseed flour (DeRuiter 1978).

Table 8.3
Formula and Procedure for Preparing Bread from Cassava
Plus Soy[a]

Formula	Parts by weight
Cassava starch	80
Defatted soy flour	20
Compressed yeast	2
Salt	2
GMS (10% emulsion)	13.5
CSL	0.15
Water (depends on water absorption of flour)	60–80

Procedure	Time (min)
Mixing	10
Bulk fermentation	40
Remixing (depends on size of mixer)	5–10
Dividing	0
Final proof (depends on CO2)	Variable
Baking	30

[a]Source: DeRuiter (1978).

Development and Application of Composite Flours Containing Wheat Flour

To maintain a good bread quality with high proportions of nonwheat flour requires modification of the bread-making procedures. Solutions to these production problems have been developed in The Netherlands (ICFB-TNO), in England (Tropical Products Institute), and in the United States (Kansas State University). A large variety of wheat breads with 30% nonwheat flours have been made and are generally satisfactory. Given a good-quality wheat flour with proper additives, it may be possible to mix up to 50% nonwheat flours into acceptable doughs by mechanical dough development using a high-speed mixer (Dendy *et al.* 1970). A wide range of nonwheat components have been used such as cassava starch, sorghum, millet, rice flour, and high protein components such as soy flour.

The Interpan Project implemented by IIT in Colombia involved application of mechanical dough development using sheeting rolls. The dough was prepared manually or in a low speed mixer from a composite flour of wheat/rice/soy flours (70/27/3), to which 0.5 parts of CSL had been added (DeRuiter 1978).

Pomeranz *et al.* (1969) examined the effect of using synthetic sucroglycerides in making bread from a wheat/soy flour and found that at 16% soy flour sucroesters improved loaf volume, crumb grain, and bread softness. Considerable research has been done at Kansas State University on use of composite flours in bread making. It was here that the usefulness of CSL and SSL was discovered for maintaining loaf volumes in high-protein breads. For nonwheat components, Tsen and Hoover (1973) and Tsen *et al.* (1975) used defatted soy flour, defatted corn germ flour, full-fat soy flour, and extruded soy products from whole and dehulled soybeans cooked at various temperatures. A soy product, used at 12% level and moderately toasted with NSI above 40, was best.

Ranhotra and Loewe (1974) experimented with 15 commercial soy protein products for making bread with a composite flour at 10, 15, and 20% protein. The soy products included concentrates, isolates, and defatted and full-fat flours. The poorest results in terms of loaf volumes at 20% protein were obtained with the soy protein isolates, followed by the two concentrates and the defatted soy flours. The best volumes at high protein content were obtained with either full-fat or high-fat soy flours. The authors thought that phospholipids in the full-fat flours may have been responsible, because total fat in the breads was maintained constant.

When the performance of extruded soy products was compared with that of nonextruded soy flours, loaf volume decreased less with the extruded soy products (Bookwalter *et al.* 1971). Fellers *et al.* (1976) specified a number of requirements for chemical composition, moisture content, and bromate level in order to obtain a satisfactory loaf volume with a standard straight dough method for 6 and 12% soy fortified bread flours. For example, the NSI of the defatted soy flour must be between 55 and 75, the Falling Number (a measure of α-amylase activity) of the wheat flour between 200 and 300, and that of the composite flour between 225 and 350.

An acceptable product was obtained with 10% soy flour added to wheat flour for production of *balady* bread (a white Arabic bread) (Hallab *et al.* 1974). A composite flour of wheat/barley/defatted soy (75/20/5) was acceptable for a number of bakery products (Kwon *et al.* 1976), as was wheat/cassava/defatted soy (70/25/5) (Kim and DeRuiter 1969).

The addition of substantial levels of soy flour increases dramatically the protein nutritional values of the baked goods. The PER for white bread is about 0.7 and increases to 0.8 with 3% added soy flour. If the soy flour is increased to 6%, the PER increases to 1.3, and at 12% added soy the PER is 1.95. Furthermore with 12% soy flour, there is 50% more protein in the fortified bread.

The bread with added soy flour has an extended shelf-life. For breads with soy flour and SSL incorporated, staling is delayed. This is probably due to the increased absorption and retention of moisture by the added soy flour and to the known enhancement of softness by SSL. On the minus side, soy-fortified breads develop mold growth faster than regular breads because moisture content is higher (but only slightly). Addition of the mold inhibitor calcium propionate may be necessary to preserve bread if long periods are anticipated before consumption (Hoover 1979).

In addition to bread formulas, pasta doughs with added soy have been tried. Golden Elbow macaroni, formulated from a flour mixture containing wheat/corn/soy (30/40/30) has been used in school feeding programs in Brazil. Good results were also obtained with wheat/corn/soy (32/60/18) and cassava, corn, or rice/wheat/soy (50/25/25) in making macaroni in Guatemala and Colombia, respectively (DeRuiter 1978).

CEREAL BLENDS CONTAINING SOYBEANS

The guidelines for cereal blends prepared by USDA, USAID, and NIH include recommendations on composition and nutritional contents. At the outset, it was recognized that a lipid content greater than 2% to increase calorie density would be desirable in a weaning food. Subsequent studies have shown that lipid can be increased to 6% with retention of storage stability for at least one year at 25°C (77°F). Current USDA purchase specifications for the blended cereal product corn–soy–milk (CSM), require a minimum fat content of 6%.

Table 8.4 lists recommendations on composition, including moisture, protein, lipid, and crude fiber. Table 8.5 shows minimum recommended levels of selected essential amino acids, most frequently low in cereal grains and oil seed meals, per 100 g of the final product. Protein quality is further specified to have a minimum apparent NPU of 60, and PER of 2.1 or greater. Recommendations on

Table 8.4

Recommended Minimal Composition(s) for
Cereal Blends per 100 g[a]

Moisture	10
Protein	19
Lipid	6
Crude fiber	2

Source: Bookwalter 1981

minerals and vitamins are shown in Table 8.6. In general, the levels are such that
100 g of formulated product or one serving should supply at least half of the
National Academy of Sciences–National Research Council (NAS-NRC) RDA of
these nutrients for 6- to 8-year-old children and two-thirds of the allowance for 1
to 2 year old children. The level for vitamin B_{12} was set at the full NAS-NRC
RDA, since this vitamin occurs mainly in animal products and therefore may not
be available.

General guidelines on the ingredients, processing, and properties of the formulated products were also developed. The products had to be precooked, so that
they would be ready for serving after an additional boiling for 1 or 2 min. The
precooked product is important in developing countries where quick-cooking
saves scarce fuel and simplifies preparation. A partial cooking can also serve to
destroy enzymes such as lipases and lipoxygenases, which if active could cause
off-flavor development. A precooked product that requires minimal further cooking before serving has the advantage that water and other added ingredients are
pasteurized in the process.

Other properties required of the cereal blends are a bland flavor, low bran
content, smooth texture, and a total plate count of 50,000 bacteria/g or less. Also
the blends must be *Salmonella* free. The texture is controlled mainly by the
particle size achieved by dry milling and sieving. For CSM, a minimum of 99%

Table 8.5

Guidelines for Essential Amino Acids and Protein Quality
for Cereal Blends

Essential amino acid	Minimum (per 100 g)
Lysine (g)	0.95
Methionine (g)	0.30
Total sulfur-containing amino acids (g)	0.60
Tryptophan (g)	0.22
Threonine (g)	0.65
Apparent NPU	60
PER	2.1

[a]Source: Senti (1969).

Table 8.6

Guidelines for Vitamins and Minerals in Cereal Blends[a]

Vitamin or mineral	Minimum (per 100 g)	Maximum (per 100 g)
Vitamin A (int. units)	1500	2250
Vitamin D (USP units)	200	250
Vitamin C (mg)	20	
α-Tocopherol acetate (IU)	1.5	
Thiamin (mg)	0.5	
Riboflavin (mg)	0.5	
Niacin (mg)	6	
Pantothenic acid (mg)	3	
Pyridoxine (mg)	0.33	
Vitamin B_{12} (μg)	3	
Folacin (mg)	0.1	
Calcium (g)	0.5	0.6
Phosphorous (g)	0.42	0.6
Sodium (g)	0.3	0.45
Iron (mg)	5	15
Copper (mg)	0.5	1
Zinc (mg)	2.5	5
Iodine (μg)	60	100

[a]Source: Senti (1969).

must pass a U.S. standard screen no. 6 and a maximum of 60% can pass a U.S. standard screen no. 60. These limits are also useful for good wetting characteristics.

Some of the specific blends are described in more detail in the following sections. One of the first blends, which is of interest now as a part of the history of the concept, was named Ceplapro. It was a blend of corn meal, wheat flour, soy flour, and nonfat dry milk (58/10/25/5) with vitamins and minerals added. The cost of this product was relatively high and this was attributed to extrusion cooking. Hence subsequent products were made without extrusion cooking.

It is interesting to note that some countries are now processing their own cereal blends to avoid dependence on supplies from the United States. With many of these indigenous cereal blends, the cooking method is low-cost extrusion. Thus it is possible to make economical cereal blends using extrusion cooking.

CSM

The second cereal blend developed was the one we know today as CSM. The development was done by the American Corn Millers Federation under the direction and guidelines of the USDA. The specifications for amounts of ingredients are shown in Table 8.7.

The required fat content may be achieved by alternative ingredients such as a

Table 8.7

Specifications for CSM[a]

Ingredient	%
Corn meal, processed	63.8
Soy meal, defatted	24.2
Nonfat dry milk	5
Soybean oil, refined	5
Mineral and vitamin premix	2
Total protein, minimum	19
Total lipid, minimum	6
Crude fiber, maximum	2
Moisture, maximum	10

[a]Source: Senti (1969).

corn germ fraction, full-fat soy flour, or soy oil. By specifying limits on composition and on granulation, density, and consistency, it has been possible to assure uniformity in the final product. This product uniformity is an important part of achieving and maintaining consumer confidence in the product.

CSM has a PER of 2.5, and so it is comparable to casein in protein quality. Experience in clinical feeding of CSM to both healthy and malnourished 1- to 3-year-old children has proven that CSM will maintain nitrogen balance when fed as the primary nitrogen source. Furthermore CSM and WSB (mentioned below) have not only supported normal growth but have allowed "catch up" growth in malnourished children and infants (Bookwalter 1981).

CSM has had good acceptance even though its major component, cornmeal, is not familiar to the children of many recipient countries. Two modifications of the basic formulation for CSM have been introduced to enhance acceptance in special feeding situations. One, designated Instant CSM, is a product that requires no cooking before serving, thus saving fuel and enhancing convenience. The other is Instant Sweetened CSM, which contains 15% sucrose. The sugar was added to improve acceptance of the product by adults and by older children in emergency feeding situations.

Wheat Soy Blend (WSB)

Although CSM has had wide acceptance, it was desirable to develop high-protein formulated foods with other cereal components. Wheat is widely grown around the world and is known as a desirable food grain. It is also a major U.S. grain. A formulated food using wheat as an alternative ingredient to corn could be introduced as a prototype into countries where wheat is grown, with the expectation that the product would eventually be made from indigenous wheat.

The composition of WSB is shown in Table 8.8. For good flavor and storage

Table 8.8

Specifications for WSB[a]

Ingredient	%	%
Wheat fractions (total)		73.4
(a) straight grade flour	38.4	
and wheat protein concentrate, or	35.0	
(b) bulgur flour	53.4	
and wheat protein concentrate	20.0	
Soy flour, defatted		20
Soybean oil, refined, stabilized		4
Minerals and vitamins		2.6
Moisture, maximum		11
Protein, minimum		20
Lysine, minimum		0.95
Lipid, minimum		6
Crude fiber, maximum		2.5
Ash, maximum		6.6

[a]Source: Senti (1969).

stability, it is important to cook all components. To ensure that the final product has received sufficient heat to inactivate enzymes, a test for peroxidase activity as an indicator enzyme has been specified. Specifications also ensure that sufficient cooking has been achieved to remove the raw wheat or starch flavor, and that excessive color development, which is indicative of lysine destruction due to overheating, has not occurred.

Chemical composition of WSB (protein, fat, crude fiber, ash) is comparable to CSM. WSB is made with a smaller particle size than CSM, and this gives more fluidity to the cooked WSB. For this reason, WSB may be useful in the right concentration as a beverage rather than as a porridge or gruel. As noted above, WSB has been tested clinically and is as useful as CSM in treating malnourished infants and children. The PER for WSB is lower than for CSM averaging 2.1.

Other Soy Fortified Cereal Products

Many children, as well as adults, who need additional proteins, vitamins, and minerals are not reached by the distribution of CSM or WSB through the World Food Program, various voluntary agencies, and recipient governments. These people may receive commodities such as cornmeal, bulgur wheat, rolled oats, and wheat flour under some type of donation program. To provide additional and better quality protein these cereal products in the U.S. Food for Peace Program have been fortified with soy flour, flakes, or grits. Table 8.9 lists these fortified cereal products, the form in which soy protein is added, and the proportions of the ingredients. Cornmeal and wheat flour are fortified not only with protein but also with vitamins and minerals. An emulsifier, SSL, is added to wheat flour to

Table 8.9

Soy-Fortified Cereal Products Distributed by United
States Overseas Food Assistance Programs[a]

Commodity	Soy component	Proportion
Corn meal	Flour	85:15
Bulgur	Grits	85:15
Rolled oats	Flakes	85:15
Wheat flour	Flour	94:6
Wheat flour	Flour	88:12
Sorghum grits	Grits	85:15

[a]Source: Senti (1974).

improve loaf volume and crumb texture of baked products, particularly at the 12% level of soy fortification.

REFERENCES

Anon. (1983). PAG/UNU Guideline No. 6: Preclinical testing of novel sources of food. *Food Nutrition Bull.* **5**(1): 60.

Berg, A. (1973). "The Nutrition Factor." The Brookings Institute, Washington, D.C.

Bookwalter, G. N. (1981). Requirements for foods containing soy protein in the Food for Peace Program. *J. Am. Oil Chem. Soc.* **58**:455.

Bookwalter, G. N., G. C. Mustakas, W. F. Kwolek, J. E. McGhee, and W. J. Albrecht (1971). Full-fat soy flour extrusion cooked: Properties and food uses. *J. Food Sci.* **36**:5.

Bressani, R. (1969). Formulation and testing of weaning and supplementary food containing oilseed proteins. *In* "Protein Enriched Cereal Foods for World Needs" (M. Milner, ed.). American Association of Cereal Chemists, St. Paul, MN.

Bressani, R., and L. G. Elias (1966). All-vegetable protein mixtures for human feeding. The development of INCAP vegetable mixture 14 based on soybean flour. *J. Food Sci.* **31**:626.

Cheigh, H. W. (1978). "Development of High Nutrition—Low Cost Supplementary Foods and Production System for Wonseung County Comprehensive Nutrition Program" Korea Institute of Science and Technology, Seoul.

Del Valle, F. R., E. Montmayor, and H. Bourges (1976). Industrial production of soy-enriched tortilla flours by lime cooking of whole raw corn-soybean mixtures. *J. Food Sci.* **41**:349.

DeMayer, E. M. (1969). FAO/WHO/UNICEF guidelines for safety evaluation and human testing of supplementary food mixtures. *In* "Protein-Enriched Cereal Foods for World Needs" (M. Milner, ed.). American Association of Cereal Chemists, St. Paul, MN.

Dendy, D. A. V., P. A. Clarke, and A. W. James (1970). The use of blends of wheat and nonwheat flours in breadmaking. *Trop. Sci.* **12**:131.

DeRuiter, D. (1978). Composite flours. *Adv. Cereal Sci. Technol.* **2**:349.

Fellers, D. A., D. K. Mecham, M. M. Mean, and M. M. Hanamoto (1976). Soy-fortified wheat flour breads. 1. Composition and properties. *Cereal Foods World* **21**:75.

Franz, K. (1975). Tortillas fortified with whole soybeans prepared by different methods. *J. Food Sci.* **40**:1275.

Hallab, A. H., H. A. Khatchadourian, and I. Jabr (1974). The nutritive value and organoleptic properties of white arabic bread supplemented with soybean and chickpea. *Cereal Chem.* **51**:106.

Harper, J. M., D. A. Cumings, J. D. Kellerby, R. E. Tribelhorn, G. R. Jansen, and J. A. Maga (1978). "Evaluation of Low-Cost Extrusion Cooker for Use in LDC's." Annual Report, Colorado State University, Fort Collins.

Hoover, W. (1979). Use of soy proteins in baked foods. *J. Am. Oil Chem. Soc.* **56**:301.

Jongh, G. (1961). The formation of dough and bread structures. 1. The ability of starch to form structures and the improving effect of glycerlmonostearate. *Cereal Chem.* **38**:140.

Kapsiotis, G. D. (1969). History and status of specific protein-rich foods. FAO/WHO/UNICEF Protein food programs and products. *In* "Protein-Enriched Cereal Foods for World Needs" (M. Milner, ed.). American Association of Cereal Chemists, St. Paul, MN.

Kim, J. C., and D. DeRuiter (1968). Bread from nonwheat flours. *Food Technol.* **22**:867.

Kim, J. C., and D. DeRuiter (1969). Bread from nonwheat flours. *In* "Protein-Enriched Cereal Foods for World Needs" (M. Milner, ed.). American Association of Cereal Chemists, St. Paul, MN.

Kwon, T. W., H. S. Cheigh, C. H. Ryu, J. S. Jo, Y. R. Pyun, and H. E. Snyder (1976). "Development of Composite Flour Using Local Resources and Use of the Flour to Produce High Nutrition–Low Cost Food Products." Korea Institute of Science and Technology, Seoul.

Liener, I. E. (1978). Nutritional value of food protein products. *In* "Soybeans: Chemistry and Technology," Vol. 1, Proteins, 2nd ed. (A. K. Smith and S. J. Circle, eds.). AVI Publ. Co., Westport, CT.

Pomeranz, Y., M. D. Shogren, and K. F. Finney (1969). Improving breadmaking properties with glycolipids. 1. Improving soy products with sucroesters. *Cereal Chem.* **46**:503.

Ranhotra, G. S., and R. J. Loewe (1974). Breadmaking characteristics of wheat flour fortified with various commercial soy protein products. *Cereal Chem.* **51**:629.

Rotsch, A. (1954). Chemical and baking technological investigations on artificial doughs. *Brot. Gebaeck* **8**:129.

Senti, F. R. (1969). Formulated cereal foods in the U.S. food for peace program. *In* "Protein-Enriched Cereal Foods for World Needs" (M. Milner, ed.). American Association of Cereal Chemists, St. Paul, MN.

Senti, F. R. (1974). Soy protein foods in U.S. assistance programs. *J. Am. Oil Chem. Soc.* **51**:138A.

Smith, O. B. (1969). History and status of specific protein-rich foods. Extrusion processed cereal foods. *In* "Protein-Enriched Cereal Foods for World Needs" (M. Milner, ed.). American Association of Cereal Chemists, St. Paul, MN.

Sterner, M. (1976). Exploration of potential for low-cost extrusion cooker—worldwide. *In* "Low-Cost Extrusion Cookers. International Workshop Proceedings." (J. M. Harper, G. R. Jansen, D. Wilson, and P. Stumf, eds.). Colorado State University, Fort Collins.

Tsen, C. C., and W. J. Hoover (1973). High-protein bread from wheat flour fortified with full-fat soy flour. *Cereal Chem.* **50**:7.

Tsen, C. C., E. P. Farrell, W. J. Hoover, and P. E. Crowley (1975). Extruded soy products from whole and dehulled soybeans cooked at various temperatures for bread and cookie fortification. *Cereal Foods World* **20**:413.

Wang, H. L., G. C. Mustakas, W. J. Wolf, L. C. Wang, C. W. Hesseltine, and E. B. Bagley (1979). "Soybeans as Human Food—Unprocessed and Simply Processed." *Util. Res. Rep.* **5**, USDA, NRRC, Peoria IL.

9

Soy Protein Food Products

In addition to the use of soybeans in traditional Oriental foods and in cereal blends for international use, both the protein and oil fractions are important food ingredients. We shall discuss in this chapter the use of the protein fraction and in Chapter 10 the use in foods of soybean oil.

In cereal blends, soybeans are useful for their nutritive qualities—particularly their protein quantity and quality. However, the uses of soybeans in Western foods are based on the functional properties of soybeans discussed in Chapter 5.

Table 9.1 reviews the functional properties of soy proteins and gives examples of food products making use of these functional properties. The greatest amounts of soy proteins used in Western foods are in baked goods.

BAKED GOODS

Attempts to introduce soy flour to the baking industry before World War II were an unfortunate experience for both the soy processing and the baking industries. Following the war, early mistakes of the processing industry were corrected, and new and improved procedures for processing soy flour for bakery uses were developed. A better understanding of the functional uses of soy flour in

Table 9.1

Functional Properties of Soy Proteins and Their Food Applications

Functional property	Mode of action	Major food system
Emulsification	Emulsion formation	Sausages, baked goods, whipped toppings, soups
	Emulsion stabilization	Sausages, soups
Fat absorption	Promotion: reduce cooking loss and improve dimensional stability	Sausages, meat patties, simulated meats
	Prevention	Doughnuts, pancakes
Water absorption	Promotion: improve doughs, extend shelf life of baked goods	Breads, confections
	Promotion: reduce cooking loss and develop texture in meats	Sausages, meat patties, simulated meats
Texture development	Thickening	Soups, gravies
	Gelation	Comminuted meat products
	Cohesion and adhesion	Sausages, simulated meats, baked goods
	Elasticity	Simulated meats, fruits, nuts
	Structural (chips, chunks, fibers)	
	Film formation	Special sausages
Other properties	Flavor absorption and entrapping	Simulated meats
	Color bleaching	Breads
	Browning	Breads, pancakes
	Foaming	Whipped toppings, chiffons

bakery products was developed, and now there is an ever-increasing number of successful uses for soy flour in the baking industry.

Although concentrates and isolates can be used in some baked goods, defatted and full-fat soy flours are the primary soy protein products used in baking. The flour is low in price and is often used at low levels for functional effects. At low levels, off-flavor is not a serious problem in the final products. Many types of soy flours are used in a wide range of baked goods at levels up to 5% of the weight of wheat flour, and for specialty products at levels up to 10% and more in some instances. The advantages of using soy flour in baked goods are increased mixing tolerance, easier machining, improved moisture retention, improved crust color, and longer shelf-life. At the same time, the need for nonfat dry milk and shortening is reduced.

In most bread formulations, soy flours replace a predetermined portion of wheat flour. In exactly balanced formulations where full-fat soy flour is to be replaced with defatted flour, 0.8 part of defatted flour is used per part of the full-fat product. A partial list of bakery products that use soy products includes all types of breads, bran muffins, coffee cake, devils' food cake, pound cake, pie crust, pancake mix, cookies, fruit bars, and doughnuts. Generally no formula changes are necessary when soy flour is used in such products. but there will be increased water absorption.

The soy industry supplies a wide range of products for the baking industry, and these products have a variety of functional uses in different bakery foods. Although the technology for soy use in bakery items is well established and reasonably simple, the functional properties and flavor are continually being improved.

Specifications for soy flours in bakery food applications are protein content, protein solubility, lipid content, urease and lipoxygenase activities, and particle size. Of these, protein solubility affects functionality the most. A lightly toasted soy flour has a PDI of 60–80, whereas a heavily toasted soy flour has a PDI of 10–20. The enzyme activity of soy flour is related to protein solubility. If lipoxygenase-active flour is desired, a relatively high PDI flour should be used.

A baking test is essential in determining the suitability of a soy product for a specific bakery application. The type of soy products available to the baking industry and a summary of bakery food applications are shown in Table 9.2.

Initially, defatted soy flours excessively softened bread doughs, which was corrected by the toasting process. However, excessive toasting decreased bread volume and produced inferior texture. Most flours for bread have a PDI of 50–75. Defatted soy flours are permitted in standardized bakery items at a maximal level of 3%, flour weight basis, in the United States (Anon. 1979). No maximum has been established for other bakery food items. Fat containing soy products (high-fat, full-fat, and lecithinated soy flours) are often used at 3–5% in heavier cakes, such as sponge and pound cakes, to give increased richness and emulsification. In addition, high-fat or lecithinated soy flour may decrease the need for

Table 9.2
Applications of Soy Products in Baked Goods[a]

| Soy product | Bread and rolls | | Cakes | Doughnuts | | Sweet goods | Cookies |
	White	Specialty		Cake	Yeast-raised		
Defatted soy flour	X	X	X	X	X	X	X
Enzyme-active soy flour	X						
Low-fat soy flour			X	X		X	
High-fat soy flour			X	X		X	
Full-fat soy flour			X	X		X	
Lecithinated soy flour			X	X			X
Soy grits		X					
Soy concentrates		X					
Soy isolates		X	X	X	X		
Soy fiber		X					

[a]Source: Dubois and Hoover (1981).

Table 9.3

Functions of Soy Flours in Bakery Products

Facilitate greater water incorporation
Improve dough handling and machineability
Improve moisture retention
Improve crust color development
Prolong freshness and storage stability
Bleach bread color to produce white bread
Retard fat absorption by doughnuts
Improve cake tenderness and crumb structure
Improve nutritional quality

eggs and shortening. Heavily toasted soy grits with PDIs of 20–30 are used at 2–4% in whole grain, multigrain, and natural grain breads to add color and improve flavor. The functions of soy flours in bakery products are summarized in Table 9.3.

Breads

Many of the white breads in the United States now contain 1.5–2 parts of soy flour per 100 parts of wheat flour to replace nonfat dry milk. Many nonfat milk replacers for baking applications have been developed. Some consist of cheese whey–soy flour blends and some contain soy protein isolate combined with casein. Whey protein gives desirable flavor and crust color and adds lysine. Table 9.4 shows a typical American bread formula.

Soy flour places an added stress on the gluten when added to bread dough. If too much is added, it will modify the structure of the bread and reduce loaf volume. High-protein wheat flours will tolerate a larger addition of soy flour than

Table 9.4

Typical American White Bread Formula

Ingredient	Amount by weight
Enriched flour	100
Water	65
Yeast	2.5
Yeast food[a]	0.5
Salt	2.25
Sugar	8
Shortening	3
Mono- and diglycerides	0.5
Nonfat dry milk or soy flour and dry whey	2
Calcium propionate	0.125

[a]Yeast food contributes 15 ppm bromate plus ammonium and calcium salts.

low-protein or weak flours. With bakers' patent flour, bromate in the range of 1–3 mg% results in satisfactory dough handling properties and normal loaf volume with 5% added soy flour (Ofelt *et al.* 1954). An increase of 1% water absorption is desirable for each 1% of soy flour added.

When high-extraction wheat flour (90%) is used in bread making (straight dough method) with defatted soy flour at the 5% level, the deleterious effects of soy flour on dough handling and loaf volume are practically eliminated. Optimum amounts of oxidizing agents give a bread equal or superior to the control loaf (Ofelt and Smith 1955).

European-type bread refers to bread produced from flour, yeast, salt, and water without nonfat dry milk, sugar, or shortening. In many European countries, bread improvers, such as bromate, SSL, CSL, and sucrose esters are not allowed or are not used. Jakubczyk and Haberowa (1974) reported that with wheat flour, rye flour, and mixtures of the two, water absorption, dough development time, dough stability, and dough gasing power were increased, while dough softening was decreased with 5% added soy flour. A shorter proofing time and more intensive mixing with the sponge dough process produced loaves only slightly smaller in volume than the control with 3% soy, and the bread was still acceptable at 5% added soy.

In general, increasing the protein in the soy flour leads to increased water absorption capacity, and increasing the water-soluble protein improves the grain and crumb color of the bread. The PER of bread containing 3% soy flour was 1.02, and for bread containing the same level of nonfat drymilk, PER was 0.97. Thus, at the 3% level, soy flour improved the protein quality of bread to the same extent as nonfat dry milk.

Soy flours are useful in controlling color in bakery products in two ways: (1) they may act as bleaching agents or (2) they may promote color formation in the final products. When added at a level of 0.5–1% to a bread dough made from unbleached wheat flour, the enzyme-active or unheated soy flour will bleach the color of the bread. This color improvement is effected by the lipoxygenase in the soy flour bleaching carotene in the wheat flour. British bakeries add 0.7% of the enzyme-active soy flour to about 90% of their bread to improve both color and flavor. Another enzyme present in raw soy flour, α-amylase, affects dough adversely by decreasing its viscosity. If 2–3% raw soy flour is used in bread, the resulting dough becomes slack or sticky. The crust color of bread is enhanced by the reactions between enzyme-inactive soy protein and wheat flour carbohydrates.

Doughnuts

Full-fat or lecithinated soy flour (5–15% lecithin) with a high NSI reduces fat absorption and enhances the eating quality of doughnuts and pancakes. Data for the effect of soy flour on fat absorption in doughnuts are shown in Table 9.5. Soy

Table 9.5
Control of Fat Absorption in Doughnuts with Soy Flour[a]

	NSI of soy flour	Fat absorption during frying (g fat/100 g dry mix)
Wheat flour mix	—	27.6
Wheat flour mix plus 4% soy flour[b]	60	22.8
Wheat flour mix plus 4% soy flour[b]	80	11.1

[a]Source: Wolf and Cowan (1975).
[b]Based on weight of wheat flour.

flours with a high NSI impart a beany flavor to doughnuts and are not used by bakers. Instead, a compromise is made between functionality and flavor, and flours with NSI values of 50–65 are used. The decreased fat absorption may be caused by heat denaturation of the proteins, forming a fat-resistant barrier at the doughnut surface, since a high-NSI flour performs better than a low-NSI flour.

MEAT PRODUCTS

When soy flours were first used in processed meats, abuse by many processors caused unfavorable customer reaction. This led to the enactment of meat regulations that either restricted or prohibited the use of soy proteins in meat processing. Despite these restrictions, the use of soy protein products increased in comminuted meat products. This growth can be attributed to the following:

(1) The soy protein products have been improved in functionality and eating qualities.

(2) Processors have learned the proper use of soy protein products.

(3) Economic benefits came from extending the supply of meat products without a loss in quality.

(4) Consumer interest in nutrition is satisfied by the protein quality of soy products.

Currently, the second largest use of soy proteins is in meat products as extenders and as functional ingredients. The traditional meat extender has been bread crumbs. The primary purpose of this addition was to lower the cost of the finished products, but at high levels, the bread crumbs cause undesirable texture in the meat products. This self-limiting aspect is also true for soy protein products in meats and is noticeable mainly in undesirable mouth feel and flavor.

Soy protein concentrates and isolates have found increasing usage in the meat industry because they have less off-flavor and because prices for nonfat dry milk are increasing. Both concentrates and isolates are high in protein, and their addition does not lower the protein content of finished meat products as do cereal products. Instead the appearance and eating qualities of the meat products with added soy products are improved.

Soy proteins are most successful as a partial replacement of the animal proteins without changing the traditional meat characteristics and quality. Experience has shown that quality must be maintained for consumer acceptance of added soy proteins.

A reformulation experiment illustrates this concept and the upper limits of soy protein application (Fig. 9.1). Waggle *et al.* (1981) experimented with a bologna containing lean beef, lean pork, and pork backfat reformulated with graded levels of two soy proteins, A and B. The fat and protein contents of the final products were equal to those of an all-meat control. Color of the reformulated products was maintained by using low levels of beef blood as a source of heme pigment.

A trained 16-member panel compared the soy products to the all-meat control for differences in color, flavor, odor, texture, and overall eating quality. A linear relationship between the amount of meat replacement and overall quality for the final products prepared from both the soy products A ($R = 0.997$) and B ($R = 0.996$) was found to exist. About 16% of the meat could be replaced with soy product A before the soy bologna became significantly different than the all-meat control, whereas only 9% of the meat could be replaced with soy product B before the quality difference became significant. Because of the color correction, the differences were primarily due to flavor and texture. These data clearly demonstrate that soy protein products are not the same in their ability to maintain quality.

Soy proteins are also useful in the meat industry in new products, where quality standards are not established, and soy protein products contribute to the overall appeal of the new products. The successful introduction of soy proteins into poultry meats is an example. Many new products, such as poultry rolls and poultry convenience foods, contain soy ingredients.

Meat is consumed in the form of comminuted products and intact muscle. Soy protein products are used in both forms. Emulsified meats (frankfurters, for example) and coarse ground meats (such as ground beef patties) are two important classes of comminuted products. Whole cuts of meat also can be augmented with soy protein products. A slurry of isolated soy protein with other ingredients can be injected into the muscle, or the slurry can be massaged into the muscle. The major functions of soy protein products in meats are shown in Table 9.6.

The use of textured soy proteins in the U.S. National School Lunch Program beginning in 1971 and the introduction of ground beef–soy blends in supermarkets in 1973 have contributed to increased usage of soy proteins. Both developments resulted from the increased cost of meats. The increases of soy protein products in meat applications are not limited to the United States but are found wherever the balance between quality and price of the meat products favors soy usage.

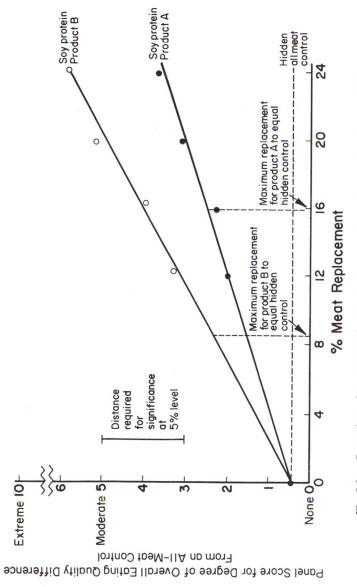

Fig. 9.1. Comparison of two soy protein products as meat replacers. Source: Waggle *et al.* (1981).

Table 9.6

Functions of Soy Proteins in Meat-Based Products

Improve uniform emulsion formation and stabilization
Reduce cooking shrinkage and drip
Prevent fat and water separations
Enhance binding of meat particles without stickiness
Improve moisture holding and mouthfeel
Improve firmness, pliability, and texture
Improve nutritional quality

Comminuted Meat Products

The most popular comminuted meat products are sausages. Next in popularity are the loaf products, many of which are canned, although in the United States coarse ground meat products such as hamburger patties and chili products are more popular than the loaf products. Soy flours, grits, concentrates, and isolates are used extensively in comminuted meat products. The use of soy protein products in comminuted meats is primarily functional, but in many meat products the soy protein level is high enough to have a substantial effect on nutritional value. Table 9.7 shows comminuted meat products in which soy protein products are used and gives the permitted level according to the present federal U.S. regulations.

Of the various types of soy flours available, the toasted products are preferred in cooked sausages, since the inherent enzymes in raw soy flours can cause problems in emulsion systems. Lecithinated or fatted soy flours have no advantage in meat systems.

Soy grits are used in sausage products but to a lesser degree than soy flour, and like soy flours toasted grits are preferred. Soy grits have greater utility in coarse ground meats such as hamburger than in sausage emulsions. In the United States

Table 9.7

Permitted Levels of Soy Protein Products for Meat Products in the United States[a]

Meat products	Soy products	Permitted levels (%)
Sausages, fresh and cooked	Flours, grits, concentrates	3.5
	Isolates	2.0
Chili con carne	All products	8.0
Spaghetti with meat balls, salisbury steak	All products	12.0
Imitation sausage, nonspecific loaves, soups, other related items	All products	Sufficient for the purpose

[a]Source: Wolf and Cowan (1975).

the name "hamburger" is not permitted for products with soy grits. Hamburger by definition is only meat; it may or may not contain seasoning. At present the only name permitted for hamburgerlike products is "pattie." Present usage of soy grits in ground meat patties is about 6%. Rakosky (1974) pointed out that the poor mouth feel in products containing soy flour is less of a problem in products containing soy grits.

Soy protein concentrates are used to a greater extent than grits in coarse ground meat products. The advantages over grits are less off-flavor and a higher protein content. Thus the soy product can be hydrated to a greater degree and used at higher levels. In supplementing ground meat to make a pattie product, additions of concentrates can be made to about 20% without flavor changes. In applications of this type, a coarse form of the concentrate is used rather than a fine flour.

The advantages in using soy products in patties are less shrinkage (about 10%), good size stability, better taste, and juicier products. For optimum results, the soy product is presoaked with water for a short time before it is added to the meat. Finely ground soy protein concentrates are used primarily in emulsion systems similar to soy flour, but because of their higher protein content and bland flavor, concentrates are preferred.

Isolated soy proteins are available in either the isoelectric (acid) form or as a sodium proteinate. They act as both emulsifiers and binders. In a frankfurter in which the red meat portion is high in myosin, there is little need for an extender. Therefore, use of 2% isolate is little more than insurance against processing defects. In formulations where myosin is lacking, the isolate has more utility. Using 2% soy protein isolate has the functional effect of using 10% meat. Isolates have greatest utility in nonspecific loaves, where there is no restriction on the amount of water used. The ranges of amounts of soy protein products usually used in meat products are shown in Table 9.8.

Textured soy protein should be handled like meat when preparing meat products. First, the lean and fat meat portions are cut into pieces. Next, textured soy,

Table 9.8
Ranges of Soy Protein Products in Meat Products

Meat product	Soy protein product	Range (%)
Breakfast sausages, links, strips	Isolate	3–9
Luncheon meats	Concentrates and isolates	1–6
Red meat patties, meat loaves	Concentrates, isolates, TVP	2–8
Pizza topping meats	Concentrates, isolates, TVP	1–17
Casserole dishes	Concentrates, TVP	1–5
Ham salads, paste	Concentrates, TVP	3–5
Poultry rolls, patties	Concentrates, isolates	2–7

water, and spices are weighed, and the spices or flavorings are added to the water used to hydrate the textured soy. All are mixed together thoroughly before coarse grinding. Then, the mixture is ground to the desired degree of texture before forming into patties, meat balls, loaves, or other products. To minimize the soy flavor, spices and seasonings should be added to the hydration water and the mixture allowed to remain at refrigerated temperature for flavor penetration into the soy.

Many of the problems of sausage making are due to the wide variation in the raw materials used. Variations are the results of age, diet, weight, sex of the animal, anatomical origin of the trimmings, and whether the raw materials are fresh or frozen. The proper use of soy proteins in the formulations can compensate for these variations.

The functionality of soy flours, concentrates, and isolates in comminuted meat depends to a degree on the amount of soluble protein. Soy protein isolate, which is very soluble at a pH of 6.9–7.2 and which is more than 90% protein, is the most effective and most expensive of the several soy products used. There is the possibility that a sausage emulsion will break, and the separated white fat will create an undesirable appearance, requiring that the product be reprocessed. Thus, it is important to use an emulsifier–binder as insurance against a processing failure. The soy protein supplements myosin and actomyosin as an emulsifying and encapsulating agent. Soy protein helps to prevent fat separation and to retain meat juices during cooking.

The dry sausage industry has traditionally used nonfat dry milk as a binder in its products, but the costs of milk proteins are rising. At maximum usage levels (3.5% nonfat dry milk vs. 2% soy protein isolate), a soy protein isolate costs less than nonfat dry milk. In Italian dry salami, the moisture:protein ratio is 1.9:1 and in pepperoni the ratio is 1.6:1. In formulating salami, a higher yield can be obtained at a moisture:protein ration of 1.9:1 with soy isolate than with nonfat dry milk.

A good example of both the functional and nutritional use of soy protein is in making meat patties with soy flour, grits, or concentrates. In making patties, water is needed at 2–3 times the weight of the soy protein. The primary function of the protein is to improve size stability of the patties and to decrease cooking losses during cooking. Patties correctly formulated with soy will be tastier, have higher protein and lower fat, and be better balanced nutritionally than without soy. As little as 2% soy grits in patties can improve tenderness scores (Huffman and Powell 1970).

In the usual method of making chili con carne, much of the fat is cooked out of the meat, but it can be retained by the addition of corn meal or soy protein. A soy product increases the protein content more than corn meal. If soy grits are used, they give a pleasant grainy texture. Other meat products that benefit from added soy protein are stews, scrapple, tamales, meat pies, and beef and pork barbecue sauce.

Whole Meat Cuts

Special isolated soy protein products and the refinement of cured meat technology have permitted use of the isolate in whole meat cuts. Cured meat products are prepared with solution containing soy protein isolate, sodium chloride, nitrite, ascorbate, phosphates, sugars, and flavorings. The protein must be thoroughly hydrated before other ingredients are added. A multineedle injector is used to pump the protein brine directly into the muscle tissue for faster and more even distribution. Using this technique, it is possible to incorporate soy protein isolate into cured meats such as hams, pork shoulders, pork loins, corned beef, and uncured products such as roast beef and poultry. The soy protein addition reduces product cost while increasing the supply of products.

Two protein brines for use in cooked hams are given in Table 9.9. The protein brine for 40% injection contains polyphosphates and the 30% brine does not. The amounts of soy protein that will be present in the final products with 40 and 30% levels are 2.4 and 1.5%, respectively, based on 94% cook yield. The amount injected varies depending on the products and can range from 15 to 50%. After injection, the meats are massaged or tumbled to facilitate diffusion of the salt and curing agents throughout the meat and to solubilize meat proteins for binding pieces of meat together during cooking. The meat pieces move horizontally in a massager and vertically in a tumbler. After massaging or tumbling, there may be a resting period. The treated meats are put in molds, stuffed into casings, or canned. These products are heat processed and smoked. Injection provides for fast distribution of the brine ingredients. This system can be used for large pieces of meat and provides a reliable cure throughout the muscle (Desmyter and Wagner 1979).

The injection, massaging–tumbling technique is also used for extending whole cuts of meat. In one method a nonprotein brine is injected into the meat cuts, the

Table 9.9

Typical Composition of Two Protein Brines
Designed for 40% and 30% Injection Level in
Hams[a]

Ingredient	Injection level	
	40%	30%
Water	79.6	80.3
Soy protein isolate	8.0	6.0
Sodium chloride	9.5	11.7
Polyphosphate	1.3	0
Sodium erythorbate	Normal level	
Sodium nitrite	Normal level	
Sugar	1.6	2.0

[a]Source: Desmyter and Wagner (1979).

meat placed into a massager or tumbler, and the soy protein isolate added in hydrated form. The soy dispersion is about 12.5% protein. Higher concentrations may result in high viscosity or a protein gel that would be difficult to absorb. As a result of the massaging–tumbling action, the added soy protein is absorbed.

Uncured roast beef also can be extended by an injection method. A brine of 10% soy protein isolate, 5% salt, and 85% water is added at a 30% level. After massaging, the product is stuffed into plastic containers and cooked to the desired temperature. Typically, a yield of 110% (based on original meat weight) will be obtained. The finished product is of excellent quality and comparable to the usual roast beef.

Poultry products, such as turkey breasts and poultry roasts, can be extended with protein isolate by the injection and massaging–tumbling methods.

Seafood Products

In comparison to usage in red meats, utilization of soy protein products in seafood is limited. However, textured soy protein products recently have been used as seafood extenders, usually in pink or off-white products.

The soy proteins are used in several ways. In one use, the textured material is hydrated and then mixed with ground or minced seafood and a matrix-forming material (cereal flours, gums, and spices or flavorings). The mix is extruded or molded into shapes, such as stick, shrimp, or fish shapes, and then battered, breaded, fried, and frozen.

Another use is directed toward canned seafood items such as tuna or salmon. Hydrated textured soy protein or analog products are used in preparing tuna salad or salmon patties. Soy protein isolate also is a useful ingredient for such applications. The integrity of fish cakes or patties is improved by incorporating isolate.

The water absorption and retention properties of the soy protein are used also to hold moisture in fish blocks, to bind fish pieces in minced fish blocks, and to retain moisture lost during processing. Frozen minced fish blocks containing soy protein are becoming popular with primary processors, as this commodity allows the incorporation of deboned fish flesh. Fresh filleted fish cuts or deboned fish are combined with soy protein concentrate or textured soy protein, water, and flavor enhancer, mixed, and formed into standard blocks. The blocks are frozen for storage or further processing. Table 9.10 shows formulations for such fish blocks. Further processing of formulated fish blocks would be identical to that of conventional blocks.

Fish muscle can be extended by injecting the muscle with a soy protein isolate slurry. Whole eviscerated fish, fish fillets, and frozen/tempered fillets have been treated in this way. The level of injection ranges from 10 to 25%, and the brine is usually made up by hydrating soy protein isolate (9–10%) before adding the salt and phosphate. Flavorings, such as smoke or lemon, may be added to the brine. A boneless or bacon model stitch pump is recommended for injecting. Care

Table 9.10

Fish Block Formulations[a]

Ingredient	%
Fish flesh	69
Textured soy flour	10
Water	20
Fish block flavor enhancer	1
Fish flesh	79
Soy protein concentrate	5
Water	15
Fish block flavor enhancer	1

[a]Source: Sipos *et al.* (1979).

should be taken not to tear or break the fish muscle. The fish products can be smoked, frozen, or further processed. The finished products are more tender and juicy with added soy isolate.

In Japan, a new seafood product consisting of fish paste and soy protein has been developed and marketed for industrial and institutional uses. The soy protein is isolate or concentrate used as a powder, as a paste prepared by mixing the soy products with water, or as an emulsifier paste prepared by mixing soy products with water and fat or oil. Because of inherent weakness in texture and flavor, the soy protein is usually below 30% based on fish paste. Fish paste also is widely used as an ingredient in ham, sausage, retorted foods, and other ready-to-eat food products.

Japanese fish sausage, which contains fish paste, has been successfully formulated with isolated soy proteins. The emulsion formation and stabilization properties of soy protein are effective at 7% level of the isolate. Fish sausage is popular in Japan, and flavor of the products can be varied depending upon the source of minced fish and spices used (Sipos *et al.* 1979).

Simulated Meat Products

Simulated meats, also called meat analogs, are products processed from vegetable proteins to resemble meat products in texture, flavor, and color. Methods of texturizing soy proteins are by spinning soy isolates into fibers, shaping, and flavoring, or by thermoplastic extrusion of soy flour or protein concentrates into different sizes and shapes to give a chewy product of the desired flavor. A third method is heat gelation of a dispersion of soy protein isolate under the proper conditions for forming sausagelike products.

Extruded soy flour products have a cost advantage over spun isolates, but extruded products contain residual flavors and oligosaccharides. Bacon-flavored textured products are sold in retail stores and are marketed to hotels, restaurants, and institutions. A bacon strip analog has been developed. It is made by binding

Table 9.11

Typical Formulation for the Bacon Strip Analog with Red and White Strips

Ingredient	Red strip (%)	White strip (%)
Spun fiber	18	1.5
Egg albumen	10	8.2
Tapioca starch	7.5	5.8
Water	42.5	42.3
Corn oil	6.7	25.7
Soy protein isolate	3.4	2.6
Carrageenan	0.5	0.2
Sodium caseinate	0	5.2
Colors, flavors, seasonings	11.4	8.5

alternating layers of red and uncolored soy fibers to simulate the lean and fat portions of bacon. After shaping, heat setting, and slicing, the product is frozen. For consumption, all that is needed is to heat the product. A typical formulation is given in Table 9.11.

Meat analog production is primarily for the foodservice industries. Scarcity of skilled help in the foodservice industry places a premium on convenience in food preparation. Meat analogs and prepared foods containing them meet this need for convenience.

Spun-Fiber Meat Analogs.   To simulate the texture of muscle in meat, fish, and poultry, it is helpful to incorporate spun fibers into analogs. Spun fibers also can be used to extend meat at levels up to 45%. Fabricated protein products containing spun fibers most closely simulate the structural characteristics of animal protein. Spun fibers are not needed in products that are not analogs or are not going to be used as meat extenders at levels in the 45% range.

There are many ingredients used in spun soy fiber products, and the soy fiber is only one ingredient though the most critical for texture. However, texture also is affected by the other ingredients and the processing techniques for the finished products. The other ingredients are binders, colors, nutrients, oils, emulsifiers, and water. Such ingredients are mixed and blended with fibers and vegetable oil at roughly 2 parts water, 1 part fiber, and ½ part vegetable oil by weight. The final blended mixtures go into dry-, frozen-, or canned-product lines. The dry-product line consists of extenders, such as bits, cubes, and granules. The frozen and canned lines are meat analogs.

The spun-fiber foods usually contain one- to two-thirds of their weight in monofilament fiber on a dry basis (Odell 1967). Although the composition of the final product can vary widely, a typical example contains approximately 40% protein fiber, 10% binder, 20% fat, and 30% flavor, color, and supplemental nutrients (Thulin and Kuramoto 1967). Another analysis on a dry basis was reported as 60% protein, 20% fat, 17% carbohydrates, and 3% ash. Such products have a moisture range of 50–70%.

The final simulated products usually are more tender than the real meat products and may be refrigerated, canned, frozen, or dried. Several products made from spun fibers are shown in Fig. 9.2. These products are boneless, cooked, and completely edible, but they have not been a success in the marketplace. Nutritional tests indicate that the PER of such analogs is about 90% of casein (Odell 1967).

Extruded Meat Analogs. Several meat analogs have been developed commercially by continuous extrusion under heat and pressure of a soy flour formulation. The process makes small chunks, which when hydrated have a chewy meatlike texture. The advantage of this method is the use of low-cost soy flour rather than expensive protein isolate. A premix of soy flour with added fat, flavoring, carbohydrates, coloring, and other necessary ingredients is made and extruded under proper conditions of moisture, temperature, time, and pressure.

Fig. 9.2. Spun-fiber meat analogs are available in a variety of tastes and textures. Courtesy of Worthington Foods, Inc.

Table 9.12

Ingredients for Soy Frankfurters and Soy Bologna Sausages

Ingredients	Soy frankfurters (g)	Soy bologna (g)
Water	596	640
Trisodium phosphate	24	16
Protein isolate	220	100
Fat with emulsifier	120	100
Hydrolyzed vegetable protein	24	24
Smoke flavor	8	12
Spices	8	8
Certified color	Optional	Optional

When the product leaves the extruder, it expands and is shaped by the die at the extruder outlet and by the speed of the revolving knife. Extruded meatlike products are also produced from protein concentrates.

Gelled Meat Analogs.   A meat analog can be obtained by adjusting concentration and pH of an aqueous protein system, shaping, and applying heat to the system to produce a chewy gel. Frank and Circle (1959) prepared meat analogs simulating bologna and frankfurters from isolated soy protein. In the process, isoelectric protein isolate was added to a hot solution of trisodium phosphate in a mixer. While mixing, fat, emulsifier, smoke flavor, spices, and color were added, and the pH was adjusted to 6.3. The mixture was stuffed into cellulose frankfurter casings, linked, tied, and steamed at 10–15 psig for about 10 min. With this heat treatment, the protein dispersion is changed from a viscous sol to a gel that binds the fat, water, and other components. Formulations for soy frankfurters and bologna are given in Table 9.12. The cooked frankfurters have 18.6% protein, 16.4% fat, and 57% moisture.

For inspection and regulatory purposes, it is necessary to know the amount of soy protein incorporated in meat products. This subject will be discussed in Chapter 11.

DAIRY PRODUCTS

In recent years, food manufacturers have developed dairylike products that contain vegetable fats and sodium caseinate. With the increasing cost of caseinates, attempts are made to use soy protein isolate to replace milk protein in such products. One stimulus for the development of nondairy products is religious groups seeking to conform to dietary laws. Also, dietary and ethnic constraints have become important in the formulation and marketing of these products. Furthermore, due to its excellent hydration capacity and ability to emulsify fats, soy protein is useful in simulated dairy products. Examples of dairylike products containing soy proteins are simulated milk products, infant formulas, simulated cheese products, whipped toppings, coffee whiteners, simulated yogurt products, and frozen desserts.

Simulated Milk Products

Simulated or imitation milks include the category of filled milk, a product made by replacing the butterfat in cow's milk with a vegetable oil. Filled milk cannot be marketed either interstate or within more than 30 states in the United States, and the marketing of any product containing milk or nonfat milk solids is governed by various state and federal dairy regulations. Imitation milks usually contain soy protein isolate, vegetable oil, corn syrup solids, vitamins, and minerals. Nondairy milks, now being marketed in several parts of the world, have the distinct advantage of digestibility by those groups with low tolerance for the lactose in cow's milk.

The milklike product prepared from soybeans with or without any added ingredients is soymilk. Soymilk is prepared in households and commercially. For home preparation whole soybeans are used, while both whole soybeans and full-fat soy flour can be used as the starting materials for commercial production of soymilk.

As already described in Chapter 7, soybeans have been converted into a number of foods in the Orient from prehistoric days. Soymilk is one of these basic products and can be consumed as is or used as the starting material for other processed foods such as soy curd.

An example of an imitation milk with soy protein has been produced commercially in Mexico as part of an industry–government nutrition improvement program (Del Valle 1982). The process consists of dispersion of an extruded full fat flour, sucrose, vegetable oil, and stabilizers in water followed by mixing, batch pasteurization [30 min at 90°C (194°F)], homogenization (2000 psi), cooling, addition of colorings and flavorings, and packing in Tetra-Pak cartons. Beverage flavorings are chocolate, vanilla, and strawberry. The product has been tailored to approximate cow's milk in fat content (4%). Since full-fat soy flour does not have enough fat, vegetable oil must be added. Stabilizers used are carboxymethyl cellulose and mono- and diglycerides.

It is reported that the imitation milk produced has good sensory quality, with little or no detectable soybean flavor, which is attributed to the use of extruded soy flour. Due to high shear forces and temperatures developed within the extruder, cotyledon cells are ruptured, releasing protein and oil. Thus, the protein is roasted in oil, and a pleasant toasted almond flavor develops. Also, the abrupt temperature and pressure drop as the product leaves the extruder causes many undesirable flavor components to be volatilized. The heat treatment applied is sufficient to inactivate lipoxygenase, thus contributing to flavor stability. Another important factor contributing to good sensory quality is adequate masking of soy flavor by added flavorings.

Simulated Cheese Products

A simulated cheese product can be prepared from soymilk by fermenting with *Streptococcus thermophilus*. Autoclaved soymilk is inoculated with the starter organism and incubated at 32°C (87°F) for 15 hr. The resulting product is converted into soy cheese by precipitating with calcium sulfate or food acid, cutting, cooking, and pressing. The final product remained fresh and elastic (Hang and Jackson 1967). Imitation cream cheese also can be prepared with soy protein isolate as the protein base.

Infant Formulas

Infant formulas based on soy protein are designed primarily for infants who are allergic to cow's milk. Such infant formulas containing soy protein have been available commercially for about 50 years. The early formulas contained full-fat soy flours, were dark in color, and had a beany flavor. The presence of soluble carbohydrates from the soybean were the cause of flatus in the infant and of foul-smelling stools. The technology that yielded the high-quality soy protein isolate of today enabled the infant formula industry to make great improvements over the first generation products made from soy flours.

In the 1970s, formulas based on soy protein isolate replaced the soy flour formulas. These hypoallergenic formulas normally are produced from soy protein isolate, corn syrup solids, sucrose, and vegetable oils and supplemented with vitamins and minerals. The protein content will be 2–2.5% when reconstituted. In some applications, soy protein isolates are used for their functional properties, and the nutritional quality of the protein is not of primary importance, but in dietary applications for infants it is of the utmost importance. The isolates usually have PERs of 1.8–1.9 compared to 2.5 for casein. If a PER of 2.5 is essential, one can supplement the soy protein isolates with up to 1% methionine. A well-processed product with methionine added and providing 1.6 g of protein per 100 kcal should be nutritionally equivalent in protein quality to cow's or human milk.

The latest formulas are milklike in color, are bland in taste, and yield normal stools. Formulas are available in powdered and liquid forms. The powder products require reconstitution with water. The two liquid forms are ready-to-feed and a concentrate. The latter requires dilution with an equal volume of water before feeding. For some infants, the formula is the sole source of nutrition for weeks or months of their lives. Consequently, it is imperative that the formula be adequate nutritionally, that it be microbiologically safe, and that it be free from toxic or antinutritional factors.

Table 9.13

Approximate Analysis of a Typical Soy Protein Infant Formula[a]

Approximate analysis (wt/liter)		Vitamins per liter	
Protein	20.0 g	Vitamin A	2,500 I.U.
Fat	36.0 g	Vitamin D	400 I.U.
Carbohydrate	68.0 g	Vitamin E	15 I.U.
Minerals	3.8 g	Vitamin C	55 mg
Calcium	0.70 g	Vitamin B_1	0.04 mg
Phosphorous	0.50 g	Vitamin B_2	0.60 mg
Sodium	0.30 g	Vitamin B_6	0.40 mg
Potassium	0.71 g	Niacin (mg equiv)	9.0 mg
Chloride	0.53 g	Folic acid	0.10 mg
Magnesium	50 mg	Vitamin B_{12}	3.0 μg
Iron	12 mg	Pantothenic acid	5.0 mg
Zinc	5.0 mg	Biotin	0.15 mg
Copper	0.5 mg	Vitamin K_1	0.15 mg
Manganese	0.20 mg		
Iodine	0.15 mg		
Water	901.6 g		
Calories per fluid oz.	20		
Calories per 100 ml	68		

[a]Isomil, Ross Laboratories, United States. Source: Thomson (1979).

Some governments have standards for infant formulas. Others use the standards of the Codex Committee on Foods for Special Dietary Use (Anon. 1976). In the United States, the standards of the FDA must be adhered to (Anon. 1977). The modern infant formula meets the requirements of the growing infant for protein, fat, carbohydrate, vitamins, and minerals. Table 9.13 shows the approximate analysis of a typical soy protein infant formula available in the United States.

Although the details of processing infant formulas are proprietary, it is likely that formulas are made by first preparing the aqueous portion containing protein, carbohydrate, and minerals and a fat portion with emulsifiers. The separated portions are mixed, homogenized, analyzed, and ingredient adjustments made if necessary. Ingredients that are most heat and oxygen sensitive, such as vitamin C and B complex, are added later with water to standardize the formula. The product is either heat treated and spray dried or filled into glass or metal containers and sterilized. The finished products are analyzed chemically and microbiologically to ensure product specifications are met.

The infant formula manufacturer is aware that toxic or antinutritional factors such as pesticides or heavy metals might enter the formula. These undesirable agents may be introduced with the protein, other ingredients, or process water. Therefore, all ingredients are subject to stringent analyses before acceptance.

In addition to hypoallergenic formulations, maternalized or humanized milks are appearing. The purpose of such milks is to provide nutrition to infants who can not receive maternal milk. In such milks, the aim is to reconstitute maternal milk with respect to lipids, proteins, lactose, minerals, and vitamins, and it is possible to use soy protein isolates for the protein portion.

Whipped Toppings

Soy proteins are surface active and will form a stable foam when whipped. Surface-active substances have hydrophobic and hydrophilic groups in the same molecule. Consequently, these molecules tend to concentrate at the interface between the two immiscible phases of oil and water, or solid and liquid, or liquid and gas. Through specific orientation and intermolecular bonding, stable three-dimensional networks develop that increase the stability of the dispersion. For gas–liquid dispersions, foams, the surface-active agents are called whipping or aerating agents.

Traditionally, whipped topping has been prepared by mechanically whipping dairy cream. This resulted in finished products of variable quality since composition of the cream varies. Because of this, gums and other stabilizers have been added to minimize the variation. Many nontraditional whipped toppings have been developed and are widely used in the United States. Nonfat dry milk, sodium caseinate, egg albumen, and gelatin have been used as a source of protein and vegetable oil as a source of fat. With these ingredients the quality of the topping is more uniform and higher overruns (volumes) can be achieved. These toppings have excellent flavor, are lower in fat content, and are more convenient than the traditional topping.

Today, isolated soy proteins are replacing milk proteins in the manufacture of aerosol, liquid, frozen, and frozen prewhipped toppings. The major purpose of protein in whipped toppings is to emulsify, to incorporate air during whipping, and to produce a protein–fat film that will hold the incorporated air until the product is consumed. For frozen, prewhipped toppings, the film must be stable enough to undergo several freeze–thaw cycles. In most formulations, 0.4–1.2% isolated soy protein is used. This is less than sodium caseinate concentrations because soy protein has higher viscosity and excellent emulsifying properties.

Enzyme-modified proteins are used for producing stable foams by whipping and can be prepared by several methods from a number of different starting materials (Gunther 1979). For example, soy protein isolate can be subjected to enzyme hydrolysis for 12–24 hr using pepsin, papain, ficin, trypsin, or bacterial proteases. These enzymes are most effective in producing proteins with maximum whipping functionality. Often more than one enzyme is used, with the several enzymes acting sequentially. The hydrolyzate is separated by centrifugation, concentrated, and spray dried.

In another method, defatted soy flakes are washed to remove soluble salts and sugars, and then hydrolyzed enzymatically without first isolating the soy protein. After enzymatic digestion, the purification and concentration are the same as already described. The resulting dried products are bland, light colored, soluble in hot or cold water, and functional over a wide pH range.

Depending upon the source and process, products have a protein content ranging from 50 to 85% and a whipping power ranging from equal to twice that of egg albumen. These products differ from egg albumen in not coagulating appreciably when heated.

Figure 9.3 compares the rate of whip in the first stage whip of a conventional corn syrup, sucrose confectionery frappe for three protein products. The enzyme-modified product (labeled soy albumen) whips faster and to a greater volume than egg albumen. Also the hydrolyzed soy protein does not exhibit an increase in density, which is characteristic of egg albumen in this system (Gunther 1979). Note the improvement in whipping achieved by enzymatic hydrolysis of soy protein.

The range of products using protein-based whipping agents includes sugar confectionery (nougat, fondant cream, marshmallow) biscuits, cookies, snack items, foam in soft drinks, specialty ice cream products, and other frozen confections.

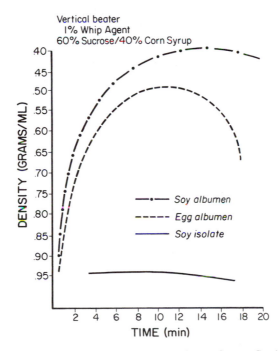

Fig. 9.3. Comparative whip rates in confectionary frappe. Source: Gunther (1979).

Coffee Whiteners

Some people regularly add milk or cream to coffee. The added dairy product lightens the color and moderates the acidity and bitterness of the coffee. Dairy product substitutes can be made that contain corn syrup, vegetable oil, protein (sodium caseinate or isolated soy protein), emulsifiers, and stabilizers. A typical formula for a liquid coffee creamer or whitener with soy protein isolate is given in Table 9.14. These products are popular because of their low cost and convenience. They are grouped into three categories: liquid, frozen, and dry. A whitener with superior lightening power must maintain a stable emulsion in coffee. The primary function of the protein is to emulsify the oil and to stabilize the emulsion in coffee. The emulsified oil globules are 0.2–0.4 μm in diameter, and this small size contributes to whitening power, body, and viscosity. The whitening is due to light scattering at the surface of the emulsified oil.

The method of processing will influence the performance of coffee whiteners. The finely divided oil globules are achieved by homogenization, and rapid cooling immediately after homogenization helps stabilize the emulsion.

For some purposes, it is desirable to have frozen rather than liquid coffee whiteners but frozen coffee whitener emulsions are difficult to stabilize. The frozen products may undergo several freeze–thaw cycles before being used, and the emulsion breaks easily under these conditions. Soy protein isolate at a concentration of 0.8–1% can stabilize the emulsion to the stress of freezing and thawing. Dry coffee whiteners undergo even more stress than frozen or liquid products due to dehydration and rehydration.

Coffee whiteners made with soy protein have whitening power equal to those with caseinate based on Hunter color difference meter L values (Kolar *et al.* 1979).

Simulated Yogurt Products

Yogurt is a popular item in the U.S. food market, and production of yogurt in various forms also has been increasing worldwide. Yogurts are usually eaten as

Table 9.14
Formulation for Liquid Coffee Whitener[a]

Ingredient	Weight (%)
Corn syrup solids	15
Vegetable oil	10
Soy protein isolate	0.8
Mono- and diglycerides	0.5
Sodium stearoyl-2-lactylate	0.2
Polysorbate 60	0.2
Dipotassium phosphate	0.2
Water	73

[a]Source: Kolar *et al.* (1979).

such but can be used as ingredients in dessert products and candy bars, and as a frozen soft-serve product.

Yogurt probably originated in the Balkans. It is produced by culturing milk or milk with reduced fat levels with *Lactobacillus bulgaricus* and *Streptococcus thermophilus*, but cultures can vary from country to country. Yogurt products have a high viscosity and generally set up in a weak gel. To enhance gel formation and to increase viscosity of the finished product, a high solids content is desirable. Often concentrated milk or nonfat dry milk is added to increase the solids content (4% nonfat dry milk added to low fat milk).

Stabilizers such as gelatin, carrageenan, and xanthan are used for improved product texture. Efforts have been made to use soy protein to replace such stabilizers and sodium caseinate. Isolated soy protein contributes to increased viscosity and gel strength. In addition, the isolate is effective in reducing syneresis of the yogurt and increases protein for nutrition. Experiments have shown that soy protein isolates are effective in improving gel strength and viscosity per unit of added protein when compared with nonfat dry milk or sodium caseinate (Kolar *et al.* 1979).

Frozen Desserts

Ice cream is a sensitive test system in which to evaluate the flavor and functional properties of soy protein. Sensory panel data have shown that properly processed soy protein isolate can be used in place of nonfat dry milk in making ice cream with no discernible differences (Wilding 1979).

OTHER FOODS CONTAINING SOY PROTEIN

There are many foods that contain small amounts of soy protein or that contain appreciable soy protein and are eaten in small amounts. Data on such products are presented in the following sections.

Snack Foods

Basic raw materials for snack items are corn meal, wheat flour, potato flour, oat flour, tapioca, milo, and some modified starches. These starchy products can be extruded into a wide variety of textures from light, fragile, highly puffed foods to dense crisp products. Soy flour or other soy protein products can be added to snack formulas up to about 15%. The moisture content will need to be adjusted with added soy protein. Added soy makes the dough pliable and easy to handle, darkens crust color, increases shelf-life, and upgrades the nutritive quality of the high starch products. Added soy also toughens the texture of starchy products.

Roasted soybeans are available as a snack food from several companies. They are normally deep fried in fat and salted to compete with peanuts.

Infant Foods

Examination of labels of products on supermarket shelves reveals that many infant foods contain soy protein in various forms. A high-protein cereal containing 35% protein has defatted soy flour as the major ingredient. Soy protein products such as soymilk, soy flours, soy concentrates, and isolates are used depending on the final products. Cookies, vegetables, and meat foods containing soy flour are available for infants.

Breakfast Cereals

Soy flour can be added to breakfast cereals to improve their amino acid balance and to increase their protein content. One oat flake cereal with added soy flour contains 18% protein. A number of high-protein cereals are on the market, but their protein quality often leaves much to be desired because of processing or protein composition of the raw materials. Four commercial oat cereals with soy flour added and with protein contents of 15.5–23% have PER values ranging from 2.18 to 2.79, whereas a rice cereal with 21.3% protein and a wheat cereal with 19.5% protein have PER values of only 0.4 and 0.19, respectively. Supplementing breakfast cereals with soy flours or concentrates offers potential for increased use of soy proteins.

Macaroni Products

Wheat flour is used in making macaroni and noodles and can be supplemented with soy flour or other soy protein products in the range of 12.5–25%. These products will retain additional moisture, and the color will darken somewhat. When soy flour is added to macaroni, U.S. regulations require a minimum of 12.5%. A combination of soy flour with semolina increased the firmness of spaghetti subjected to long cooking periods and improved its nutritive quality (Paulsen 1961).

Dietary Foods

These specialty items command relatively small shelf space in the supermarkets, but a number of them contain soy proteins. Cookies and candies containing soy flour are available. A low-cholesterol egg powder containing soy protein isolate is being marketed.

Other Products

Soy flours are used as emulsifiers in soups. In creamed soups, full-fat soy flour supplies additional fat in a fine dispersion that remains stable during canning, freezing, storage, and on subsequent reheating for serving. Soy flours also improve body of soups, gravies, and sauces by increasing their viscosities.

Soy flakes and grits have been used since the 1930s as adjuncts in brewing beer. Soy flours or flakes having a high PDI are used in the normal mashing operation to provide nutrients for the yeast, and addition of hydrolyzed soy protein directly to the beer improves foam stability, flavor, and body.

Soy protein products have applications in candy and confectionery products, simulated nutmeats, spreads similar to peanut butter, high-protein puddings, and as foam-mat drying adjuncts in producing orange powders.

Soy proteins can be acid hydrolyzed to produce meaty flavors. The hydrolysis with hydrochloric acid proceeds until 35–58% of the protein of a soy concentrate is converted to amino nitrogen. The hydrolyzates are neutralized to pH 4.5–7 with sodium hydroxide and spray dried.

Other uses for soy proteins in foods are as carriers for artificial spices, reduction of stickiness in manufacture of confectionery, and application to baking pans to aid browning of breads and cakes where they contact the pans.

REFERENCES

Anon. (1976). "Report of 11th meeting of Codex Alimentarius, FAO, and WHO." FAO, Rome.

Anon. (1977). Code of Federal Regulations, Title 21, Section 105.65. Washington, D.C.

Anon. (1979). Code of Federal Regulations, Title 21, Section 136.110(c). Washington, D.C.

Del Valle, F. R. (1982). Industry–government nutritional improvement programs. *Food Technol.* **36**:120.

Desmyter, E. A., and T. J. Wagner (1979). Utilization of vegetable proteins in meats of large cross sectional area. *J. Am. Oil Chem. Soc.* **56**:334.

Dubois, D. K., and W. J. Hoover (1981). Soy protein products in cereal grain foods. *J. Am. Oil Chem. Soc.* **58**:343.

Frank, S. S., and S. J. Circle (1959). The use of isolated soybean protein for non-meat simulated sausage products, frankfurter and bologna types. *Food Technol.* **13**:307.

Gunther, R. C. (1979). Chemistry and characteristics of enzyme-modified whipping proteins. *J. Am. Oil Chem. Soc.* **56**:345.

Hang, Y. D., and H. Jackson (1967). Preparation of soybean cheese using lactic starter organism. 1. General characteristics of the finished cheese. *Food Technol.* **21**:1033.

Huffman, D. L., and W. E. Powell (1970). Fat content and soy level effect on the tenderness of ground beef patties. *Food Technol.* **24**:1418.

Jakubczyk, T., and H. Haberowa (1974). Soy flour in European-type bread. *J. Am. Oil Chem. Soc.* **51**:120A.

Kolar, C. W., I. C. Cho, and W. L. Watrous (1979). Vegetable protein application in yogurt, coffee whiteners and whip toppings. *J. Am. Oil Chem. Soc.* **56**:389.

Odell, A. D. (1967). Meat analogs from modified vegetable tissue. *Proc. Int. Conf. Soybean Protein Foods, USDA-ARS* **71-35**:163.

Ofelt, C. W., and A. K. Smith (1955). Importance of oxidation and the use of soy flour with high extraction wheat flours. *Trans. Am. Assoc. Cereal Chemists* **23:**122.

Ofelt, C. W., A. K. Smith, and R. E. Derges (1954). Baking behavior and oxidation requirements of soy flour. I. Commercial full fat soy flours. *Cereal Chem.* **31:**15.

Paulsen, T. M. (1961). A study of macaroni products containing soy flour. Food Technol. 15:118.

Rakosky, J., Jr. (1974). Soy grits, flour, concentrates, and isolates in meat products. *J. Am. Oil Chem. Soc.* **51:**123A.

Sipos, E. F., J. G. Endres, P. T. Tybor, and Y. Nakajima (1979). Use of vegetable protein in processed seafood products. *J. Am. Oil Chem. Soc.* **56:**320.

Thomson, W. A. B. (1979). Infant formulas and the use of vegetable protein. *J. Am. Oil Chem. Soc.* **56:**386.

Thulin, W. W., and S. Kuramoto (1967). Bontrae—A new meat-like ingredient for convenience foods. *Food Technol.* **21:**168.

Waggle, K. H., C. D. Decker, and C. W. Kolar (1981). Soya products in meat, poultry and seafood. *J. Am. Oil Chem. Soc.* **58:**341.

Wilding, M. D. (1979). Vegetable protein application in whey soy drink mix and ice cream. *J. Am. Oil Chem. Soc.* **56:**392.

Wolf, W. J., and J. C. Cowan (1975). "Soybeans as a Food Source," rev. ed. CRC Press, Cleveland, OH.

10

Soybean Oil Food Products

In the early 1940s, soybean oil was not considered a quality edible oil. Since that time, research, technology, and production skills have made soybean oil a resounding success and the major edible oil of the United States. Furthermore, its popularity continues to increase worldwide.

The first processing of soybeans into oil and meal in the United States was in 1911 using hydraulic pressing on beans imported from Manchuria. After many unsuccessful attempts by various companies to press the oil from soybeans, an effective expeller plant was established in 1922 by A. E. Staley Manufacturing Co. in Decatur, Illinois. Solvent extraction of soybean oil was also started in the 1920s. In 1945, soybean oil surpassed cottonseed oil in production to become the leading edible oil in the United States, a position it has maintained. In part, increased consumption of soybean oil was due to the exigencies of World War II shortages in the United States and in part due to low prices.

Despite its expanding market, the flavor of soybean oil was singled out in 1945 as the principal problem of the soybean industry. Overcoming the flavor problem as described in Chapter 4 was a major factor in the continued success of soybean oil utilization in food products.

Today, soybean oil is not only the leading edible oil in the United States but also in Western Europe and in many other countries. Soybean oil is used in the edible oil products shown in Table 10.1. The oil is used in numerous other food

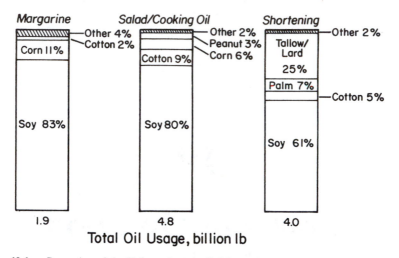

Total Oil Usage, billion lb

Fig. 10.1.   Proportion of the U.S. market supplied by soybean oil in 1978 for production of margarines, shortenings, and salad/cooking oils. Source: Brekke (1980a).

applications, and sometimes in very inconspicuous ways such as spraying pieces of refrigerated dough so they will separate readily. However, the major use of soybean oil is in salad and cooking oils, shortenings, and margarines. As shown in Fig. 10.1, soybean oil supplied 80, 61, and 83% of the total lipid required for the 1978 production of salad and cooking oils, shortenings, and margarines, respectively (Brekke 1980a).

These products are widely used by households, by foodservice industries, and by food-processing industries. They are important in bakery and confectionery items, fried snack foods (potato and corn chips), frozen fried foods, canned and dehydrated foods, imitation dairy products, and processed meat products.

Table 10.1

Edible Oil Products in Which Soybean Oil is Predominant

Product	Oil treatment[a]	Oil content (%)
Salad and cooking oils	PH, W, D	100
Mayonnaises	PH, W, D, B	77–82
Imitation mayonnaises		14–40
Salad dressings	PH, W, D, B	35–65
Imitation dressings		2.5–11
Shortenings	PH, D, S	100
Margarines	PH, D, S	80
Imitation margarines		40
Vegetable spreads		60

[a]PH, partially hydrogenated; W, winterized; D, deodorized; S, solidified; B, bleached.

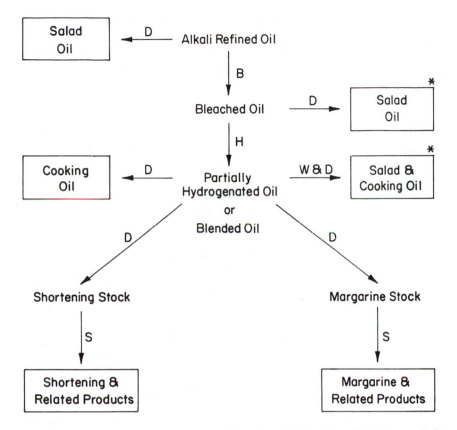

Fig. 10.2. Processing of edible soybean oil products. B, Bleaching; H, hydrogenated; D, deodorization; W, winterization; S, solidification; *, also used to manufacture salad dressing, mayonnaise, shortening, and margarine products.

In some oil-containing food products, soybean oil is the sole edible oil, and in others it is combined with one or more other oils and fats. Soybean oils, depending on the intended use, often contain additives, such as metal chelators, antioxidants, emulsifiers, coloring and flavoring agents, crystallization inhibitors, and antifoaming agents. These additives improve the stability and the functionality of the oils. In the United States the additives must be either generally recognized as safe (GRAS) or have prior approval by FDA.

Figure 10.2 outlines the various processing steps used to manufacture edible oil products from alkali-refined soybean oil. Salad oil, which does not crystallize on chilling, is distinguished from unwinterized oil or cooking oil. Salad oil may be obtained directly from alkali-refined oil or from bleached oil after deodorization. Cooking oil may be obtained from partially hydrogenated oil after deodorization, but usually salad and cooking oil are obtained from partially hydrogenated oil after winterization and deodorization. The partially hydrogenated oil alone or blended with other vegetable oils or animal fats is used for shortening and margarine after deodorization and solidification.

Blends are used because they have improved mouth feel, plasticity range, and heat stability that are not readily attained with any single oil or fat. Tub margarines, for example, are blends of a liquid oil and a hydrogenated oil. Blending usually is done just before deodorization but may be done afterwards. It is now possible to design an integrated base stock program in which a limited number of oil and fat base stocks are used in shortening and margarine formulations. This program simplifies scheduling of overall operations (Latondress 1981).

SALAD AND COOKING OILS, MAYONNAISE, AND PREPARED SALAD DRESSINGS

Salad and Cooking Oils

The processed edible oil products that remain liquid at room temperature have a number of different names: salad oil, salad and cooking oil, cooking oil, frying oil, cooking and frying oil, baking oil, or table oil. We shall use only the terms "salad oil" and "salad and cooking oil" to avoid some of the confusion. Salad oil should remain clear at low temperatures without turbidity due to the solidification of saturated fat. Therefore, a salad oil prepared from unhydrogenated soybean oil with a relatively high iodine value may perform well as a salad oil but may not be suitable for cooking because of less stability at high temperatures. A salad oil prepared from hydrogenated and winterized soybean oil with a relatively low iodine value is reasonably stable even at cooking temperatures and may serve well as a cooking oil.

Salad oil may be prepared directly from alkali-refined oils by deodorization (Fig. 10.2). Without bleaching, this oil meets the U.S. government specifications for shipment under the Public Law 480 Food for Peace Program.

A salad oil that has been alkali refined, bleached, and deodorized is referred to as an RBD oil. Bleaching usually improves color and flavor stability of an oil. Supporting data for this observation based on taste panel and other analyses are given in Table 10.2. This bleached oil also can be used to produce mayonnaise, prepared salad dressing, and margarine, or as an ingredient in bread and cakes.

Partially hydrogenated, winterized soybean oil (HWSB) is the most popular dual-purpose oil sold to consumers today as a salad and cooking oil. The alkali-refined and bleached oil is hydrogenated to an iodine value of about 104 and winterized to remove fats that crystallize at 0°C (32°F). The crystals (vegetable stearine) are removed by plate and frame filter presses, and the oil is deodorized. When properly stabilized, the HWSB oil is an improved product compared to unhydrogenated soybean oil. HWSB oil is also used in the preparation of fluid margarines and shortenings.

Table 10.2

Effect of Bleaching on Flavor Stability of Alkali-Refined and
Citric Acid Treated Soybean Oils[a]

	Bleached	Unbleached
Initial flavor score[b]	7.9	6.7
Aged flavor score	6.4	5.2
Increase in PV on aging	1.0	3.5
Increase in PV by AOM[c]	2.2	20.0

[a]Source: Wolf and Cowan (1975).

[b]Based on a flavor scale of 1 to 10 (highest quality).

[c]Active-oxygen method: 8 hr at 100°C with bubbling O_2.

The typical fatty acid compositions and other data on RBD salad oil and
HWSB salad and cooking oil from soybeans are given in Table 10.3. Among the
salad and cooking oils now marketed in the United States and throughout the
world, those of soybean origin are most popular. In the United States and other
countries, most brands contain a single type of oil, but blends of two types are
available. Blending soybean oil with an oil containing little or no linolenic acid,
such as cottonseed oil or sunflower oil, will lower the linolenic acid content.
Partial and selective hydrogenation of soybean oil to lower its linolenic acid
content to about 3% (HWSB oil) improves the flavor and oxidative stability of
the oil. Packaging the oil under nitrogen and minimizing light exposure also are
important in maintaining quality during storage. Some salad oils contain oxy-
stearin or polyglycerol esters as crystallization inhibitors (Krishnamurthy 1982),
and some dual-purpose salad and cooking oils contain a silicone compound as an
antifoam agent. Both RBD and HWSB oils will remain clear after immersion for
5.5 hr in an ice bath held at 0°C (32°F) and thus meet the criteria for a salad oil
(Brekke 1980a). Well winterized oils remain clear for periods much longer than
5.5 hr.

Table 10.3

Some Data on Representative Soybean Salad and Cooking Oils[a]

	Salad oil	Salad and cooking oil
Iodine value	131	109
Cold test	Pass	Pass
P/S ratio	4	2.3
Palmitic acid (%)	10	10
Stearic acid (%)	5	4
Oleic acid (%)	26	47
Linoleic acid (%)	52	36
Linolenic acid (%)	8	3

[a]Source: Wolf and Cowan (1975).

A product derived from soybean oil is extensively used for deep fat frying of foods (Wolf and Cowan 1975). This cooking oil is prepared by hydrogenating to an IV of about 95–100 to minimize linolenic acid and still maintain fluidity at room temperature. This cooking oil may be cloudy but will pour readily.

Soybean oils without hydrogenation or partially hydrogenated are widely used in canned foods (tomato or chicken soups, etc.), pudding mixes, pancake and waffle mixes, macaroni and cheese mixes, spaghetti sauces, dry breakfast cereals, frozen fried seafoods, meat pattie extenders, pizza mixes, and others. Partially hydrogenated soybean oil is also used in some whipped toppings, filled evaporated milks, and imitation cheeses.

Mayonnaise

Mayonnaise is defined under the U.S. FDA Standards of Identity as an emulsified semisolid food prepared from edible vegetable oil (not less than 65% by weight), acetic or citric acid (not less than 2.5%), and egg yolk. Optional ingredients permitted are salt, natural sweeteners, spices or spice oils, monosodium glutamate, and any suitable flavor from natural sources. These ingredients have limitations imposed on them. For example, the added seasonings cannot simulate the color of added egg yolk. EDTA salts are permitted as metal chelators at levels up to 75 ppm to protect the oil from oxidizing or reverting in flavor and to protect the mayonnaise from loss in color.

Mayonnaise may be whipped with and packed under nitrogen or carbon dioxide. Nitrogen is preferred and addition of nitrogen can lower the specific gravity from 0.94 to 0.88–0.90. Since mayonnaise is sold by volume rather than weight, lower specific gravities result in higher yields.

Commercial mayonnaise usually fits into the formula range given in Table

Table 10.4

Typical Example of Commercial Mayonnaise Composition[a]

Ingredient	Weight (%)
Salad oil	77–82
Fluid egg yolk[b]	5.3–5.8
Vinegar	2.8–4.5
Salt	1.2–1.8
Sugar	1.0–2.5
Mustard flour[c]	0.2–0.8
Oleoresin, paprika, and other ingredients	Optional
Water to make 100%	

[a]Source: Weiss (1983, p. 211).

[b]Egg solids 43%. May substitute whole or fortified egg, fluid or dry, on a total solid basis.

[c]Spice oils or oleoresins may be substituted.

10.4. The quantity of each ingredient is carefully controlled by the manufacturer to achieve its specific function and to give the desired characteristics. The desired body, viscosity, and texture are achieved through the proportion of oil to egg and through the operation of the emulsifying equipment available. Salt, sugar, vinegar, and spices are used to achieve the desired flavor, but flavor is also influenced by the emulsion. A strong emulsion results in less flavor, whereas a weak emulsion emphasizes the sweetness, tartness, and saltiness and makes a poorly balanced flavor apparent.

Salad oils are used in making mayonnaise. Until recently, cottonseed oil was preferred because of its flavor stability. However, the cottonseed oil forms weak emulsions that may break at refrigerator temperatures or from mechanical shock. RBD soybean oil forms strong emulsions and is now the choice of most manufacturers due to improved flavor stability, low cost, and ready availability. Any oil that crystallizes at refrigerator temperatures will break the mayonnaise emulsion and is avoided for that reason.

Despite the high proportion of oil in mayonnaise, it is an oil-in-water emulsion. Oil is dispersed in a continuous aqueous phase. The texture of the emulsion depends on the size of the oil droplets and how closely they are aligned. The viscosity of the emulsion increases as the amount of oil increases. The Standards of Identity require a minimum of 65% oil, but the resulting product would be too thin for most consumers. Hence, the 77–82% oil content (Table 10.4) is used by most commercial manufacturers. A thicker product containing 80–84% oil is often used as a heavy-bodied mayonnaise in institutional food preparations. It has the advantages of not soaking into bread so readily and staying on salads rather than flowing.

Imitation mayonnaises are manufactured in the United States, and they contain 14–40% oil and considerably more water than regular mayonnaise. Starch is used as a thickener in such products.

Mayonnaise is usually packed in glass containers with minimum head space for protection from oxidation. Once the container is opened, the mayonnaise will rapidly oxidize unless it is stored in the refrigerator. Polyethylene jars can be used for mayonnaise, but they are permeable to oxygen and only provide a shelf-life of about two months. Therefore, the use of polyethylene is restricted to institutional products with a high turnover rate.

Prepared Salad Dressings

The prepared salad dressings may be classified as spoonable and pourable, and some of them are covered by the Standards of Identity, while some are nonstandard dressings. The spoonable or pourable classification is based on the consistency or fluidity of the dressings. For example, prepared dressings such as Thousand Island, Russian, or Italian can be either spoonable or pourable, but French dressing is defined to be pourable by the Standards of Identity. However, nonstandard dressings could be in any form with any suitable name. Soybean oil is now used almost exclusively in all prepared salad dressings in the United States.

Spoonable salad dressing initially was a low-cost substitute for mayonnaise. Prepared dressings such as Russian and Thousand Island at one time used mayonnaise as the base and included one or more additional ingredients. However, spoonable dressings are now accepted as different products with distinct characteristics. The choice between mayonnaise and spoonable salad dressing is one of personal taste rather than of quality or cost. Spoonable salad dressings under the Standards of Identity resemble mayonnaise in that they are oil-in-water emulsions using egg as an emulsifier and starch as a thickener. The Standards of Identity require at least 30% vegetable oil and at least 4% liquid egg yolk by weight. The oil, egg, vinegar, and other ingredients are described by the Standards of Identity in identical language for salad dressings and for mayonnaise. There is, however, no limitation on the level of total acid in vinegar or citrus juice used as an ingredient for salad dressing.

While the oil is the major source of mayonnaise viscosity and body, starch has the same function in salad dressing. Oil modifies the mouth feel of the starch paste making it smooth and rich. Most spoonable salad dressings are formulated within the ranges given in Table 10.5. As with mayonnaise, each of the components of salad dressing has a specific function. Many of the ingredients are common to both products.

Pourable salad dressings, of which French dressing is an example, contain oil, vinegar, lemon or lime juice, and other ingredients. These dressings may be of the emulsified type, or they may require shaking before use to mix the oil and

Table 10.5

Typical Formulation of Spoonable Salad Dressing[a]

Ingredient	Weight (%)
Salad oil	35–50
Fluid egg yolk[b]	4.0–4.5
Salt	1.5–2.5
Sugar	9.0–12.0
Mustard flour	0.2–0.8
Starch	5.0–6.5
Vinegar	9–14
Spices[c]	Optional
Water to make 100%	

[a]Source: Weiss (1983, p. 211).

[b]Egg solids 43%. May substitute whole or fortified egg, fluid or dry, on a total solids basis.

[c]Spice oils or oleoresins may be substituted.

aqueous layers. Various types of gums as well as egg yolk are used as emulsifiers.

French dressing is manufactured under the Standards of Identity, while other pourable salad dressings are not covered by such standards (Weiss 1983, p. 241). French dressing is described as liquid or emulsified viscous food prepared from not less than 35% vegetable oil by weight and acetic or citric acid. Although 35% is the minimum oil level, 55–65% is more usual. This level of oil improves the mouth feel, especially when emulsified. French dressing has a reddish color supplied by paprika. EDTA salts are added to help stabilize the oil and the color. Garlic is one of the major flavorings. Sugar, salt, and vinegar levels are a matter of taste and are determined by each processor, but sufficient vinegar is needed for preservation.

Many emulsified or pourable dressings, spoonable dressings, and heat-stable dressings are available as imitation or low-calorie products, which are not covered by the Standards of Identity. The nonstandard pourable dressings resemble French dressing in many respects but depart from the standards mainly in the amount of oil used. Low-calorie emulsified pourable dressings may contain 2.5–11% oil. Viscosity is achieved by increasing the gum content, but the satisfying mouth feel of regular dressings is sacrificed. Some of these dressings are sold as Italian, Russian, or blue cheese, or by other exotic names to describe flavors and to attract consumers.

SHORTENINGS

Shortening refers to the change in texture due to incorporation of fat into baked goods. A short product is one that breaks easily. For example, shortening as lard made tender, flaky pie crusts. Now the term "shortening" includes a cake-baking ingredient as well as frying fats and is used to differentiate shortenings from other oil-derived products (margarine, mayonnaise, salad oil, etc.) that contain various nonfat materials. Thus, shortenings may be described as edible fats used in baked goods, icings, and fillings, and in frying.

Lard and butter have a long history as shortenings, and the first manufactured shortenings were an attempt to duplicate the plasticity and performance of lard. Manufactured shortenings vary both in physical state and performance, through proper selection of vegetable oils and animal fats, controlled hydrogenation, appropriate blending, plasticizing, and tempering, and through the incorporation of plasticizers, emulsifiers, and antifoams as needed. The various forms include conventional plastic shortenings, pourable shortenings, and dry (powdered and pelletized) shortenings. Shortenings may be classified in several ways depending upon the criteria used (Table 10.6).

Soybean oil is excellent for the base stock for a number of shortenings. Since soybean oil normally contains only about 11% palmitic acid, and since palmitic

Table 10.6
Classification of Shortenings[a]

Classification basis	Type
Physical state	Plastic
	Pourable (fluid and liquid)
	Dry (powdered and pelleted)
Functional usage	All purpose
	High stability
End usage	Baking
	Frying
	Household
Package form	Bulk
	Cubed
	Sheeted
	Printed (sticks and bricks)

[a]Source: Brekke (1980a).

acid is important in generating β′ crystal structure, soybean oil needs to be combined with other oils and fats for shortening manufacture. Cottonseed oil contains 20–23% palmitic acid, lard 20–28%, and tallow 24–37%. Palmitic esters melt at lower temperatures than stearate esters; hence wider plasticity ranges are achieved with mixtures of soybean oil and other fats than with only soybean oil. Wide plasticity ranges also can be obtained with mixtures of soybean oil and hydrogenated hard fat. Soybean hard fat is used in pourable shortenings and in plastic shortenings in amounts up to 50% of the total oil.

Plastic Shortenings

Solid or plastic shortenings are the most widely used and most varied of the shortening types. They can be formulated in many ways based on raw materials and end usage. Fully hydrogenated fats are used to extend the plasticity range in shortening formulations. The widest plastic range is obtained by chilling liquid oil with hard fat. Shortenings formulated by blending soybean oil that was selectively hydrogenated to an IV of 70 with 5% hard fat would have high stability but a narrow plasticity range. All-purpose shortenings prepared from a blend of partially hydrogenated soybean oil (88 IV) and 10–15% hard fat would have moderate plasticity. The SFI values, defined as percentage solid at given temperatures, for shortenings with different plastic ranges are given in Table 10.7. The SFI values correlate positively with firmness of a shortening since consistency increases as the amount of solid fat in a product increases.

Shortenings are plastic and workable for SFI values in the range of 15–22. Above a SFI of 22, shortenings are brittle, and below 15 they are too fluid. Hard fats affect SFI values of shortening oils more at high temperatures than at low

Table 10.7

Properties of All-Purpose and High-Stability Shortenings Made from All Hydrogenated Vegetable Oils and Blends with Animal Fats[a]

Property	High-stability shortening		All-purpose shortening	
	All vegetable	Blend with animal fats	All vegetable	Blend with animal fats
Iodine value	69	58	91	60
Linoleic acid (%)	1.5	3.5	30	8
SFI 10°C	40	39	16	29
21.1°C	27	26	15	22
26.7°C	22	22	14	20
33.3°C	11	14	13	17
40°C	5	9	11	13
Melting point (°C)	42.8	46.1	50.6	—

[a]Source: Wolf and Cowan (1975).

temperatures. A general rule is that increases in SFI are slightly greater than the percentage of added hard fat at 40°C (104°F), equal at 33.3°C (92°F), and about 60% of the added amount at 10°C (40°F) (Weiss 1983, p. 122).

Shortenings are formulated from one or more partially hydrogenated vegetable oil basestocks or from mixtures of such basestocks with animal fat basestocks. Optional ingredients include emulsifiers, antioxidants, metal chelators, antifoams, colorings, and flavors.

The crystal structure of shortenings is important. The β' crystals are desirable in plastic shortenings because they are small (no grainy mouth feel), and they tightly knit the shortening structure. Such shortenings provide optimal aeration for cakes and icings. Shortenings with α crystals (large crystals giving a grainy texture) are poor aerators but function well in making pie crusts.

Table 10.8

Composition of Representative Samples of Plastic Shortening of High and Low Polyunsaturated Fatty Acids[a]

	High polyunsaturated		Low polyunsaturated	
	Oil 1	Oil 2	Oil 1	Oil 2
Iodine value	95	86	76	76
Palmitic acid (%)	13	17	13	15
Stearic acid (%)	11	11	13	10
Oleic acid (%)	45	41	62	62
Linoleic acid (%)	27	27	13	12
Linolenic acid (%)	3	2	0	1
P/S ratio	1	1	0.4	0.3

[a]Source: Wolf and Cowan (1975).

The basic steps in shortening manufacture (Brekke 1980a) are (1) preparing the basestocks and hard fats; (2) formulating the fat blend with other ingredients; (3) solidifying and plasticizing the blend; (4) packaging; and (5) tempering the shortening, generally only for plastic shortenings for baking. Tempering consists of holding the shortening for 24–72 hr at 27–32°C (80–90°F). Container size and type of shortening will determine the holding time. The purpose of tempering is to stabilize the crystal structure, which continues to develop slowly during tempering.

Fatty acid compositions and some other data on typical plastic shortenings are given in Table 10.8.

Pourable Shortenings

Pourable shortenings include fluid shortenings that have suspended or emulsified fats and liquid shortenings that are clear.

Development of the process for continuous bread making created a demand for fluid shortenings that could be pumped, metered accurately, and conveniently handled and stored. With the development and introduction of appropriate emulsifiers the baking industry has found it possible to switch from animal fats and plastic shortenings to refined vegetable oils. The combination of oil and surfactants provides a satisfactory and economical pourable shortening for continuous bread or conventional production (Brekke 1980a).

In plastic shortenings crystal structure is important in the development of a fine-grained, cellular structure in baked goods, but in pourable shortenings, the starch, protein, and emulsifiers take over this function. For icings and fillings, however, it has not been possible to substitute pourable shortenings and still obtain the desirable body or stiffness.

A liquid shortening can be prepared by winterizing soybean oil that is selectively hydrogenated to an IV of 81–84. Winterizing gives a yield of only 48–54% and increases the IV to 86–88. The liquid shortening remains free of solid fat at 16°C (60°F), and less than 1% solid fat is dispersed at 10°C (50°F) (Brekke 1980a).

Fluid shortenings are an opaque mixture of a base oil with added hard fats. Emulsifiers are added for baking applications but are to be avoided in frying applications, because they bring about excessive foaming and smoking. The size of added solid particles is critical. Large particles settle and cause inhomogeneity. Smaller particles such as β' crystals settle more slowly but eventually pack into agglomerates that settle. The ideal size and density of hard-fat particles come from β crystals.

Fluid shortenings can be prepared by rapidly chilling the base stock and the hard fat in a scraped surface heat exchanger and then allowing β crystals to form during a holding period. The length of the holding period depends on the temperature, whether or not emulsifiers are present, and the heat of crystallization.

The important change during holding is formation of stable β crystals. The crystallization may result in a soft solid, but fluidity is regained by stirring. A second method is to grind the hard fat to fine particles and suspend the particles in oil. This mixture is then homogenized with any type of suitable homogenizing equipment.

Fluid shortening products are used in cake production and may contain hard fat flakes, emulsifier combinations, or both. Specific fluid shortenings perform well in specific types of cakes, but there is no one combination that serves as an all-purpose shortening. Some bakers obtain the flexibility they need by using a vegetable oil base, usually soybean, and metering in the appropriate emulsifiers and additives. Cakes made by the soybean oil–emulsifier system can have excellent quality with respect to crumb softness and moisture retention (Brekke 1980a).

Dry Shortenings

To serve the dry cake mix industry it is necessary to provide shortening in a dry form. Dry shortenings can be prepared by spraying and cooling droplets of warm molten shortening in a stream of refrigerated air. The solidified, powdered shortening is collected at the bottom of the spray tower. A second method is to chill molten shortening on a cold metal flaking roll and then cool the flakes further, so that they can be ground into a free-flowing powder (Brekke 1980a).

MARGARINES AND RELATED PRODUCTS

Although margarine was invented in France as a butter substitute in 1869, today it is recognized as a high-quality, nutritious product in its own right. Regular margarine can be prepared from partially hydrogenated oils and is a water-in-oil emulsion in contrast to mayonnaise. Some specialty margarines such as bakery, puff paste, and roll-in margarines may include animal fats.

In the United States, margarine is manufactured under a Standard of Identity that defines it as a liquid or plastic food mixture of fat and water. The minimum amount of fat is 80% by weight. The Standard of Identity allows cow's milk, water, or water and edible proteins (nonfat dry milk solids or soybean protein) as the aqueous phase. Other ingredients in margarine are 2–3% salt (except in salt-free products), emulsifiers (mono- and diglycerides and/or lecithin), preservatives, flavorings, and colorings (usually β-carotene). Each 0.45 kg of margarine has to be fortified with 15,000 IU of vitamin A and vitamin D fortification is optional. Butter, sweeteners, and antioxidants may be included as optional ingredients. Vitamin E and marine oils are specifically excluded as margarine ingredients in the United States.

A number of different margarines (stick, liquid, and soft tub types) and man-

Table 10.9

Composition of Typical Margarine Products[a]

	Liquid	Soft tub	Regular	Imitation[b]	Vegetable oil spread[c]
Iodine value	109	101	83	94	76
Palmitic (%)	11	12	11	11	13
Stearic (%)	6	7	7	7	6
Oleic (%)	35	38	59	48	64
Linoleic (%)	41	34	17	28	12
Linolenic (%)	3	4	1	2	tr
P/S ratio	2.7	1.9	1.0	1.6	0.6

[a]Source: Brekke (1980a).
[b]About 40% oil.
[c]About 60% oil.

ufactured sandwich spreads are available. The latter include vegetable oil spreads having 60% fat and diet imitation margarine containing 40% fat. These margarines were developed to meet consumer demands for products that are readily spreadable at refrigerated temperature, have a high polyunsaturated to saturated fatty acid ratio, or have a low fat content. The fatty acid compositions of typical margarines and related products are given in Table 10.9.

Regular Margarines (Stick) and Soft Tub Products

Regular margarine is formulated from blends of one or more oil stocks, at least one of which is sufficiently hydrogenated to give margarine its semisolid form. The stocks may come from a range of different vegetable oils, but soybean oil is predominant.

The several steps in margarine manufacture include (1) blending the vegetable oils to be used; (2) blending ingredients in the aqueous phase; (3) emulsifying the oil and aqueous phases; (4) chilling and working the emulsion for control of plasticity; (5) packaging and tempering if necessary.

The physical characteristics of margarines are a result of the SFI values of the oil used. The SFI values for oil components in several U.S. margarines are shown in Table 10.10. Values over 30 at 10°C (50°F) give a brittle margarine that is difficult to spread. An SFI below 28 at 10°C (50°F) will promote spreadability. The SFI at 33°C (50°F) and 37.8°C (100°F) determine mouth feel. For good mouth feel, fat should melt completely at 37.8°C (100°F).

Special soft printed (stick) margarines can be produced by blending a hard and soft component each with an IV 5 units greater than used for table margarine.

Interest in increasing the polyunsaturated fatty acid content of margarines for nutritional purposes led to a formulation incorporating unhydrogenated oils. The

Table 10.10

Typical Solid-Fat Index Values for the Oil Component in Several U.S. Margarines[a]

Margarine type	SFI values at designated temperatures				
	10°C (50°F)	21.1°C (70°F)	26.6°C (80°F)	33.3°C (92°F)	37.8°C (100°F)
Stick (three components)	28	17	12	2–3	0
Stick (80% liquid oil)	15	11	9	5	2
Soft tub	13	8	6	2	0
Liquid (5% hard fat)	7	6	6	5.4	4.8
Puff paste	26	24	22	21	19
Roll in	25	20	18	15	12
Bakers'	30	19	16	8	4
Table	28	15	9	3	0

[a]Source: Brekke 1980A.

blends included 50–70% liquid vegetable oil and 30–50% hydrogenated soybean oil with an IV of about 60. The liquid oil could be corn, cottonseed, or safflower depending on which was to be promoted.

Soft tub margarines use larger proportions of liquid oil than printed margarines. The hardened component may be hydrogenated to an IV of 60–65, and the margarine will be 74–85% liquid. Other soft margarines may be produced from an ordinary margarine blend of hard and soft hydrogenated components with 50% liquid oil added (Weiss 1983, p. 197).

Liquid and Fluid Margarines

These products are in limited production but have found acceptance by both home consumers and commercial users because of the convenience of controlling amounts to be dispersed. As with their shortening counterparts, liquid margarines remain clear at refrigerator temperatures, whereas fluid margarines have suspended solid fats and are opaque. These margarines can be used for pan frying, or they can be spread on frozen foods in preparation for later frying.

A liquid margarine can be prepared by blending oil with a small amount (5% or less) of hard fat. The blend is chilled and held for several hours to produce β' crystals. The chilled fat is then combined with the aqueous phase of the margarine and emulsified. This emulsion remains a liquid and retains its stability at temperatures higher than refrigeration temperatures (Brekke 1980a).

The margarine emulsion does not break if there are sufficient fat crystals present. Also, the SFI of hard fat suspended in liquid oil changes very little between 3 and 33°C (40 and 92°F), and thus fluidity is maintained as the temperature is lowered.

Bakery Margarines

Bakery margarines are margarinelike products with special uses in commercial baking. They resemble shortenings except for the aqueous phase but are rarely aerated and contain insufficient emulsifiers for usual cake preparation.

One example of bakery margarine is table margarine packed in large containers and used primarily in the preparation of cream icings. Margarine is preferred to shortening because of improved flavor, color, and melting characteristics.

Regular bakery margarine is essentially table margarine but with 4–8% hard fat added as a plasticizer. Cookies, pound cakes, and pastry are made with bakery margarine rather than shortening, and bakery margarine can be used for Danish pastry. More frequently, special roll-in margarines are used for Danish pastries. These are prepared from blends of unhydrogenated tallow and soybean oil with about 8% hard fat added. Plasticity is all important in Danish pastry. If the margarine is too fluid, it will be lost as the pastry is refolded, and if the margarine is too hard, it will tear the dough.

SOYBEAN LECITHIN PRODUCTS

Production and Types

The process for recovery of soybean lecithin from soybean oil was described in Chapter 3 (see Fig. 3.30). As soybean lecithin leaves the degumming centrifuge, it contains 40–50% moisture and must be dried. Drying is often done continuously in an agitated-film evaporator.

If the moisture is removed without addition of any fluidizing agents, a plastic lecithin results. This product has functional properties and is a regularly produced type of lecithin, but it lacks convenience because heating and mixing are required to fluidize it. If soybean oil, fatty acids, or calcium chloride is added to the soybean lecithin before drying, the final product is fluid.

Both plastic and fluid lecithins are available as natural color, bleached, or doubly bleached products. The bleaching is done with hydrogen peroxide, and in addition to lightening the final lecithin color, hydrogen peroxide can control bacterial growth, which may be a problem before the lecithin is dried.

The 30–35% portion of plastic and fluid lecithins that is not lecithin (acetone insoluble) is soybean oil. Some lecithin products are produced by deoiling, using warm acetone, resulting in a waxy solid that is sold as a free-flowing powder in various particle sizes.

Lecithin can be fractionated into two products using ethanol solubility. As shown in Table 10.11, the lecithin fraction soluble in ethanol is enriched in phosphatidyl choline and is particularly useful as an emulsifier for oil-in-water

Table 10.11

Approximate Composition of Commercially Refined Lecithin Fractions[a]

Fraction	Oil-free lecithin	Alcohol-soluble lecithin	Alcohol-insoluble lecithin
Phosphatidyl choline	29	60	4
Phosphatidyl ethanolamine	29	30	29
Inositol and other phosphatides including glycolipids	32	2	55
Soybean oil	3	4	4
Other constituents[b]	7	4	8
Emulsion type favored	Either oil-in-water or water-in-oil	Oil-in-water	Water-in-oil

[a]Source: Brekke (1980B).
[b]Includes sucrose, raffinose, stachyose, and about 1% moisture.

emulsions. The ethanol-insoluble fraction is enriched in phosphatidyl inositol and is used for water-in-oil emulsions.

It is possible to modify the properties of lecithins further by various chemical treatments or by mixing with other emulsifying agents. For example, lecithin can be reacted with hydrogen peroxide in the presence of a weak acid such as lactic or acetic. This product is known as hydroxylated lecithin, and it has excellent water solubility and dispersibility in contrast to normal lecithin. However, the flavor tends to be soapy (Szuhaj 1983). Hydroxylated lecithin is approved for food use in the United States.

Functions

Lecithin products have industrial as well as food uses. For example, the cosmetic industry makes use of their emulsifying capability, but we shall concentrate only on the functions of lecithin products in foods.

The functionality of lecithin is a result of its structure. The lipophilic portion of the molecule comes from the hydrocarbon chains of the fatty-acid molecules, and the hydrophilic portion comes from the phosphate group and whatever groups are combined with the phosphate. This bipolar nature of lecithin makes it an excellent surface-active agent, which in turn allows it to function as an emulsifier, antispatter agent, wetting agent, release agent, to modify viscosity, or to control crystallization (Szuhaj 1980).

Lecithin is one of the emulsifiers used to improve properties of margarines. At levels ranging from 0.1 to 0.5%, it can prevent the loss of moisture from the water-in-oil emulsion system, it can minimize spattering during frying (again through control of moisture loss), and the antioxidative properties of lecithin can minimize vitamin A loss due to oxidation. Lecithin improves the shortening power of margarines used in baking and facilitates shortening incorporation into doughs.

In the candy industry, lecithin is used to decrease the viscosity of chocolate at levels of 0.25–0.35%. This is helpful in applying uniform coatings, and mixing ingredients into chocolate products, and again antioxidant activity helps to stabilize chocolate candies to oxidation.

Manufacturers of peanut butter make use of lecithin (1–2%) to improve smooth spreading and to prevent oil separation.

As a release agent, lecithin prevents sticking of baked bread to baking pans, and it is a useful home product as a nonstick pan coating for frying.

The wetting properties of lecithin are particularly useful to manufacturers of instant products such as dry milk and chocolate drink mixes. Along with agglomeration of particles, lecithin greatly facilitates dispersion in water of any dry product.

REFERENCES

Brekke, O. L. (1980a). Soybean oil food products—Their preparation and uses. *In* "Handbook of Soy Oil Processing and Utilization" (D. R. Erickson, E. H. Pryde, O. L. Brekke, T. L. Mounts, and R. A. Falb, eds.). American Soybean Association, St. Louis, MO, and American Oil Chemists Society, Champaign, IL.

Brekke, O. L. (1980b). Oil degumming and soybean lecithin. *In* "Handbook of Soy Oil Processing and Utilization" (D. R. Erickson, E. H. Pryde, O. L. Brekke, T. L. Mounts, and R. A. Falb, eds.). American Soybean Association, St. Louis, MO, and American Oil Chemists Society, Champaign, IL.

Krishnamurthy, G. N. (1982). Cooking oils, salad oils, and salad dressings. *In* "Bailey's Industrial Oil and Fat Products," Vol. 2, 4th ed. (D. Swern, ed.). Wiley, New York.

Latondress, E. G. (1981). Formulation of products from soybean oil. *J. Am. Oil Chem. Soc.* **58**:185.

Szuhaj, B. F. (1980). Food and industrial uses of soybean lecithins. *In* "World Soybean Research Conference II: Proceedings" (F. T. Corbin, ed.). Westview Press, Boulder, CO.

Szuhaj, B. F. (1983). Lecithin production and utilization. *J. Am. Oil Chem. Soc.* **60**:258A.

Weiss, T. J. (1983). "Food Oils and Their Uses," 2nd ed. AVI Publ. Co., Westport, CT.

Wolf, W. J., and J. C. Cowan (1975). "Soybeans as a Food Source," rev. ed. CRC Press, Cleveland, OH.

11

Grades, Standards, and Specifications for Soybeans and Their Primary Products

Soybeans and the products derived from them are not uniform and identical. As with any natural product, variations in quality and quantity have an influence on the economic value and also can influence the ultimate usage. Some of the sources of variability in soybeans are growing conditions, cultivars, storage conditions, and production practices. For example, abnormally wet conditions during the normal period of soybean harvest can lead to problems of mold growth, sprouting, abnormally high free fatty-acid content of the oil, and possible mycotoxin development in the meal.

With respect to soybean products, quality changes may originate with the raw material as noted above or may be due to processing, storage, or packaging considerations. If an inexperienced processor inadvertently allows air to enter the soybean oil each time it is transferred, the oil will lose oxidative stability and flavor quality.

To take into account the many aspects of quality when soybeans or soybean products are bought and sold, it is necessary to have well recognized standards. The purpose of this chapter is to state those standards and any pertinent background information about them.

It is not an easy task to establish standards that are considered fair by the parties involved in buying and selling agricultural products. Nor is it easy to establish the methods of measurement used to determine whether or not standards are being met. Often an organization of interested parties plays an important role in establishing and maintaining standards, and frequently federal or local governments are involved.

The grading of soybeans in the United States is the responsibility of the Federal Grain Inspection Service of the USDA, and the standards for oil and meal are established by the National Soybean Processors Association. Equally important to the actual standards are the methods used for making measurements. Several professional societies develop and compile methods used for measuring attributes of soybeans or soybean products. The most important of these are the Association of Official Analytical Chemists, American Oil Chemists' Society, and American Association of Cereal Chemists.

GRADES OF SOYBEANS

Sampling

As the first step in establishing a suitable grade for soybeans the barge, truck, or railway car must be sampled to insure that the grade established is representative of the entire lot. This is easy to state but difficult to accomplish. Some guidelines have been established by the various agencies that have an interest in impartial grading.

The size of the sample from which the grade determination is made must be at least 2 qt (about 1.9 L). If the soybeans to be sampled are being loaded or unloaded, the proper sampling procedure is to obtain cuts from the stream by means of a "pelican" sampling device. If the soybeans to be sampled are already in the container in which they are being shipped, that container needs to be sampled in at least five places by means of a double-tube trier that can remove amounts of soybeans from several depths in the container. When soybeans are being shipped in sacks rather than in bulk, a sufficient number of sacks needs to be sampled, and that number is at least the square root of the total number being shipped.

After obtaining the initial sample, that sample needs to be protected from any changes that may be caused by exposure to the atmosphere that may change the grade before the actual grading is done. For example, protection from insect contamination is necessary by using a suitable container. Also, after obtaining the initial sample, it is necessary to reduce the size of the sample while retaining the unique characteristics of that sample. If any one of the individual samples from the pelican or trier appears to be substantially inferior to the bulk of the sample, those individual samples should be kept separate and graded separately. The remainder of the samples may be pooled and reduced to the 2 qt size by means of a Boener sampler.

Grading

United States. In the United States soybeans are considered as grains for trading purposes, and they are regulated under the United States Grain Standards Act, administered by the Federal Grain Inspection Service of the USDA. Soybeans are subdivided into classes based on color, and yellow soybeans are the major commercial class.

Grades are based on test weight, moisture content, and percentages of splits, damaged kernels, and foreign material. As of September 1985, moisture content was no longer a grade-determining factor, because it will be treated as a condition of the grain rather than as a fixed measure of quality. This is reasonable since freshly harvested soybeans may grade no. 1 in all aspects except moisture. They then would change grade as they were dried. Moisture will still be determined and will be shown on all official certificates of grade, because it is an important piece of information about the soybeans being offered for sale, but it will no longer be part of the grading criteria.

Table 11.1 shows the limits for various numerical grades of soybeans. Following are some of the definitions and principles governing the grading system

Table 11.1

Specifications for the U.S. Numerical Grades and U.S. Sample Grade[a] of Soybeans

		Maximum limits (%)				
			Damaged kernels			Brown, black, or bicolored soybeans
Grade	Minimum test weight (lb/bu)	Splits	Total	Heat damaged	Foreign material	
1	56	10	2	0.2	1	1
2	54	20	3	0.5	2	2
3[b]	52	30	5	1	3	5
4[c]	49	40	8	3	5	10

[a]U.S. sample grade shall be soybeans that do not meet the requirements for any of the grades from U.S. no. 1 to U.S. no. 4; or that are musty, sour or heating; or that have any commercially objectionable foreign odor; or that contain stones; or that are otherwise of distinctly low quality.

[b]Soybeans that are purple mottled or stained shall be graded not higher than U.S. no. 3.

[c]Soybeans that are materially weathered shall be graded not higher than no. 4.

(Anon. 1984a). The numbers refer to the pertinent sections of the U.S. Code of Federal Regulations.

26.601 (a) Soybeans.   Soybeans shall be any grain which consists of 50% or more of whole or broken soybeans which will not pass readily through an $^8/_{64}$ sieve and not more than 10.0% of other grains for which standards have been established under the United States Grain Standards Act.

26.601 (b) Classes.   Soybeans shall be divided into the following five classes: yellow soybeans, green soybeans, brown soybeans, black soybeans, and mixed soybeans.

26.601 (c) Yellow soybeans.   Yellow soybeans shall be any soybeans which have yellow or green seed coats and which in cross section are yellow or have a yellow tinge and may include not more than 10.0% of soybeans of other classes.

26.601 (h) Grades.   Grades shall be the U.S. numerical grades, U.S. sample grade and special grades provided for in 26.603 (Table 11.1).

26.601 (j) Splits.   Splits shall be pieces of soybeans that are not damaged.

26.601 (k) Damaged kernels.   Damaged kernels shall be soybeans and pieces of soybeans which are heat-damaged, moldy, diseased, stinkbug-stung, or otherwise materially damaged. Stinkbug-stung kernels shall be considered damaged kernels at the rate of one-fourth of the actual percentage of the stung kernels.

26.601 (m) Foreign material.   Foreign material shall be all matter, including soybeans and pieces of soybeans, which will pass readily through an $^8/_{64}$ sieve and all matter other than soybeans remaining on such sieve after sieving.

26.602 (a) Basis of determination.   Each determination of class, splits, damaged kernels, and heat-damaged kernels, and of black, brown and/or bicolored soybeans in Yellow or Green Soybeans, shall be upon the basis of the grain when free from foreign material. All other determinations shall be upon the basis of the grain as a whole.

26.602 (b) Percentages.   All percentages shall be upon the basis of weight. The percentages of splits shall be expressed in terms of whole percents. All other percentages shall be expressed in terms of whole and tenths percents.

26.602 (c) Moisture.   Moisture shall be ascertained by the air-oven prescribed by the USDA, as described in Service and Regulatory Announcement No. 147, issued by the Agricultural Marketing Service, or ascertained by any method which gives equivalent results.

26.602 (d) Test weight per bushel.   Test weight per bushel shall be the weight per Winchester bushel as determined by the method prescribed by the USDA, as described in Circular No. 921 issued June 1953, or as determined by any method which gives equivalent results.

26.603 (d) Special grades, special grade requirements, and special grade designations for soybeans.   (1) Garlicky soybeans—(i) Requirements.   Garlicky soybeans shall be soybeans which contain 5 or more garlic bulblets in 1,000 grams. (ii) Grade designation.   Garlicky soybeans shall be graded and designated according to the grade requirements of the standards applicable to such soybeans if they were not garlicky and there shall be added to and made a part of the grade designation the word "garlicky."

(2) Weevily soybeans—(i) Requirements.   Weevily soybeans shall be soybeans which are infested with live weevils or other live insects injurious to stored grain. (ii) Grade designation.   Weevily soybeans shall be graded and designated according to the grade requirements of the standards applicable to such soybeans if they were not weevily, and there shall be added to and made a part of the grade designation the word "weevily."

The specifications of these standards do not excuse failure to comply with the provisions of the Federal Food Drug and Cosmetic Act. While it is apparent that increasing amounts of moisture, damaged kernels, and foreign material would reduce the value of soybeans and should be considered in the determination of grades, the significance of splits is not so obvious.

The reason for assessing penalties in grade for increasing quantities of splits is that splits tend to deteriorate more rapidly in storage than whole soybeans, especially as a result of increased free fatty-acid content of the oil. An additional objection to the occurrence of excessive splits is that their presence indicates the presence of an equivalent amount of separated hulls. In handling soybeans through elevators and processing plants, the hulls tend to segregate and impair uniformity of flow, especially when storage bins are being emptied. This segregation may sometimes cause irregularity in the composition of the soybeans unless careful control is maintained.

Brazil.   Brazil, the second largest producer of soybeans in the world, has not concentrated a lot of attention on grades and standards. The Ministry of Agriculture has issued a directive, no. 228 of July 29, 1980, dealing with the standardization, classification, and trade of soybeans, and from this document we can learn which characteristics are considered important.

It defines soybeans as the mature, clean, dry, perfectly developed and uniformly colored seeds of any of the cultivars of *Glycine max*. Furthermore, soybeans should be free of mechanical and physiological damage, weed seed, diseases, insects, and substances damaging to health.

Characteristics considered in classification of soybeans are moisture, foreign material, bean damage, broken beans, and green beans; and any beans that have more than 14% moisture, 30% brokens, 3% foreign material, 8% damage, or 10% green beans are considered below grade. Other attributes that would cause beans to be below grade are poorly preserved beans, moldy or fermented beans, sour or acid odors or odors of any kind that would stop utilization, or presence of castor beans or other poisonous seeds.

Data are also given on sampling procedures, packaging rules, methods of analysis, and issuing of certificates of grade.

Japan.   Although her own production of soybeans is limited (109,500 MT in 1976), Japan maintains unique grade specifications for home-produced soybeans to help assure fair trading practices in markets (Anon. 1977). The Japan Agricultural Standards (JAS) specify soybeans produced in the country based on the Agricultural Products Inspection Act (no. 144, April 10, 1951). There have been a number of revisions, but the pertinent portions of the present standards (as of 1977) may be summarized as follows (Watanabe *et al.* 1971, Anon. 1977).

Kinds.   Soybeans are classified into two kinds, ordinary soybeans (soybeans for common uses) and seed soybeans. They are further classified into four categories based on the size of the soybeans: large (7.9 mm diameter), medium (7.3 mm diameter), small (5.5 mm diameter), and extra small (4.9 mm diameter).

Grades.   Two grade specifications have been established for ordinary soybeans, and one grade specification for seed soybeans. Grade specification no. 1 is

Table 11.2

Grade Specification No. 2 for Ordinary Soybeans in Japan[a]

| | Minimum limits (%) | | | Maximum limits (%) | | | |
| | | | | | Damaged and foreign | | |
Grade	Whole[b] sound kernels	Size uniformity[c]	Visible quality[d]	Moisture	Total	Foreign kernels	Foreign material
1	90	70	Grade 1 standard	15	10	0	0
2	85	70	Grade 2 standard	15	15	0	0
3	65	70	Grade 3 standard	15	35	1	0

[a]Sample grade: soybeans that do not satisfy the specifications for any of the grades from 1–3 and do not contain more than 50% of foreign kernels and materials.

[b]Proportions of the whole, sound kernels without damaged, immature, and foreign kernels and foreign materials.

[c]Proportions of the soybean kernels remaining on the sieve used to classify into four size categories.

[d]Quality of soybeans judged on the bases of thickness of hulls, maturity, shape of kernels, color, and uniformity of size.

Source: Anon. 1977.

set for ordinary soybeans as produced by farmers, while specification no. 2 is for soybeans with quality judged on market requirements. As an example, grade specification no. 2 for ordinary soybeans is given in Table 11.2. The grade specification for seed soybeans requires the minimum limits of whole and sound kernel proportions and ability of germination to be 90%, and the maximum limits on moisture content to be 15%.

Classification of soybeans into several categories based on their size is convenient for those who wish to choose soybeans for specific uses. For example, the large soybeans are suitable for cooked fresh beans and confectionary uses; the medium soybeans are used commonly for the production of soy curd and soybean paste; the small soybeans are known to be suitable for the production of soybean paste and natto; and the extra small soybeans are used commonly in natto fermentation and for soybean sprouts (Anon. 1977).

SPECIFICATIONS FOR SOYBEAN MEALS AND FLOURS

In contrast to soybeans, the protein products resulting from oil extraction are judged for quality by the National Soybean Processors' Association (NSPA), the

Association of American Feed Control Officials (AAFCO), and the International Feed Numbers (IFN). The criteria used to separate the several products involved are particle size and protein content. Most of these products are used for feed, but the products labeled soy flour may also be used for food.

Definitions for several of these products are specified by NSPA as follows (Anon. 1983-84).

Soybean cake or *soybean chips* is the product after the extraction of part of the oil by pressure or solvents from soybeans. A name descriptive of the process of manufacture, such as expeller, hydraulic, or solvent extracted shall be used in the brand name. It shall be designated and sold according to its protein content.

Soybean meal is ground soybean cake, ground soybean chips, or ground soybean flakes. A name descriptive of the process of manufacture, such as expeller, hydraulic, or solvent extracted shall be used in the brand name. It shall be designated and sold according to its protein content.

Solvent-extracted soybean flakes is the product obtained after extracting part of the oil from soybeans by the use of hexane or homologous hydrocarbon solvents. It shall be designated and sold according to its protein content.

Soybean mill feed is the by-product resulting from the manufacture of soybean flour or grits and is composed of soybean hulls and the offal from the tail of the mill. A typical analysis is 13% crude protein, 32% crude fiber, and 13% moisture.

Soybean mill run is the product resulting from the manufacture of dehulled soybean meal and is composed of soybean hulls and such bean meats that adhere to the hull in normal milling operations. A typical analysis is 11% crude protein, 35% crude fiber, and 13% moisture.

Soybean hulls is the product consisting primarily of the outer covering of the soybean. A typical analysis is 13% moisture.

The proximate composition of several of these products is given in Table 11.3. All of the products have been heated sufficiently to give PER values of 2.3–2.4,

Table 11.3

Proximate Composition of Several Soy Protein Products Used as Animal Feedstuffs

Product	Bulk density (lb/ft³)	Protein minimum	Fat minimum	Fiber maximum	Moisture maximum
Soybean cake, soybean chips, soybean meal	36–40	41	3.5	6.5	12
Soybean flakes, 44% protein soybean meal	35–38	44	0.5	7.0	12
Dehulled soybean meal	41–42	48–50	0.5	3–3.5	12
Soybean mill feed	25–27	13–15	1–2	32–34	9–11
Soybean mill run	25–27	10–12	1–2	35–40	9–11

[a]Source: National Soybean Processors Association (Anon. 1983–84).

Table 11.4

Soy Flour Standards (%) in the United States[a,b]

Component	Full-fat soy flour	Low-fat soy flour	Defatted soy flour
Protein (N × 6.25)[c]	40.0 min.	45.0 min.	50.0 min.
Protein (N × 5.7)[c]	36.5 min.	41.0 min.	45.0 min.
Fat (ether extract)[c]	18.0 min.	4.5–9.0 max.	2.0 max.
Fiber[c]	3.0 max.	3.3 max.	3.5 max.
Moisture	8.0 max.	8.0 max.	8.0 max.
Ash[c]	5.5 max.	6.5 max.	6.5 max.

[a]Source: American Soybean Association, Blue Book (Anon. 1984a).
[b]Particle size: all flours must have 97% passing a no. 100 U.S. standard screen.
[c]Moisture-free basis.

to give urease activity equivalent to a pH rise of 0.05–0.2, and to have PDI or NSI between 15 and 30%. Particle sizes for all of these products have at least 95% passing a U.S. screen no. 10 and no more than 6% passing a U.S. screen no. 80.

Any of the products listed in Table 11.3 may contain a nonnutritive, inert, nontoxic conditioning agent to reduce caking and improve flowability in an amount not to exceed that necessary to accomplish its intended effect and in no case to exceed 0.5%. The name of the conditioning agent must be shown as an added ingredient. Most frequently calcium carbonate is used for this purpose.

In contrast to soy meals for feed use, soy flours are ground much finer, with 97% passing a U.S. screen no. 100. Proximate composition for three soy flours differing in fat concentration is given in Table 11.4. Soy flour is defined as "the products... processed from high quality, sound, clean, dehulled yellow soybeans as defined in the United States Grain Standards Act. The soybeans shall be subjected to a thorough initial cleaning operation, that shall substantially remove all foreign material. Disagreeable flavors and odors shall be removed by subjecting the soy material to adequate processing. Soy flour shall be prepared and packaged under modern sanitary conditions. The soy flour shall be free from burnt, musty, rancid or other undesirable flavors or odors; free from burnt, scorched or grayish color and free from insects, insect webbing, dirt or other extraneous matter" (Anon. 1984a).

TRADING SPECIFICATIONS FOR SOYBEAN OILS

Just as for successful trading of soybean meals, it is necessary to have a set of rules and standards for use by buyers and sellers of soybean oil. NSPA has provided such standards, and they are shown in Table 11.5. Not only are these

rules useful for trading, but also they are useful for distinguishing the various stages of refining of soybean oils that are traded.

Note in Table 11.5 that there are three grades of crude oil: prime, no. 2, and sample grade. The criterion used to distinguish these three grades is color. Prime crude oil has the least green color and a 3.5 red based on the Lovibond system. For crude degummed oil and for once-refined oil there are no grades. Note also that if a sample of oil does not fit the criteria established, often the oil can be sold anyway but with a discount in price. The discount varies, depending on the specification and on the amount of variation. In addition to discounts in price, there are price premiums provided for oils with minimum refining losses.

Samples for determining grade and specifications are taken from the bulk oil. Uniformity of samples is less of a problem with a liquid product than with a solid product such as a meal or flour, but there still can be discontinuities in the oil such as settling out of phospholipids in crude nondegummed oil. The trading rules call for the taking of two 1 qt (1 L) samples and one 2 qt (2 L) sample. The 1 qt samples go to the buyer and the seller. The 2 qt sample is held for a referee in case of a dispute about any of the trading rules.

There is a problem of determining if adulteration of an oil shipment has taken place by addition of an oil other than that which was specified in the transaction. This problem is not covered by the trading rules and is a difficult problem to solve. For some oils, the sterol content is a good indication of the type of oil being shipped. For example, there is no cholesterol in vegetable oils, and so addition of animal fats to vegetable oils might be detected by finding cholesterol in the vegetable oil sample. In some other instances the types and amounts of tocopherols might be a good indication of the kind of vegetable oil being shipped. There are real difficulties of detection, however, when only a small amount of a different kind of oil is mixed with the specified oil.

Several sets of specifications for soybean salad oils are shown in Table 11.6. All salad oils are to be free from sediment and foreign material and are to be clear and brilliant at 21–29°C (70–85°F). As with crude oils, there is provision for selling at a discount if the oils do not meet some of the specifications (moisture, color, or peroxide value).

For hydrogenated soybean oil, the IV should be 105–115, and there should be less than 3% linolenic acid. Generally, additives as preservatives are permitted at the levels specified by the FDA. If packaging gas is to be used, it should be pure nitrogen or a mixture of nitrogen and 10% carbon dioxide, but it should not contain more than 0.05% oxygen.

Those analyses that are concerned with flavor or free fatty acids have to be made within seven days of packaging.

The specifications state that the salad oils should be bland with respect to flavor and odor and free of beany, rancid, painty, musty, soapy, fishy, metallic, and other undesirable or foreign flavors after heating to 177°C (350°F). At the

Table 11.5

Trading Specifications for Crude, Degummed, and Once-Refined Soybean Oil[a]

Factor	Prime crude[b]	No. 2 grade crude[b]	Crude degummed	Once refined
Moisture and volatiles (% max) (0.15 discount)	0.3 (0.5 discount)	0.3 (0.5 discount)	0.3	0.1
Flashpoint (min) °F	250	250	250	250
°C	121	121	121	121
Free fatty acids as oleic (% max)	—	—	0.75 (1.25 discount)	0.10 (0.15 discount)
Unsaponifiables (% max)	1.5	1.5	1.5	1.5
Green color	<Std. A	>Std. A, <Std. B	—	—
Phosphorous	—	—	0.02 (0.025 discount)	—
Refining loss (% max)	<5 premium (5.1–7.5 discount)	<5 premium (5.1–7.5 discount)	—	—
Refined bleached color (max red)	3.5 (6.0, 1.5% discount)	<2.8, 1% discount 5 max, 3.3% discount	—	3.5

[a]Source: Brekke (1980A).

[b]Crude oil not meeting specifications for prime or no. 2 is sold as sample grade.

Table 11.6

Trading Specifications for Soybean Salad Oils[a]

Factor/source of rule:	Refined, deodorized U.S. General Services Administration	Fully refined/ NSPA	Fully refined/ USDA	Refined, hydrogenated, winterized U.S. General Services Administration
Moisture and volatiles (% max)	0.06	0.10	0.10	0.06
Unsaponifiables (% max)	—	1.5	—	—
Flashpoint (min) °F	—	—	550	—
°C	—	—	288	—
Free fatty acids as oleic (% max)	0.05	0.05	0.05	0.05
Color, Lovibond, max red	4	2 (2.6 discount)	2	2
yellow	35	20	20	20
Peroxide value (meq/kg max)	1.0	2.0	0.5 (1.0 discount)	1.0
Fat stability by AOM peroxide value 8 hr				
hours to 100	15	35	35	25
Cold test (min hr)	5.5	5.5	5.5	5.5

[a]Source: Brekke (1980a).

same time a footnote indicates that no common vegetable oils can pass this test at the present time.

SPECIFICATIONS FOR LECITHINS

Lecithin specifications by the NSPA for six commercial grades of lecithin are shown in Table 11.7. The lecithins are classified by plastic or fluid consistency and are further subdivided as natural color, bleached, or double bleached.

Phosphatide content is equivalent to the percentage of acetone-insoluble matter. "Benzene insolubles" refers to the contamination by soybean meal or filter aids fines.

Acidity of the phosphatides, acidity of the oil, and fatty acids are given by the acid value, expressed as potassium hydroxide (mg) required to neutralize the acids in 1 g of the lecithin sample.

Table 11.7
NSPA Trading Specifications for Soybean Lecithin

Analysis	Grade		
	Fluid natural-color lecithin	Fluid bleached lecithin	Fluid double-bleached lecithin
Acetone insolubles (% min)	62	62	62
Moisture (% max)[b]	1	1	1
Benzene insolubles (% max)	0.3	0.3	0.3
Acid value (max)	32	32	32
Color, Gardner (max)[c]	10	7	4
Viscosity (poise, 25°C max)	150	150	150
	Plastic natural-color lecithin	Plastic bleached lecithin	Plastic double-bleached lecithin
Acetone insolubles (% min)	65	65	65
Moisture (% max)[b]	1	1	1
Benzene insolubles (% max)	0.3	0.3	0.3
Acid value (max)	30	30	30
Color, Gardner (max)[c]	10	7	4
Penetration (max mm)[d]	22	22	22

[a]Source: Brekke (1980b).
[b]By toluene distillation for 2 hr or less.
[c]As a 5% solution in colorless mineral oil.
[d]By specified cone penetrometer test.

STANDARDS FOR THE USE OF SOY PROTEIN PRODUCTS IN OTHER FOODS

In contrast to standards and specifications for soybeans, soy meals, and soy oils, which are relatively well established, standards for soy protein products are still evolving. Of course, most countries have well-established food laws based on safety of the products and freedom from any misleading labeling or advertising. Usually it is possible to introduce a new food or a new food ingredient under the existing law. For soy protein products in Western countries this has not always been possible.

In countries where soybeans have been part of the traditional diet for centuries, there is less of a problem in introducing soy protein products. In fact, Japan is a leader in introducing new soy protein foods due in large part to minimal controversy about their legal status (Ward 1979).

There are many reasons for the failure of existing food laws to supply the basis for introduction of soy protein products. One of the more significant reasons is an increased awareness on the part of consumers of nutrition, diet, and new food products. Consumers and consumer advocates insist on a degree of product safety and a standard of product labeling that go beyond what conventional food law requires. Another significant reason is the economic loss feared by conventional food producers and manufacturers if a significant portion of meat, eggs, or dairy products were to be replaced by soy protein.

The development of regulations for soy protein products has proceeded differently in the United States, Canada, Western Europe, and Japan, and so we treat each area separately.

United States

It became apparent by the late 1960s that the development of soy protein concentrates and soy protein isolates had led to a situation in which it was possible to substitute appreciable amounts of soy protein for meat protein. At the same time technologies for texturizing the soy products made new meat–soy mixtures feasible. The resulting products had nutritional and sensory qualities that were sufficiently good (along with economic advantages) to lead to their use in the school lunch program.

In regulations issued in September 1970, the USDA made it possible to substitute up to 30% of vegetable protein (hydrated to 60–65%) for the meat and meat alternate portion of the type A school lunch. The meat and meat alternate portions of a type A school lunch are 2 oz (56.8 g) of lean meat, poultry, or fish; 2 oz (56.8 g) of cheese; one egg; 1/2 cup of cooked dry beans or peas; 4 Tbsp of peanut butter; or an equivalent quantity of a mixture of these foods (Hagg 1981).

Even though the regulatory authority exists for use of soy protein in the school lunch program, there are no good data to show that individual school districts are making use of this authority.

The Department of Defense has chosen to approve the use of soy protein concentrate as an extender for ground beef. The final product is a blend of 80 parts beef to 20 parts hydrated soy protein concentrate and cannot be more than 22% fat. The extended ground beef must meet standards for flavor.

These developments in the wide use of soy protein products made it obvious that the regulatory agencies should adopt rules for products available to the general public. The emphasis for new regulations came from the products in which considerable quantities of soy protein have been substituted for the meat or dairy protein. This use needs to be contrasted with the use of small amounts of nonmeat proteins in sausage products. Casein and soy proteins have been used for some time for their functional properties in sausages. For example, 2–3% added soy protein can improve a sausage product by increasing water binding and by stabilizing the emulsion. The regulations for this type of soy product use have been established by the USDA, and there is little controversy over this kind of usage in the United States.

The more difficult problem of regulating soy–meat blends has been the concern of the FDA, which in 1970 proposed to regulate these new foods with a Standard of Identity. It has been a traditional approach in U.S. food law to prescribe the ingredients that can be used in a food product with a Standard of Identity. The new food products were to be named "textured protein products." Based on comments received and the FDA's own evaluation it was decided to abandon this approach (Roberts 1979).

In 1974 the FDA published proposed regulations based on a common or usual name approach, in which the information concerning the product was contained in its name. This approach was maintained and extended in the 1978 tentative final regulation.

The regulatory philosophy of this approach by the FDA is that people recognize a food and its nutritional role in their diet by its common name. If substitute foods can be made with soy protein that are nutritionally equivalent to the original products, then the new foods can be marketed as substitutes. For example, if a macaroni and cheese product is made with a soy protein product replacing the cheese, it can be named "macaroni casserole made with soy protein product cheese substitute." If the soy protein is not nutritionally comparable to the component it replaces, then the common name is still used, but the word "imitation" has to be used instead of "substitute."

The rules establish six classes of common foods: (1) breakfast meats (bacon, sausage) and lunch meats; (2) seafood, poultry, and meats other than those in class 1; (3) eggs; (4) cream cheese; (5) cottage cheese; and (6) cheeses other than those in classes 4 and 5. For each class a level of 10 vitamins, 7 minerals, and protein was established that would satisfy the nutritional equivalency criterion.

In addition to the proposed regulations on nutritional equivalency to allow use of the term "substitute" rather than "imitation," the proposed regulations have defined primary vegetable protein products (Table 11.8). The proposal is based on the terms that have been established for soy protein products and extends these terms to any vegetable protein product. There are three categories: protein content less than 65%; protein content 65–90%; and protein content greater than 90%. The word protein is not allowed in conjunction with the term "flour" used for all products with protein content less than 65%. The justification is that other flours (wheat or corn) cannot use the term "protein," although the difference in protein content between wheat flour (about 14% protein) and soy flour (about 44–50% protein) is considerable.

A special category of glutens is included because this term has become accepted for high-protein wheat and corn products. Actually the term "gluten" has a special meaning of the protein fraction developed in a wheat bread dough that gives the unique texture to wheat breads. To use this term for other high protein cereal products would be misleading in the authors' judgment.

A subject of some importance in relation to any discussion of control of vegetable protein products in food mixtures is the analysis of those vegetable protein mixtures. It is not reasonable to establish rules about the percentage vegetable protein allowed in a food product, unless means exist for determining how much vegetable protein is in a mixture. Unfortunately no foolproof analytical techniques exist for determining vegetable protein.

There are some procedures that can be helpful, however. If one has access to the records of a food company, it is possible to determine the amount of raw materials coming into a food manufacturing plant and the amount of product leaving. From these data one can make a sophisticated guess about the amount of vegetable protein being incorporated into a food product. Also, in meat plants that are under continuous USDA inspection, it is possible for the inspectors to monitor the amount of vegetable protein being used.

For a qualitative check of the presence of soy flour in a food product, microscopy is useful. Soy hulls are usually present as a contaminant of soy flour and the distinctive guard cells in soy hulls can be readily identified. Calcium oxalate crystals in the cotyledon cells of soybean are apparent in polarized light as

Table 11.8
Nomenclature for Primary Soy Protein Products[a]

Percentage protein by weight	Product name
Less than 65	Flour: "soy flour"
65 to 90	Protein concentrate: "soy protein concentrate"
90 or more	Protein isolate: "soy protein isolate"
65 to 90 (glutens)	Gluten: "wheat gluten"

[a]Source: Roberts (1979).

polygonal green bodies. Staining of the polysaccharide cell wall constituents of the soybean also may be used for quantitative purposes. However, good results are time consuming and laborious.

For soy flours and concentrates, the amount of hemicellulose in the products is determined using an alcoholic–potassium hydroxide digestion with final solubilizing in hydrochloric acid. Purified proteins, such as soy protein isolates, depend on various protein analyses. Electrophoretic methods require complete solubilization of the proteins for their subsequent separation. Soy proteins are detectable at levels of 1% in meat products. Using detergents, urea, and SH-reagents, soy proteins can be solubilized, if meat products have not been severely heat processed. Immunochemical methods, although very sensitive and specific, are only suitable if the sample temperature has not exceeded 100°C (212°F) during processing. Amino acid composition that uses computer matching of the amino acid pattern of the meat product sample with all other possible mixtures can detect added soy products (Olsman 1979).

For about 15 years in the United States, it was required that 0.1% titanium dioxide be added to soy protein isolates. The purpose of the titanium dioxide was as an aid to analysis for soy protein isolate. Since there is no good analytical method for differentiating between meat and soy protein, an analysis for titanium would allow a measure of the amount of soy protein added to a meat product. Titanium was determined spectrophotometrically in the form of yellow titanium peroxide (Smith and Circle 1978).

In 1984 the addition of titanium dioxide as a tracer for soy protein was discontinued (Anon. 1984b). It was felt that although analysis was still a problem, the meat inspectors could control amounts of soy protein used. Also, since Western Europe forbids the use of titanium dioxide in soy isolates, its removal would ease problems of using titanium dioxide for domestic markets but not for export. Finally, consumer groups in the United States are against the use of any non-nutrient materials in food, and consequently they supported the elimination of additions of titanium dioxide to soy isolates.

Canada

The Canadians have chosen to regulate meat products and poultry products with added extenders by putting the emphasis on the protein, fat, vitamin, and mineral content of the final product. For example, meat product extenders can be marketed providing that the rehydrated extender contains at least 16% protein with a nutritional quality rating of 40 or more and with at least stated amounts of 5 vitamins and 7 minerals.

Similarly, extended meat or poultry products can be marketed if they meet stated requirements for protein quantity and quality, for fat content, and for certain vitamins and minerals. These requirements vary depending on the type of extended meat product such as sausage, potted meat or ground beef.

In addition, provision has been made for the marketing of simulated meat products containing no meat ingredients. The regulations state that protein, fat, vitamin, and mineral content of the simulated products must be at least equal to those meat products that are being simulated.

Western Europe

Western Europe does not represent a unified area with respect to food law as do the United States and Canada. Nonetheless, the European Economic Community has started thinking about the regulatory aspects of vegetable protein–meat combinations for Europe as a whole.

The range of regulations for meat products in the various European countries is wide. Hence, there is potential for widely varying approaches to regulating vegetable protein–meat combinations. This subject has been reviewed (Brincker 1979), and the striking thing is how few countries have considered regulations for allowing vegetable protein–meat combinations. Fifteen countries were surveyed (Austria, Belgium, Denmark, Federal Republic of Germany, Finland, France, Ireland, Italy, Luxembourg, The Netherlands, Norway, Spain, Sweden, Switzerland, and the United Kingdom). Of these only France has specifically written regulations allowing vegetable protein as an additive in a conventional meat product or as a component of a nontraditional product.

To indicate the different philosophical approaches to regulating the food supply in Western Europe, two extremes can be cited. There is the idea that the public is best served by using strict compositional criteria for meat products. Under these rules, the name of a meat product dictates how and from what it is made, and there is little need for information on the label. This is similar to the "Standard of Identity" approach of the United States. At the other extreme is a fairly liberal approach to allowing compositional changes to a product and placing reliance on the label to inform the consumer.

Although Western Europe's regulations can be considered relatively conservative in their approach to the use of vegetable protein–meat mixtures, it is assumed that those rules will change as the nutritional and economic benefits of soy products become apparent.

Japan

As already noted, Japan seems to have the least problem with introduction of new soy protein–meat mixtures. Part of their success may be due to the Japan Vegetable Protein Food Association, which was established in 1975. This organization of producers of vegetable proteins works with the Japanese government in promoting the general use of vegetable proteins.

While new products containing soy proteins have been introduced in Japan, these new products are not a major food source. For example, it is estimated that in 1978 about 20,000 tons of soy protein were used in Japan in 570,000 tons of meat products such as hams, sausages, minced (ground) meat products, and fish paste (Kanda 1981). Also, the 20,000 tons of soy protein is a small part of the 3,500,000 tons of soybeans imported by Japan for oil and feed.

The regulation of these new products is achieved by establishing quality standards for them under the Japan Agricultural Standards (JAS) program. Once a standard is established under JAS, the product is monitored for adherence to the standard and consumers have confidence in the quality and safety of products labeled as meeting the JAS criteria.

The JAS for several forms of vegetable protein products are shown in Table 11.9. Note that moisture, protein content, and particle size are the three quality attributes controlled by JAS. Also included in the table are the permissible additives that can be included with vegetable proteins. The permitted levels of vegetable proteins in some Japanese meat products are shown in Table 11.10.

Japan does import soy protein products, and in 1984 the Japanese Government

Table 11.9

Japan Agricultural Standards on Vegetable Proteins[a]

	Powder	Paste	Textured		Structured	
			Dry	Frozen	Dry	Frozen
Moisture (%)	<10	<80	<10	<80	<10	<80
Crude protein (dry) (%)	>60	<70	>52	>52	>60	>60
Particle size (% through 350 μm mesh)	>95	—	<10	<10	—	—

Additives permitted for vegetable protein production

Food ingredients	Vegetable oils, salt, starch, lecithin	Vegetable oils, salt, starch, lecithin	Vegetable oils, salt, starch, lecithin, color, tocopherol, HVP, sucrose, spices	Same as textured
Food additives	Polyphosphate, L-ascorbic acid emulsifier,[b] acetic acid, citric acid, sodium-phosphate, sodium-bisulfite	Polyphosphate, citric acid, sodium-phosphate, sodium-bisulfite	L-ascorbic acid, mono- and di-glyceride, calcium sulfate, calcium chloride, flavors, seasonings[c]	L-ascorbic acid, calcium sulfate, calcium chloride, flavors, seasonings[c]

[a]Source: Kanda (1981).
[b]Sugar ester, sorbitan ester.
[c]Various nucleotides.

Table 11.10

Permitted Levels of Vegetable Proteins for Standardized
Processed Foods in Japan[a]

Items	Level
Pressed hams	3–5% as binder
Sausages	5 to 10%
Fish sausages	<20%
Kamaboko (boiled fish paste)	<8%
Minced (ground) meat products	20–40%

[a]Sources: Watanabe (1974); Kanda (1981).

gave a "seal of approval" to Ralston Purina Co. of the United States, which eliminates the need for inspection of each shipment. The waiving of inspection was based on inspection of production facilities in the United States. This kind of arrangement certainly facilitates trade in soy proteins.

A recent development in Japan is the use of soymilk. Although not a traditional food for the Japanese (as it is for the Chinese), soymilk use is rapidly growing as a healthful food. Again, relative freedom from regulatory problems plus a long-term familiarity with soy products seem to be helpful in the rapid adoption of this new product (Anon. 1984c).

Regulations that provide a good balance of correct information to guarantee safety and nutrition for the consumer and of sufficient leeway to encourage new product development would be the best that a country could achieve.

REFERENCES

Anon. (1977). "Soybeans in Japan." Jikyusha, Tokyo (in Japanese).

Anon. (1983-84). "Year Book and Trading Rules." National Soybean Processors Association, Washington, D.C.

Anon. (1984a). "Soya Blue Book." American Soybean Association, St. Louis, MO.

Anon. (1984b). Federal Register. No.91, May 9, 49:19621.

Anon. (1984c). Soymilk: New processing, packaging expand markets. *J. Am. Oil Chem. Soc.* **61**:1784.

Brekke, O. L. (1980a). Specifications for soybean oil. *In* "Handbook of Soy Oil Processing and Utilization" (D. R. Erickson, E. H. Pryde, O. L. Brekke, T. L. Mounts, and R. A. Falb, eds.). American Soybean Association, St. Louis, MO, and American Oil Chemists Society, Champaign, IL.

Brekke, O. L. (1980b). Oil degumming and soybean lecithin. *In* "Handbook of Soy Oil Processing and Utilization" (D. R. Erickson, E. H. Pryde, O. L. Brekke, T. L. Mounts, and R. A. Falb, eds.). American Soybean Association, St. Louis, MO, and American Oil Chemists Society, Champaign, IL.

Brincker, A. (1979). Review of European legislation on vegetable protein in meat products. *J. Am. Oil Chem. Soc.* **56**:211.

Hagg, D. D. (1981). Labeling and compliance assurance of soya protein foods. *J. Am. Oil Chem. Soc.* **58**:473.

Kanda, H. (1981). Utilization and quality standards of vegetable proteins for foods in Japan. *J. Am. Oil Chem. Soc.* **58**:441.

Olsman, W. J. (1979). Methods for detection and determination of vegetable proteins in meat products. *J. Am. Oil Chem. Soc.* **56**:285.

Roberts, H. (1979). Regulatory outlook in vegetable protein. *J. Am. Oil Chem. Soc.* **56**:206.

Smith, A. K., and S. J. Circle (1978). Protein products as food ingredients. *In* "Soybeans: Chemistry and Technology," Vol. 1, 2nd ed. (A. K. Smith and C. J. Circle, eds.). AVI Publ. Co., Westport, CT.

Ward, A. G. (1979). Basic principles underlying a legislative framework for vegetable protein. *J. Am. Oil Chem. Soc.* **56**:196.

Watanabe, T. (1974). Government role and participation in development and marketing of soy protein foods. *J. Am. Oil Chem. Soc.* **51**:111A.

Watanabe, D. J., H. O. Ebine, and D. O. Ohda (1971). "Soybean Foods." Kohrin Shoin, Tokyo (in Japanese).

Glossary of Scientific and Technical Terms

Activated earth   bleaching earth that has been treated with acid to increase its capacity for pigment adsorption from soybean oil.

Acylation   addition of an acyl group($-\overset{\text{O}}{\underset{\|}{\text{C}}}-$R) to an amino nitrogen such as the ϵ amino of lysine; can impair nutritional availability of lysine can may enhance functionality.

Albumins   solubility class of proteins that includes all water-soluble proteins.

Aleurone cells   outermost cells of the cotyledon with high protein content.

Aleurone grain   synonym for protein body.

Alkali refining   a process whereby alkali is added to crude oil to form salts of free fatty acids, which can then be removed by centrifugation; not to be confused with the more general term "refining," which refers to all processes for processing crude soybean oil.

Amyloplast   a subcellular organelle that contains starch grains.

Antioxidants   chemical compounds that retard the autoxidation of unsaturated lipids.

Autoxidation   the oxidative deterioration of fats or oils caused by a spontaneous reaction with molecular oxygen of the air.

Basket extractor   means for solvent extraction in which multiple baskets containing flakes travel vertically through a countercurrent solvent stream (see Figs. 3.16 and 3.17).

Biological value   a measure of the nutritive value of proteins in which the nitrogen ingested (I), nitrogen excreted in feces (F), and nitrogen excreted in urine (U) are measured; biological value is $I - (F + U)/(I - F)$ or nitrogen retained/nitrogen absorbed.

Bleaching   the process for removing pigments and thereby lightening the color of soybean oil; makes use of a natural earth adsorption process followed by filtration.

Bleaching earth   mined special earths or clays that when added to hot soybean oil (at about 1%) have the ability to adsorb unwanted pigments.

Bound water   water that does not freeze at $-20°C$; both proteins and carbohydrates bind water in soybeans.

Bowman–Birk trypsin inhibitor   a class of soybean trypsin inhibitors that have low molecular weight (about 8000), many disulfide links, and great stability to denaturation.

329

Browning reactions   several different processes that can cause brown pigments in food products; in dried protein products, browning is usually due to carbonyl–amino interactions also called the Maillard reaction.

Cadaverine   a polyamine compound found in the nonprotein nitrogen fraction of soybeans.

Calcium stearoyl lactylate (CSL)   one of the prominent emulsifiers used to maintain loaf volume when a composite flour is used to make bread.

Calorimetry   a technique for measuring heat changes during a chemical reaction or phase change; can be used to measure free water in a sample because only free water freezes; total water minus free water is a measure of bound water.

Campestrol   a sterol present in soybean oil (see Fig. 2.26 for structure).

Chalaza   morphological feature of the soybean seed coat; it is a small groove at the end of the hilum opposite the micropile.

Chemical score   a method of evaluating protein nutritive quality based on amino acid analysis; the amount of each essential amino acid is divided by the total amount of essential amino acids; the smallest of these values signifies the limiting amino acid, and the number (expressed as a percentage) is the chemical score.

Cholecystokinin   an intestinal hormone that provides control of the biosynthesis of trypsinogen by the pancreas.

Coffee whiteners   products substituting for dairy cream for use in coffee; often soy protein is used in these products as an emulsifier and as a freeze–thaw stabilizer.

Comminuted meat products   refers to all types of ground or emulsified meat products from frankfurters to hamburger patties.

Composite flours   mixtures of wheat flour with nonwheat flour for the purpose of minimizing use of foreign exchange to purchase wheat and for improving nutritional quality of wheat products.

Conditioning   heating and addition of moisture to soybean pieces to achieve good plasticity for flaking; conditioning minimizes fines production.

Conglycinin   a general term for the 7 S proteins from soybeans; the major protein of this group is β-conglycinin, which dimerizes when the ionic strength is decreased.

Conjugated dienes or trienes   refers to the double bonds of fatty acids found in soybean oil; conjugated double bonds are alternating single and double bonds and are found in oxidized fatty acids; normal fatty acids have methylene interrupted double bonds and are not conjugated.

Continuous loop extractor   equipment for solvent extraction of soybean flakes by a moving shallow bed of flakes and countercurrent solvent system (see Fig. 3.20).

Corn–soy–milk (CSM)   one of several cereal blends used for international relief feeding.

Cotyledon   one of the two halves of the soybean seed; the cotyledons develop into the first two leaves of the young plant.

Cracking   breaking of whole soybeans into several pieces to facilitate dehulling and flaking.

Creaming   breaking of an oil-in-water emulsion in which lipid droplets separate by rising to the top but do not coalesce into a separate oil phase.

Cyclone separator   a piece of equipment for removing fine particles from an airstream (see Fig. 3.3).

Daidzin   a soybean isoflavone; the aglycone derivative is daidzein (see Fig. 2.24 for structure).

Degumming   the removal of phospholipids from soybean oil by a water washing step.

Deodorization   the removal of volatile flavor compounds from oils by high temperature steam distillation under high vacuum.

Desolventizer–toaster   equipment for removing solvent from defatted flakes and for heating flakes sufficiently to overcome growth inhibition properties; heating medium is steam.

Dowtherm A   tradename of a heat exchange material that condenses at temperatures of 205–260°C (400–500°F) to provide heat for deodorization of soybean oil.

Emulsion capacity   a measure of the quantity of oil that can be emulsified per unit of protein or other emulsifying agent.

Emulsion stability   a measure of staying power of an emulsion after it is formed.

Estrogenic substances   compounds that mimic estrogen hormones; thought to be isoflavones in soybeans.

Expeller   equipment for expressing oil from oil seeds; consists of an augur moving through a slotted barrel through which oil can drain; also called continuous screw press.

Extruder   a jacketed augur used as an economical cooker and as a means of texturizing soy flours or concentrates; also used to treat soy flakes before solvent extraction.

Extrusion   a process for texturizing soy flours using high pressures and temperatures in an extruder.

Fat absorption   analogous to water absorption by soy proteins and measured by mixing oil and protein and centrifuging; the oil that cannot be decanted is said to be absorbed.

Fermented soy curd   produced by a brine fermentation of low-moisture soy curd; active organisms are molds (*Mucor* and *Rhizopus*) and fermentation is for several months; primarily a Chinese product called *su-fu*.

Fermented soy pulp   soy pulp is the material removed in soymilk production, and it can be fermented by *Rhizopus* (24 hr) to produce a tempehlike product; an Indonesian product.

Filled milk   a milklike product resulting from replacement of the butter fat in milk with some other oil or fat.

Fines   very small particles of soybeans produced during cracking and flaking; crushers want to minimize fines because they interfere with extraction and desolventizing the oil.

Fire point   temperature at which sustained burning of oil occurs; Used to judge residual solvent in oils.

Fish paste   a Japanese product consisting of emulsified fish muscle with added soy isolate; may be used as an ingredient in other red meat or fish products such as fish sausage.

Flaking   a process for converting soybean meats into thin flakes in preparation for solvent extraction.

Flash desolventizer   equipment for removing solvent from defatted flakes without adding appreciable heat (see Fig. 3.27).

Flash point   temperature at which a brief flame appears above the oil; used to judge residual solvent in oils.

Flatulence   intestinal gas produced by microflora acting on undigested raffinose and stachyose, two soluble carbohydrates of soybeans that cannot be digested by humans.

Flavor intensity value   a weighted sensory judgment of flavor of an oil based on overall flavor intensity and individual off-flavors such as grassy or beany.

Fluid shortening   a pourable shortening that is opaque at low temperatures, in contrast with liquid shortening, which remains clear at low temperatures.

Formulated foods   a general term for nutritious mixtures of vegetable proteins, vitamins and minerals used as supplementary foods in international relief; corn–soy–milk (CSM) and Incaparina are specific examples of formulated foods.

Fresh soybeans   vegetable-type soybeans picked green and cooked until tender; synonyms are *maotou*, China; *put-kong*, Korea; *edamame*, Japan.

Full-fat soy flour   ground whole soybeans containing all of the original oil; may be enzyme active or heated to minimize enzyme action.

Functional properties   in reference to soy proteins, the beneficial ways that they can aid in food manufacture in addition to nutritive value; includes water binding, foaming, emulsification, etc.

Genestin   a soybean isoflavone; the aglycone derivative is genestein (see Fig. 2.24 for structure).

Globoid inclusions   structural features found in protein bodies and thought to be phosphorus-containing structures.

Globulins   solubility class of proteins that includes all salt soluble proteins.

Glutelins   solubility class of proteins that includes all proteins soluble in dilute acid or base.

Glycinin   the major storage protein of soybeans synonymous with the 11 S globulin.

Glycitin   a soybean isoflavone; the aglycone derivative is glycitein (see Fig. 2.24 for structure).

Goitrogenic substances   compounds that promote formation of goiters; untoasted soy products tend to be goitrogenic but the mechanism is not known.

Hamanatto   a fermented whole soybean product from Japan; cooked soybeans are inoculated with *Aspergillus oryzae*, packed with salt, and aged for several weeks or months.

Hard soybeans   soybeans that do not imbibe water when soaked at low or moderate (ambient) temperatures.

Hemagglutinins   soybean proteins that cause red blood cells to agglutinate but do not have any growth inhibition activity; belong to the class of plant proteins known as lectins.

Hilum   morphological feature of the soybean seed coat; it is the point of attachment of the seed to the pod.

Hourglass cells   cells of the seed coat lying between the palisade and parenchyma cells; these cells are uniquely shaped like an hourglass.

Hydrogenation   process of adding gaseous hydrogen to the double bonds of unsaturated fatty acids; makes the oil more resistant to oxidation or changes an oil to a fat for shortening or margarine manufacture.

Hydroperoxide   the general term for the first stable product of unsaturated lipid oxidation.

Incaparina   a blend of cereal and oil seed ingredients with vitamins and minerals added that offers excellent nutrition at low cost; developed by the Institute of Nutrition for Central America and Panama (INCAP) in Guatemala.

Iodine value   a measure of the unsaturation of a fat or oil; percentage by weight of iodine needed to saturate an oil or fat.

Ionic strength   a way of measuring ionic effects that takes into account both the total number of ions and the charge on the ions.

Koji   Japanese term for the fungal starter culture mass used in soy sauce fermentation.

Kunitz trypsin inhibitor   a class of soybean trypsin inhibitors primarily responsible for growth inhibition from raw soybeans; has molecular weight about 20,000 and is sensitive to heat denaturation.

Lactose intolerance   inability to digest lactose in dairy products due to absence of β-galactosidase in adults; orientals and blacks are prone to this condition, which has symptoms of diarrhea and intestinal cramping.

Lecithin   has two meanings in common usage: (1) the common name for phosphatidyl choline, and (2) the common name for all phospholipids removed from an oil by degumming.

Lectins   plant proteins that have the ability to combine selectively with carbohydrates; soybean hemagglutinin is a lectin.

Legumin   the heavier (12 S) of the two major globulin fractions of storage proteins found in legumes; vicilin is the other (7 S) fraction.

Light line   a region of high refractility in the palisade cells of the soybean seed coat; some have attributed the moisture barrier properties of the seed coat to this region.

Lipid bodies   subcellular organelles of the soybean cotyledon cells with diameters of 0.2–0.5 μm; the lipid of the soybean is stored in these organelles, and they are made up of at least 80% lipid; formerly named spherosomes.

Lipoxygenase   an iron-containing enzyme in soybeans that catalyzes hydroperoxide formation in unsaturated fatty acids; synonymous with the older name lipoxidase.

Liquid shortening   a pourable shortening that remains clear at low temperatures, in contrast with fluid shortening, which is pourable but opaque.

Lovibond red or yellow   numbers used in the Lovibond Tintometer system to specify color of oil samples; the higher the number, the more of that color is present.

Lutein   the predominant pigment in soybean oil (3,3′-dihydroxy-α-carotene).

Lysinoalanine   an unusual amino acid that can form in proteins after heating or alkali treatment; in the concentrations found in soy protein, it is of no nutritional significance.

Massaging   a technique for moving whole meat cuts against one another for the purpose of even distribution of added soy protein dispersions, or for incorporating soy protein dispersions.

Maturity groups   division of soybean cultivars into 12 groups based on how flowering responds to dark periods; maturity groups are selected for the proper latitude, so that flowering is timed for maximum yield.

Meat analogs   products in which the principal protein ingredient is from soy and added fat, color, and flavor are used to produce a meatlike product; synonymous with simulated meats.

Meats   the pieces of soybean produced by cracking rolls.

Melting point   temperature at which solid fat liquefies; important physical property of fats and oils but not as precise as with pure crystalline compounds; depends on method for determination.

Mesophase   a dense highly concentrated protein phase achieved with calcium or low ionic strength and centrifugation; can be used as starting material for texturizing protein.

Micropile   morphological feature of the soybean seed coat, it is the opening through which the germ tube reached the embryo sac.

Miscella   solution of crude oil in extracting solvent.

Natto   a fermented whole-soybean product produced and used in Japan; cooked soybeans are traditionally covered with straw providing a bacterial inoculum; the organism is *Bacillus natto*.

Net protein utilization (NPU)   a measure of the nutritive value of proteins in which the nitrogen ingested (*I*), nitrogen excreted in feces (*F*), and nitrogen excreted in urine (*U*) are measured; NPU is $[I - (F + U)]/I$ or nitrogen retained/nitrogen ingested.

Nitrogen balance   a way of measuring nutritive value of protein in which nitrogen ingested is compared with nitrogen excreted; if nitrogen ingested is greater than nitrogen excreted, the test subject is in positive nitrogen balance.

Nitrogen solubility index (NSI)   a measure of the solubility of soybean protein; is important as a measure of heat treatment received by a product; contrasts with Protein Dispersibility Index (PDI) in using a slow stir method and usually lower than PDI.

Nonhydratable phospholipids   the phospholipids that are not removed by degumming or water washing of crude soy oil; may be calcium or magnesium salts and tend to increase in damaged soybeans.

Nuclear magnetic resonance   a physical method for measuring solids content based on interaction of magnetic fields on ability of nuclei to absorb radiomagnetic energy.

Palisade cells   a columnar layer of cells in the seed coat lying just under the cuticle; also, applied to the elongate cells of the cotyledon.

Pancreatic hypertrophy   enlarged pancreas resulting from feeding raw soybeans or unheated soy flours; occurs in those young animals in which the pancreas makes up a relatively large proportion of body weight.

Parenchyma cells   a layer of cells in the seed coat, loosely connected, and innermost.

Phosphatidyl choline   one of the three predominant phospholipids in crude soybean oil.

Phosphatidyl ethanolamine   one of the three predominant phospholipids in crude soybean oil.

Phosphatidyl inositol   one of the three predominant phospholipids in crude soybean oil.

Phytic acid   hexaphosphate derivative of inositol found associated with soybean protein; can have nutritional significance by limiting availability of divalent cations, especially zinc (see Fig. 2.21 for structure).

Phytin   a calcium, magnesium salt of phytic acid.

Plastic shortening   solid shortening as contrasted with dry or pourable shortenings.

Polyamines   a class of compounds found in the nonprotein nitrogen fraction of soybeans; recognized to be bad-flavored compounds, they are not present in high enough concentration to affect soybean flavor.

Press bleaching   additional bleaching that occurs during filtration of bleaching earth from soybean oil; the reason for this effect is not known.

Prolamines   solubility class of proteins that includes all proteins soluble in 50–70% ethanol; found in wheat and corn but not in soybeans.

Prooxidants   conditions or chemical compounds that speed up the autoxidation of unsaturated lipids.

Protein Advisory Group (PAG)   a United Nations advisory group to WHO, FAO, and UNICEF on nutrition and protein–calorie malnutrition; responsible for rules for testing new food mixtures; replaced in 1977 by the United Nations Advisory Group on Nutrition.

Protein bodies   subcellular organelles of the soybean cotyledon that range in size from 2 to 20 μm in diameter; they are the prominent organelles in cells and are made up of approximately 90% protein.

Protein dispersibility index (PDI)   a measure of the solubility or dispersibility of soybean protein; is important as a measure of heat treatment received by a product; contrasts with nitrogen solubility index (NSI) in using a fast-stir method and is usually higher than NSI.

Protein efficiency ratio (PER)   a measure of protein quality in nutrition given by weight increase/ weight of protein fed × 100 under carefully prescribed feeding conditions with young rats.

Protein-rich food mixtures   a general term for nutritious mixtures of vegetable proteins, vitamins, and minerals used as supplementary foods in international relief; corn–soy–milk (CSM) and Incaparina are specific examples of protein-rich food mixtures.

Putrescine   a polyamine found in the nonprotein nitrogen fraction of soybeans.

Raffinose   a soluble trisaccharide found in soybeans that is responsible for flatulence.

Refining   refers to the various processes (degumming, alkali refining, bleaching, hydrogenation, deodorization) used in processing a crude soybean oil into an edible product; not to be confused with alkali refining, which refers to the specific process of removing free fatty acids.

Refractive index   one of the physical properties of soybean oil used to detect changes in fatty-acid composition; refractive index increases as the degree of unsaturation increases.

Reversion   the development of grassy and beany flavors followed by fishy or painty flavors in soybean oil that is thought to be due to oxidation but occurs at low peroxide values; it is unique to soybean oil and is attributed to linolenic acid.

Rotocel   trade name for extraction equipment in which moving beds of flakes are contacted counter-currently by solvent (see Fig. 3.18).

Salad oil   a degummed, alkali-refined, edible oil that remains clear at low temperature storage; may or may not be partially hydrogenated, bleached, deodorized, or winterized.

Sapogenins   the aglycone portion of saponins.

Saponification   treatment of fat or oil with strong base causing hydrolysis of triglycerides to free fatty acids and glycerol.

Saponins   glycosides of various triterpine alcohols found in soybean meal; these compounds have high surface activity and can lyse red blood cells but are thought to be safe in the amounts found in soybeans.

Scalping   a cleaning process for soybeans in which all material larger than a screen size that will pass soybeans is removed.

Secondary reaction products   a general term for the chemical compounds resulting from break-down of lipid hydroperoxides; the off-flavor compounds of lipid autoxidation are secondary reaction products.

Selectivity   a measure of the degree to which hydrogenation acts faster on more highly unsaturated fatty acids than on less saturated fatty acids.

Shortening   originally the fat that caused a "short" or easily broken texture in baked goods; now a general term for a solid-fat baking or cooking ingredient.

Simulated meats   products in which the principal protein ingredient is from soy and added fat, color, and flavor are used to produce a meatlike product; synonymous with meat analogs.

Sitosterol   a sterol present in soybean oil (see Fig. 2.26 for structure).

Smoke point   temperature at which a fat or oil emanates continuous wisps of smoke; free fatty acids lower the smoke point of a fat or oil.

Sodium stearoyl lactylate (SSL)   one of the prominent emulsifiers used to maintain loaf volume when a composite flour is used to make bread.

Solidification point   similar to a freezing point but not based on crystal formation; usually a few degrees lower than the melting point due to the complexity of phase changes in oils and fats.

Solids fat index (SFI)   an approximate measure of the solids content of an oil at a given temperature; an important physical attribute of margarines and shortenings when measured for a set of five specified temperatures.

Soybean cake   material remaining after processing by a continuous screw press (expeller) to remove oil from soybean flakes.

Soybean mill feed   a by-product of soybean processing consisting of primarily hulls and other offal from the tail of the mill; has 13% protein, 32% crude fiber, and 13% moisture.

Soybean mill run   a by-product of soybean processing consisting of hulls and adhering meats; has 11% protein, 35% crude fiber, and 13% moisture.

Soy concentrate   the product achieved by extracting soluble carbohydrate from defatted soy flour; in the United States soy concentrate must have at least 70% protein on a dry-weight basis.

Soy curd   product achieved by precipitating protein from soymilk using calcium salts and removing the fluid by pressing; synonyms are *tou-fu*, China; *doo bu*, Korea; *tofu*, Japan; *tau-fu*, Indonesia and Malaysia.

Soy flour   the product achieved by defatting dehulled soybean flakes and grinding so that 97% will pass through a 100-mesh screen; contains a minimum of 50% protein.

Soy grits   soy product from defatted flakes with particle sizes larger than for soy meal or soy flour.

Soy isolate   the product achieved by separating all carbohydrate from defatted soy flour leaving mainly protein; in the United States, soy isolate must contain at least 90% protein on a dry-weight basis.

Soy meal   the soy product used for animal feeding after grinding defatted flakes; may be 44.0 or 47.5–49% protein based on whether hulls are added or not.

Soymilk   the aqueous dispersion from grinding soaked soybeans in water, heating, and filtering with many variations; synonyms are *tou-chiang*, China; *kong kook*, Korea.

Soymilk film   product achieved by boiling soymilk until a film forms, then lifting that film free and drying it; synonyms are *tou-fu-pi*, China; *yuba*, Japan; *fu chok*, Malaysia.

Soy paste   traditionally the material remaining after filtration of the fluid portion of soy sauce; now produced as a fermented, salty, soy condiment with many variations of flavor, color and texture; synonyms are *chiang*, China; *doen jank*, Korea; *miso*, Japan; *tauco*, Indonesia and Malaysia; *tao si*, Philippines.

Soy powder, toasted   dry soybeans heated until browned with a characteristic flavor; then ground into a powder and sieved and added to other foods as a flavoring adjunct; synonyms are *tou-fen*, China; *kong-ka-ru*, Korea; *kinako*, Japan; *bubuk kedele*, Indonesia.

Soy sauce   a fermented soy product from soybeans and wheat; the fluid from a prolonged bacterial and yeast brine fermentation of several months is filtered and pasteurized to yield soy sauce; synonyms are *chiang-yu*, China; *kan jank*, Korea; *shoyu*, Japan; *kecap*, Indonesia and Malaysia; *tayo*, Philippines.

Soy sprouts   product from sprouting and cooking soybeans; synonyms are *huang-tou-ya*, China; *kong na moal*, Korea; *daizu no moyashi*, Japan.

Spermidine   a polyamine found in the nonprotein nitrogen fraction of soybeans.

Spermine   a polyamine found in the nonprotein nitrogen fraction of soybeans.

Spherosomes   synonym formerly used for lipid bodies.

Spinning   a process for texturizing soy protein isolate by forcing a concentrated solution of protein through a small opening into a coagulating bath; the basis for production of meat analogs from soy protein.

Stachyose   a soluble tetrasaccharide found in soybeans that is responsible for flatulence.

Stationary basket extractor   equipment for solvent extraction of soybeans that makes use of stationary beds of flakes and a countercurrent stream of solvent (see Fig. 3.19.).

Steam texturizing   a process for texturizing soy flours in which a quantity of flour is subjected to high temperature steam and shearing action to realign protein molecules (see Fig. 3.48).

Stigmasterol   a sterol present in soybean oil (see Fig. 2.26 for structure).

Supercritical fluid   gas under high pressure that has flow and solvent characteristics of a liquid; supercritical carbon dioxide has been used to extract soybean oil in experimental systems.

Svedburg units   abbreviated S, these units are used to designate major protein fractions of soybeans as 2 S, 7 S, 11 S, and 15 S; the larger the S value the more rapid is the movement of the sedimentation boundary in the ultracentrifuge.

Synergism   a more than additive effect achieved by combining two treatments; often used to refer to the antioxidative effects of a combination of phospholipids and tocopherols.

Tempeh   a fermented whole soybean product made in Indonesia; boiled soybeans are fermented briefly with mold; it is often eaten fried.

Tempering   allowing time for moisture equilibrium after drying or wetting soybeans; also, time for holding shortenings after filling to achieve good creaming characteristics.

Thiobarbituric acid (TBA) number   a measure of autoxidation of lipid-containing foods in which the reagent TBA is reacted with secondary oxidation products (malondialdehyde) to yield a colored complex that can be measured with a spectrophotometer at 532 nm.

Thixotropy   a change in viscosity as a result of time of resting after subjecting a test fluid to shear.

Titanium dioxide   an inert inorganic compound formerly used as an indicator by adding 0.1% to protein isolate (United States only); analysis for protein isolate in a product could then be done by analyzing for titanium dioxide; use was discontinued in 1984.

Trichloroethylene   solvent used to extract oil from soybeans in the 1940s but discontinued because of toxicity in the defatted soy flakes.

Trypsin inhibitors   proteins in soybeans responsible for growth inhibition when raw soybeans are fed (see Bowman–Birk and Kunitz trypsin inhibitors).

Tumbling   a technique for moving whole meat cuts against one another for the purpose of even distribution of added soy protein dispersions, or for incorporating soy protein dispersions.

Urease   soybean enzyme that catalyzes the breakdown of urea to ammonia and carbon dioxide; its thermal denaturation is used as an indicator for proper heat treatment of defatted soy flakes for animal feeding.

Vegetable stearine   the crystalline fats removed from soybean oil by winterizing; these are predominantly saturated fats.

Vicilin   the lighter (7 S) of the two major globulin fractions of storage proteins found in legumes; legumin is the other (12 S) fraction.

Water absorption   synonymous with water-holding capacity; contrasts with water adsorption, which refers to specific interaction between a chemical compound and water leading to the bound-water concept.

Water activity (a_W)   the equilibrium relative humidity expressed as a decimal; partial pressure of water vapor in equilibrium with a foodstuff divided by partial pressure of water at the same temperature.

Water holding capacity (WHC)   degree to which a food immobilizes water and prevents it from flowing freely, but not synonymous with bound water.

Water sorption isotherm   the interaction of water vapor with a chemical compound at constant temperature; used for soy protein to show how water vapor will transfer and equilibrate in dry food mixtures.

Wheat–soy–blend (WSB)   one of several cereal blends used in international relief feeding.

Whey   the soluble protein fraction remaining after precipitation of soy protein by adjustment of pH or by addition of calcium; analogous (and the same term) to soluble milk proteins after casein precipitation.

Winterizing   process of gradual cooling of soybean oil so that saturated triglycerides crystallize and can be removed; the remaining fluid oil can then be stored at refrigerator temperatures without any crystal formation.

Index